Metallic Oxynitride Thin Films by Reactive Sputtering and Related Deposition Methods: Process, Properties and Applications

Edited by

Filipe Vaz

Minho University
Portugal

Nicolas Martin

National Engineering School ENSMM
France

&

Martin Fenker

FEM Forschungsinstitut Edelmetalle & Metallchemie
Germany

CONTENTS

About the Editors

Filipe Vaz (F. Vaz) graduated in Physics and Chemistry at the University of Minho, Portugal in 1992, where he obtained also his PhD degree in Physics in 2000. Since September 1992 he has been working at the Physics Department of University of Minho, involved in research areas related with thin films and their applications. Main research topics concern hard nanostructured thin films, with targeted applications varying from tools and machine parts, including polymers. From 2001 he is also developing new optical thin film systems, based on oxynitrides, oxycarbides, and their mixing. Recently, his research is focused on the physics and technology of magnetron sputtered thin films containing noble metal nanoparticles, namely gold and silver, revealing Surface Plasmon Resonance behaviour.

Nicolas Martin obtained a PhD in Physical Chemistry from the University of Franche-Comté in 1997 and an habilitation degree from the same University in

2005. He was a researcher at the Ecole Polytechnique Fédérale de Lausanne from 1998 to 2000 in the Physics department. He was nominated as Assistant Professor at the National Engineering School ENSMM – "Ecole Nationale Supérieure de Mécanique et des Microtechniques in Besançon in 2000, and Full Professor in 2008. His research is focused on the physics and technology of metallic and ceramic thin films prepared by reactive sputtering. He is also interested in nanostructuration of coatings prepared by Glancing Angle Deposition (GLAD). He was the head of the MIcro NAno MAterials & Surfaces team (MINAMAS) in the Micro Nano Sciences & Systems (MN2S) research department of the FEMTO-ST Institute for 2008 and 2009. He is now one of the Deputy Directors of MN2S research department.

Martin Fenker studied Physics in Ulm and Heidelberg (Germany). He obtained a PhD in Physics from the University of Heidelberg in 1998. He was a researcher at the DaimlerChrysler AG Research Institute (ULM) from 1994 to 1999, studying the deposition of diamond coatings by using CVD methods. Since 1999, he is employed at the research institute for precious metals and metals chemistry (FEM) in Schwäbisch Gmünd (Germany), heading the department Physical Surface Technology. In 2004, this department merged with the Materials Physics department. He was appointed as head of this new department and its name changed to Plasma Surface Technology and Materials Physics (POT-MPh). His research is focused on thin film deposition by PVD and PACVD and coatings characterisation. He is interested in understanding the correlation between plasma process and thin film growth. Basic investigations are performed in High Power Impulse Magnetron Sputtering (HiPIMS) and PACVD with a Plasma Beam

Source (PBS). Beneath coating deposition of nitrides, carbides/DLC, oxides and oxynitrides *etc*., surface modifications like plasma nitriding and combined processes (duplex) are explored.

FOREWORD

Due to their excellent physical, chemical and mechanical properties, metal oxynitride coatings hold considerable promise for a very wide range of applications, from electronic devices through to tribological surfaces for cutting tools. This text sets out, for the first time together in a single book, leading-edge research into the modelling, deposition and applications of oxynitride thin films. The layout of the book is well-structured, moving from modelling and simulation, through film synthesis and then to applications issues. The Chapters are all written by leading experts in this research field and the book is edited by three leading research specialists in advanced coating processes: Prof. Filipe Vaz, Prof. Nicolas Martin and Dr. Martin Fenker.

There is a need, in the case of all possible applications for oxynitride coatings, to develop improved deposition methods which allow the achievement of desired coating properties through effective control of structures and phase compositions. Therefore the papers in the book form a vital body of research which points the way forward not only for specific oxynitrides, but also provides generic information which will be applicable across all possible combinations of elements within this materials system. The Chapters are at the leading edge in their respective subject areas, and include comprehensive information on different deposition methods and how they can be used to optimise specific coating systems and achieve precise properties and to fulfil the needs of different applications.

The book is particularly useful as it includes information on the background history and previously published literature on particular coatings and processes. It therefore provides the reader with the knowledge needed to understand and develop metal oxynitride deposition processes, supported by an understanding of previous work in the field. It is invaluable to those who are just entering this research area, as well as those who have been actively researching metal oxynitrides for some time.

This book is a vital text for anyone who has an interest in the production, analysis and use of metal oxyntride thin films.

Allan Matthews
Surface Engineering
Department of Materials Science and Engineering
University of Sheffield
Sheffield, UK

PREFACE

Ceramic thin films based on metal nitrides and oxides are established materials with numerous industrial applications and still with a promising future, in addition to being materials of great interest to the academic and industrial communities. Although most of their applications are relatively recent, the metal nitrides and oxides have been known for almost a century. The industrial importance of the metal nitrides and oxides keeps growing rapidly, not only in the traditional and well-established applications based on the strength and refractory nature of these materials such as cutting tools and abrasives, but also in new and promising fields such as electronics and optoelectronics.

Recently, and using the idea of merging the benefits of the basic characteristics of both metal nitrides and oxides, a new class of materials is gaining more and more relevance in several technological applications, the metal oxynitrides, MeN_xO_y. Their relevance arises from the fact that the addition of oxygen to MeN_x allows the tailoring of film properties between those of metal nitrides, MeN_x, and those of the correspondent insulating oxides, MeO_y. Tuning the oxygen/nitrogen ratio allows playing with the band gap, the electronic conductivity, and the crystallographic order between oxide and nitride and hence the electronic and (micro)structural properties of the materials, and thus the overall set of properties. These metal oxynitrides are now investigated in several research groups and technologically developed and applied. They open up new industrial possibilities in several different fields, taking profit of the extended range of properties that arise for the mixing of the basic characteristics of both nitrides and oxides in one material.

Oxynitride thin films are either emerging or about to emerge from the research laboratory to become commercially available finding various practical applications. No textbook outlining the basic theoretical background, the methods of fabrication and applications currently exist. Thus, this book aims at presenting an in-depth overview of this topic covering a broad range of oxynitride thin films technologies including simple approaches like sputtering targets with a mixture of two reactive gases to more complex ones such as gas and power pulsing ones, presented in a systematic fashion, by the scientific leaders in the respective domains. The main purpose is to provide a systematic and an in-depth overview of the topic, on both a high and current level. It offers information on the physical basics as well as the latest results in a compact yet comprehensive manner. It covers a broad range of related topics, from physical principles to design, fabrication, characterization, and possible applications in various types of devices.

The book was written by and for scientists and engineers from all the areas related to the synthesis and application of **Metallic Oxynitride Thin Films**, which is intended to serve as a cross-disciplinary major reference work. The book will also serve as a valuable introduction to undergraduate and graduate students interested in oxynitride thin film technology. The contributions cover the physical processes involved in development of metal oxynitrides, from a growth point of view, *via* their atomic structures and the related production characteristics right up to some future industrial applications. We hope that this book will be useful to the multitude of disciplines represented by the vacuum coaters and will bring a highly pertinent tool for anyone working in the applied materials research or the modern industry. In particular, it is intended for researchers in thin film processing who want to fill knowledge gaps. Knowledge, which is specific to oxynitride compounds, from their synthesis to their resulting behaviors.

The first three chapters deal with the reactive sputtering process as an attractive way to fabricate metal oxynitride thin films. Modeling of such a complex process is especially surveyed and pulsing approaches (gas and power) appear as original experimental procedures to achieve tunable oxynitride compounds. The next five chapters provide a brief description yet pithy of different oxynitride systems: zirconium, titanium, chromium, niobium and aluminum. Each of these chapters describes the physico-chemical properties and the structural modifications induced by a systematic change of the metalloid contents in the films. The functional properties are also assessed in connection to the main characteristics of the oxynitrides. The last four chapters address a non-exhaustive overview of promising applications in various fields such as optoelectronics, biomedicals, gas barriers, decorative and wear resistance, microelectronic devices …

Combining production related issues and a selection of some successful applications; this comprehensive source will give you a unique perspective to the fascinating field where nanoscale materials and their various synthesis methods meet. Intended to be a reference work of this magnitude, it provides a broad, yet in-depth perspective of numerous facets. For all this, the editors wish to thank all the authors of the chapters for their important contribution, as well as for updating the final stage of the book publishing. Our warmest thanks go to the partners of the "Hardecoat" European project who indirectly motivated us to initiate this book. Finally, we want to thank Prof. Allan Matthews for taking his time to write the foreword.

Filipe Vaz
Minho University
Portugal

Nicolas Martin
National Engineering School ENSMM
France

&

Martin Fenker
FEM Forschungsinstitut Edelmetalle & Metallchemie
Germany

List of Contributors

Agouram Said

Department of Applied Physics and Electromagnetism, Univesitat de Valencia, C./ Dr. Moliner N 50, 46100 Burjassot, Spain

Alves Eduardo

Instituto Superior Técnico/ITN, Universidade Técnica de Lisboa, E.N. 10, 2686-953 Sacavém, Portugal

Barradas N. Pessoa

Instituto Superior Técnico/ITN, Universidade Técnica de Lisboa, E.N. 10, 2686-953 Sacavém, Portugal

Beaufort Marie-France

Institut PRIME - UPR 3346 CNRS, Université de Poitiers-ENSMA, Département de Physique et Mécanique des Matériaux, Bât. SP2MI - Téléport 2, BP 30179, 86962 Futuroscope Chasseneuil Cedex, France

Berg Sören

Solid State Electronics Division, The Ångström Laboratory, Uppsala University, Box 534, 75121 Uppsala, Sweden

Besnard Aurélien

LaBoMaP, Centre Arts et Métiers ParisTech de Cluny, Rue porte de Paris, 71250 Cluny, France

Bodart Franz

Department of Physics, LARN, FUNDP, 61, Rue de Bruxelles, 5000 Namur, Belgium

Borges Joel

Centro de Física, Universidade do Minho, 4710-057 Braga, Portugal

Camelio Sophie

Institut PRIME - UPR 3346 CNRS, Université de Poitiers-ENSMA, Département de Physique et Mécanique des Matériaux, Bât. SP2MI - Téléport 2, BP 30179, 86962 Futuroscope Chasseneuil Cedex, France

Carvalho Pedro

Centro de Física, Universidade do Minho, 4710-057 Braga, Portugal

Chappé Jean-Marie

Institut FEMTO-ST, 32, Avenue de l'observatoire, 25044 Besançon Cedex, France

Cunha Luis

Centro de Física, Universidade do Minho, 4710-057 Braga, Portugal

Depla Diederik

Department of Solid State Sciences, Ghent University, Krijgslaan, 281(S1), 9000 Ghent, Belgium

Espinos J. Pedro

Instituto de Cienca de Materiales de Sevilla, CSIC University Sevilla, Avda Américo Vespucio s/n, 41092 Sevilla, Spain

Eyidi Dominique

Institut PRIME - UPR 3346 CNRS, Université de Poitiers-ENSMA, Département de Physique et Mécanique des Matériaux, Bât. SP2MI - Téléport 2, BP 30179, 86962 Futuroscope Chasseneuil Cedex, France

Feng Li-Ping

State Key Lab of Solidification Processing, College of Materials Science and Engineering, Northwestern Polytechnical University, 710072 Xi'an, People's Republic of China

Fenker Martin

FEM – Forschungsinstitut Edelmetalle und Metallchemie, Katharinenstraβe 17, 73525 Schwäbisch Gmünd, Germany

Girardeau Thierry

Institut PRIME - UPR 3346 CNRS, Université de Poitiers-ENSMA, Département de Physique et Mécanique des Matériaux, Bât. SP2MI - Téléport 2, BP 30179, 86962 Futuroscope Chasseneuil Cedex, France

Huang Jia-Hong

Department of Engineering and System Science, National Tsing Hua University, 101, Kuang Fu Road, Section 2, Hsinchu, Taiwan, R.O.C

Lintymer Jan

Comadur SA, Le Col 33, 2400 Le Locle, Switzerland

Liu Zheng-Tang

State Key Lab of Solidification Processing, College of Materials Science and Engineering, Northwestern Polytechnical University, 710072 Xi'an, People's Republic of China

Marques Luis

Centro de Física, Universidade do Minho, 4710-057 Braga, Portugal

Martin Nicolas

Institut FEMTO-ST, Département MN2S, 32, Avenue de l'observatoire, 25044 Besançon Cedex, France

Nyberg Tomas

Solid State Electronics Division, The Ångström Laboratory, Uppsala University, Box 534, 75121 Uppsala, Sweden

Paumier Fabien

Institut PRIME - UPR 3346 CNRS, Université de Poitiers-ENSMA, Département de Physique et Mécanique des Matériaux, Bât. SP2MI - Téléport 2, BP 30179, 86962 Futuroscope Chasseneuil Cedex, France

Riviere Jean-Paul

Institut PRIME - UPR 3346 CNRS, Université de Poitiers-ENSMA, Département de Physique et Mécanique des Matériaux, Bât. SP2MI - Téléport 2, BP 30179, 86962 Futuroscope Chasseneuil Cedex, France

Sarakinos Kostas

Plasma and Coatings Physics Division, Linköping University, 58183 Linköping, Sweden

Sthal Fabrice

Institut FEMTO-ST, Département Temps-Fréquence, 26, Rue de l'épitaphe, 25030 Besançon Cedex, France

Terwagne Guy

Department of Physics, LARN, FUNDP, 61, Rue de Bruxelles, 5000 Namur, Belgium

Vaz Filipe

Centro de Física, Universidade do Minho, 4710-057 Braga, Portugal

Vetter Jörg

Sulzer Metaplas GmbH, Am Böttcherberg 30-38, 51427 Bergisch Gladbach, Germany

Wu Yung-Hsien

Department of Engineering and System Science, National Tsing Hua University, 101, Kuang Fu Road, Section 2, Hsinchu, Taiwan, R.O.C

2

CHAPTER 1

Modelling of Reactive Sputter Deposition of Oxynitrides

Sören Berg[1], Tomas Nyberg[1,*] and Diederik Depla[2]

[1]Solid State Electronics Division, The Angstrom Laboratory, Uppsala University, Box 534, 75121, Uppsala, Sweden and [2]Department of Solid State Sciences, Ghent University, Krijgslaan, 281(S1), 9000 Ghent, Belgium

Abstract: Reactive sputter deposition is frequently carried out in a mixture of argon and oxygen or nitrogen to obtain oxides and nitrides. The behavior of such "single-reactive gas" processes has been explained theoretically and verified by numerous industrial thin film deposition applications. However, mixing two reactive gases with the argon sputtering gas in order to carry out reactive sputter deposition of oxy-nitride films is a far more complicated process. A first order simple process model for such a mixed process is presented. Modelling indicates that altering the supply of one gas will not only cause a change of the partial pressure of this gas but also may significantly change the partial pressure of the other reactive gas. Moreover, different reactivities of the reactive gases result in stoichiometries that are very different from the relative reactive gas supplies. This linked behavior between the reactive gases may cause severe process control problems. In a second part of the chapter, a more advanced model is presented, which includes reactive ion implantation and knock-on implantation. First this model is tested *vs.* experimental results published in literature. Based on the good agreement in the noticed trends, the time dependence of the poisoning behavior of the target is discussed. Finally, the influence of the deposition profile on the hysteresis behavior, and the composition of the oxynitride is discussed, showing the importance of a complete model.

Keywords: Sputtering, reactive sputtering, oxynitrides, process modelling, chemical reactivity, target poisoning, process control, hysteresis.

INTRODUCTION

Reactive sputtering is a well known technique to deposit oxides and nitrides from different metal targets. Processes involving a single elemental metal target and a gas mixture of argon and one reactive gas (*e.g.* oxygen or nitrogen) have been

***Address correspondence to Tomas Nyberg:** Angström Laboratory, Uppsala University, Box 534, 751 21 Uppsala, Sweden; Tel: 46-(0)-18-471-3164; Fax: 46-(0)-18-55 50 95; E-mail: tomas.nyberg@angstrom.uu.se extensively studied and well described in the literature. Normally this type of reactive sputtering process exhibits a hysteresis behaviour that may cause complications concerning processing stability. It should be understood that despite the relative complicated processing behaviour the "one metal one reactive gas" sputtering process is the simplest reactive sputtering process to describe. A simple model has been developed that describes the main features of this process [1]. Results from this model may predict the target sputter erosion rate and the partial pressure of the reactive gas as a function of the supply of the reactive gas. Typical results are shown in Figs. **1** and **2**.

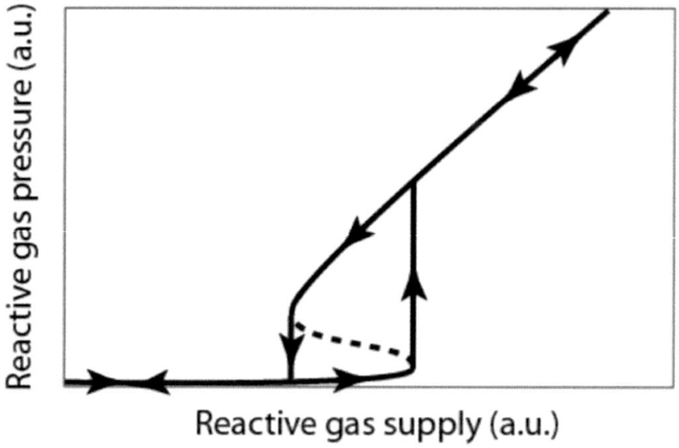

Figure 1: Schematic of processing curve for partial pressure of reactive gas *vs.* gas supply using constant sputtering current.

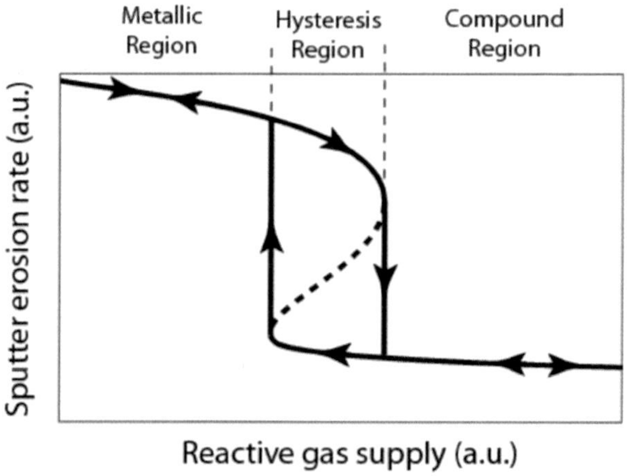

Figure 2: Schematic of processing curve for target sputter erosion *vs.* supply of reactive gas using constant sputtering current.

The S-shaped processing curves can only be obtained if the partial pressure of the reactive gas is used as control parameter in a feedback control system. The solid lines define the processing curves if the supply of the reactive gas will be used as control parameter. The dotted segment will be reached if the partial pressure of the reactive gas is used as control parameter. The width of the S-shaped curves defines the hysteresis width of the process. It should be understood that if the supply of the reactive gas is used as the input control parameter it will not be possible to reach the processing points on the S-shaped curves inside the hysteresis width. This gives a fundamental limit for what compositions that can be obtained for the deposited films.

Replacing the single element metal target with 2 different single element targets or an alloy target consisting of 2 metals significantly changes the processing conditions. Since the two metals generally have different reactivity to the reactive gas it will be necessary to adjust the supply of the reactive gas to a level where the less reactive metal will fully react with this gas. This normally forces the process into a low rate sputtering mode. This low rate may sometimes be as low as one magnitude lower than what is possible to obtain if a single element target was used.

It is also possible to use one single element target and add 2 reactive gases to the argon processing gas. By this technique it is possible to reactive sputter deposit *e.g.* oxy-nitrides. Both the nitrogen and the oxygen have to react with the sputtered metal atoms. Also for this system the difference in reactivity of these gases will give new restrictions in the processing control. In addition a complication in controlling the partial pressures of the 2 reactive gases will appear. Varying the partial pressure of one of these gases will affect the partial pressure of the other gas. Therefore quite advanced feed-back control systems are needed to obtain full control of such a reactive sputtering process [2].

BASIC MODELLING OF SPUTTERING WITH TWO REACTIVE GASES

In this chapter we will describe somewhat in detail the mechanisms responsible for the processing behaviour for a reactive sputter deposition process carried out from sputtering one single element target in a mixture of argon and 2 reactive gases (oxygen/nitrogen). The purpose of the presentation is to clarify what kind of additional complications will arise that will not be present when sputtering one single element target in a mixture of argon and one reactive gas. We believe that the simplest way to illustrate the differences may be to outline a simplified model of the process. From this model it will be possible to predict processing behaviour caused by the different involved processing parameters. As a first approximation we will follow the description earlier presented in reference [1].

We will assume a system consisting of one single element target having an area A_t. Power may be applied to this target generating an argon ion current density J evenly distributed onto the target surface. All sputtered material will be collected and evenly distributed at the collecting surface having area A_c. Two reactive gases (*e.g.* oxygen and nitrogen) will also be present in the chamber having partial pressures p_O and p_N respectively. It will be assumed that these partial pressures are so small so that they do not give rise to any contribution to the ion current bombarding the target. Thus only argon ions are responsible for the sputter erosion from the target. There will be a probability that the gases will react with the pure metal atoms at the A_t and A_c surfaces. These probabilities will be denoted α_O and α_N for oxygen and nitrogen respectively. For simplicity we will assume the same probabilities at both surfaces. It is far too complicated to include all possible chemical reactions that may occur during processing. However, we will consider one possible effect. It is well known that oxygen normally is more reactive than nitrogen. In fact it may be possible that the oxygen also may react with the nitride and convert it into oxide. This can only happen at surfaces where nitrides have been formed. The probability for this to happen will be denoted α_{ON}.

We assume the following to happen in the processing chamber. Due to the presence of oxygen in the chamber some fractions Θ_{tO} and Θ_{cO} of the surfaces A_t and A_c respectively will be covered to oxide. The fractions covered by nitride will be denoted Θ_{tN} and Θ_{cN}.

Based on the assumptions above it is possible to define a number of equations that together determine the expected processing behaviour. The outline in Fig. **3** will serve to illustrate a simplified model of this type of reactive sputtering process.

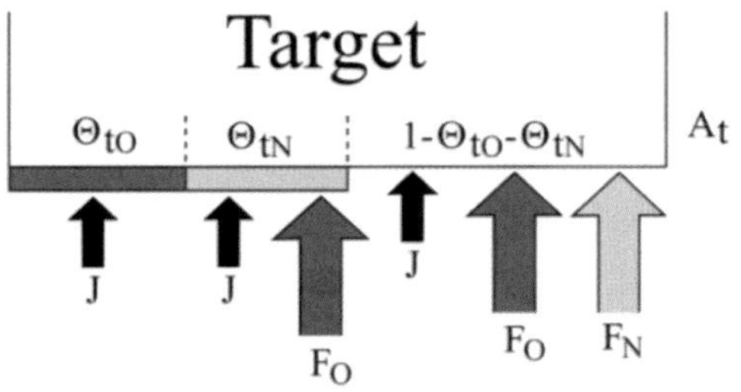

Figure 3: Schematic of incoming gas fluxes and ion current to the different area fractions at the target.

We will assume steady state conditions.

The relation between the partial pressure p and the flux of molecules F (number/ unit area and time) that this pressure generates at all surfaces will be:

$$F_O = \frac{p_O}{\sqrt{2\pi kTm_O}} \tag{1}$$

for the oxygen flux and

$$F_N = \frac{p_N}{\sqrt{2\pi kTm_N}} \tag{2}$$

for the flux of nitrogen molecules.

The flux of sputter eroded metal atoms out from the target will be denoted R^M.

$$R^M = (J/e)Y_m(1 - \Theta_{tO} - \Theta_{tN}) \tag{3}$$

Where J denotes the current density of argon ions and e denoted the elemental electronic charge and Y_m denotes the partial sputtering yield of metal by the argon ions. Oxide and nitride are formed at the target surface. We will assume that sputtering of these compounds will result in erosion of the corresponding molecules. Consequently the flux of sputter eroded oxide molecules denoted R^O will be:

$$R^O = (J/e)Y_O\Theta_{tO} \tag{4}$$

Where Y_O denotes the sputtering yield of oxide molecules. For simplicity we neglect that the oxide molecule may decompose into metal and oxygen atoms. Thus a sputtered oxide molecule will be transferred to the collecting area and be deposited as an oxide molecule. The flux of sputter eroded nitride molecules denoted R^N will be:

$$R^N = (J/e)Y_N\Theta_{tN} \quad \textbf{(5)}$$

where Y_N denotes the sputtering yield of nitride molecules. We assume that also the sputtered nitride molecules will be deposited as nitride molecules.

Since we have assumed steady state conditions it is possible to define a number of balance equations at the target and the collecting surfaces. At steady state there will be a balance for the nitride coverage at the target where the formation of nitride must be equal the removal of nitride. For simplicity we will assume that the metal nitride will have the form M_1N_1 and that the oxide molecule will have the form M_1O_1. This defines the following balance equation for the nitride coverage:

$$(J/e)Y_N\Theta_{tN} + 2\alpha_{ON}F_O\Theta_{tN} = 2\alpha_N F_N(1- \Theta_{tO} - \Theta_{tN}) \quad \textbf{(6)}$$

Where the term $2\alpha_{ON}\Theta_{tN}F_O$ defines the number of nitride molecules at the target surface being converted to oxide molecules by the flux of oxygen molecules to the fraction Θ_{tN} of the target surface already covered by nitride. The factor 2 in the gas terms originates from the fact that one gas molecule contains 2 atoms. Compounds with other stoichiometries should be compensated for by introducing proper constants in the equation.

The corresponding equation for the oxygen coverage at the target surface will be:

$$(J/e)Y_O\Theta_{tO} = 2\alpha_O F_O(1- \Theta_{tO} - \Theta_{tN}) + 2\alpha_{ON}F_O\Theta_{tN} \quad \textbf{(7)}$$

It is possible to define two balance equations at the collecting area A_c, one for the oxide formation and one for the nitride formation. An illustration for the simple model for the collecting area is shown in Fig. **4**.

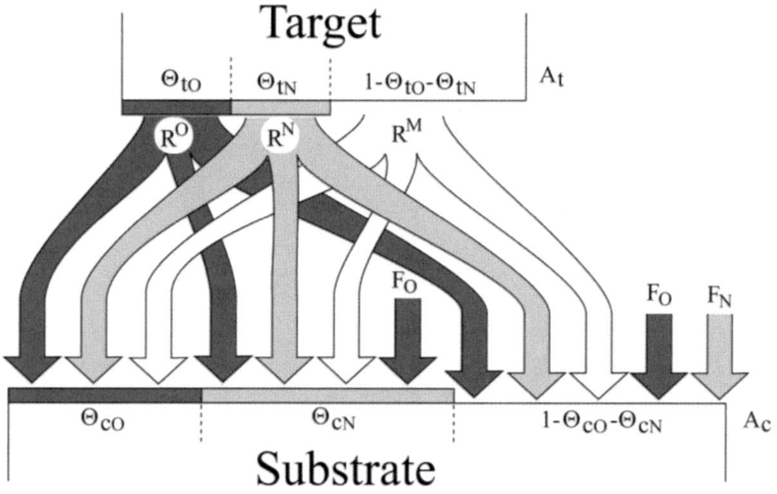

Figure 4: Schematic of fluxes of sputtered particles from target to substrate.

The balance for the nitride formation will be:

$$R^N(A_t/A_c)(1-\Theta_{cN}) + 2\alpha_N F_N(1-\Theta_{cO}-\Theta_{cN}) = (R^M + R^O)(A_t/A_c)\,\Theta_{cN} + 2\alpha_{ON}F_O\Theta_{cN} \quad \mathbf{(8)}$$

And the corresponding equation for the formation of oxide will be:

$$R^O(A_t/A_c)(1-\Theta_{cO}) + 2\alpha_O F_O(1-\Theta_{cO}-\Theta_{cN}) + 2\alpha_{ON}F_O\Theta_{cN} = (R^M + R^N)(A_t/A_c)\,\Theta_{cO} \quad \mathbf{(9)}$$

In addition to these balance equations it is possible to calculate the total consumption Q_O and Q_N of oxygen and nitrogen respectively for every value of the partial pressures of the reactive gases.

$$Q_O = Sp_O + 2\alpha_O F_O(1-\Theta_{cO}-\Theta_{cN})A_c + 2\alpha_O F_O(1-\Theta_{tO}-\Theta_{tN})\,A_t + 2\alpha_{ON}F_O(A_t\,\Theta_{tN} + A_c\,\Theta_{cN}) \quad \mathbf{(10)}$$

$$Q_N = Sp_N + 2\alpha_N F_N(1-\Theta_{cO}-\Theta_{cN})\,A_c + 2\alpha_N F_N(1-\Theta_{tO}-\Theta_{tN})\,A_t - 2\alpha_{ON}F_O(A_t\,\Theta_{tN} + A_c\,\Theta_{cN}) \quad \mathbf{(11)}$$

where the terms Sp_O and Sp_N denotes the throughput of the gases to the vacuum pump having a pumping speed S. It is also possible to obtain an expression for the target sputter erosion rate R. We will define the sum of the sputtered atoms and molecules as a measure of the target sputter erosion rate R.

$$R = R^M + R^O + R^N \quad \textbf{(12)}$$

In this simplified presentation R is a direct measure of the sputter eroded target metal atoms. From the defined balance equations above it is possible to calculate both the target sputter erosion rate R and the partial pressures p_N and p_O of the reactive gases as a function of Q_O for different constant supplies of nitrogen Q_N.

Fig. **5** shows a typical behaviour for the partial pressures p_O and p_N respectively as a function of Q_O for a constant supply of nitrogen Q_N. In this calculation we have assumed that $\alpha_{ON} = 0.3$.

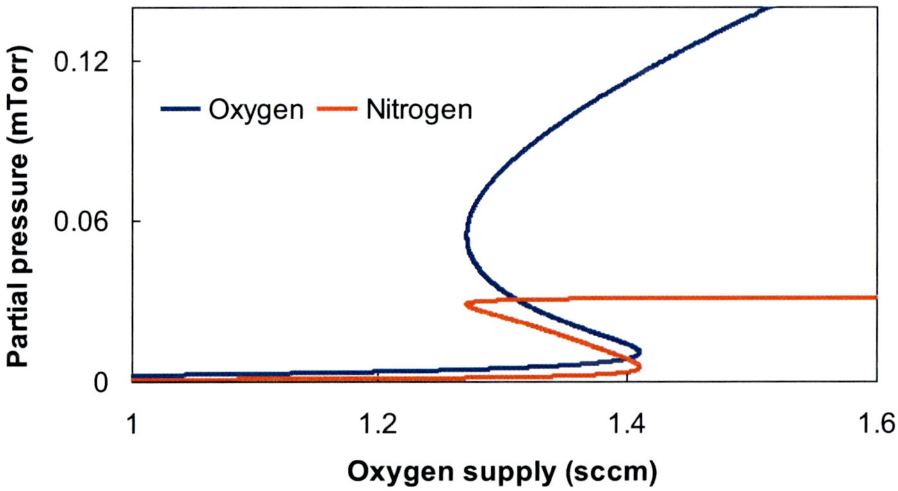

Figure 5: Calculated partial pressures of oxygen and nitrogen *vs.* supply of oxygen for a constant supply of nitrogen.

It should be noticed that the partial pressure of nitrogen p_N is significantly affected by the supply of oxygen. The reason for this is of course that the added oxygen will contribute to an increase of poisoning of the target resulting in a lower sputter erosion rate. This will of course decrease the nitrogen consumption and consequently show up as an increase of p_N.

Fig. **6** shows the sputter erosion rate R as a function of the oxygen supply Q_O for different supplies Q_N of nitrogen. Also here we assume that $\alpha_{ON} = 0.3$ but a somewhat lower Y_N than in Fig. **5**.

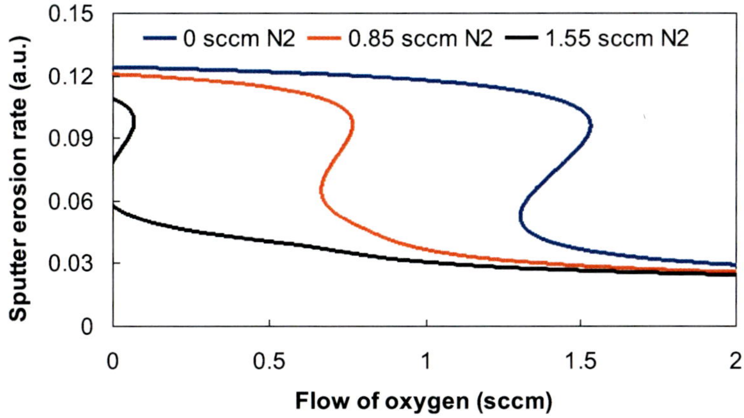

Figure 6: Calculated target sputter erosion rates *vs.* oxygen supply for three different constant supplies of nitrogen.

The results point out that it is possible to be trapped in the poisoned mode if the supply of nitrogen is sufficiently large. In this situation it is not possible to return to the starting point (zero oxygen supply) without also decreasing the supply of nitrogen. This is a serious limitation that may create significant problems for the process control system.

The relation between the sputter yields of the nitride, oxide and metal will significantly influence the process behaviour. Results from calculations with the same parameters as in Fig. **5** are shown in Fig. **7**.

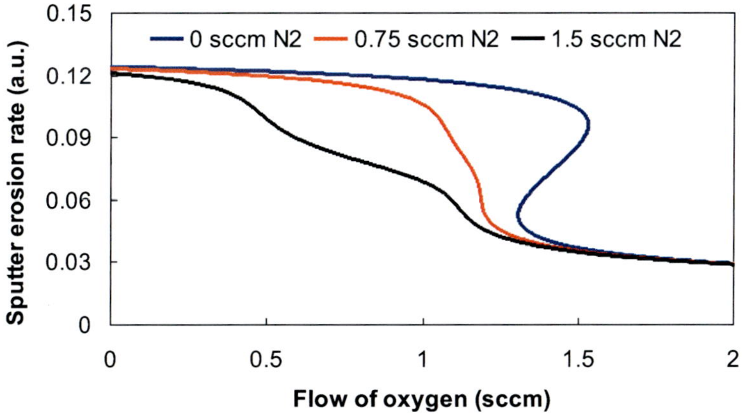

Figure 7: Calculated target sputter erosion rates for three different constant supplies of nitrogen.

It should be noticed that for this parameter combination the hysteresis will decrease as the supply of nitrogen is increased and as shown here eventually totally disappear. This prediction has been verified experimentally for reactive sputtering of Zr in a mixture of nitrogen and oxygen [3]. For metals exhibiting this behaviour it may be difficult to form oxynitrides by reactive sputtering.

It should be understood that all the above outlined calculations are based on quite crude simplifications. Despite this it has turned out that calculated predictions surprisingly well give the same general processing behaviour as observed experimentally. However, this is only true for steady state processing conditions. Investigating dynamic behaviour needs a more detailed treatment of some additional phenomena involved in the target sputter erosion mechanism.

ADVANCED STRATEGIES FOR REACTIVE SPUTTERING MODELLING

The straightforward model described in the previous section enables the fast calculation of different process parameters (such as deposition rate, reactive gas partial pressure) as a function of the reactive gas flow rates during sputter deposition of oxynitrides. In this way one can get a good, but qualitatively, idea of the process behaviour. This flexibility can however only be achieved by a simplified description of the processes occurring at the target and the substrate. For a steady state calculation, as shown above, these approximations have a limited impact on the shape of processing curves and the experimental curves can be mimicked by the model. One must however realize that the obtained parameters generally can only have a qualitative meaning. More important, to reach a more quantitative model a better description of the process kinetics is needed. For a one reactive gas/one target combination the target condition is described by the simple model as,

$$n_{o,s}\frac{d\Theta_{tO}}{dt} = 2F_O\alpha\left(1-\Theta_{tO}\right) - I_d Y_{cO}\Theta_{tO} \tag{13}$$

with $n_{o,s}$ the target atomic surface density. The equation describes the balance between compound formation by chemisorption (first term) and the compound sputtering (second term). Based on equation (13), setting F equal to zero, one can calculate a time constant for the sputter cleaning of a fully poisoned target.

$$\tau = \frac{n_{o,s}}{I_d Y_{cO}} \quad \textbf{(14)}$$

This gives a value of 0.06 s for the sputter cleaning of a fully poisoned Ta target (Y_c = 0.18, I_d 1.8×10^{17} ions.cm^{-2}.s^{-1} (0.03 A/cm^2) and $n_{o,s}$ 2×10^{15} cm^{-2}). It is common knowledge that the sputter cleaning of a poisoned target [1] takes typically a second or more, indicating that the compound layer is several monolayers thick. An example of the sputter cleaning of a poisoned tantalum target is shown in Fig. **8**.

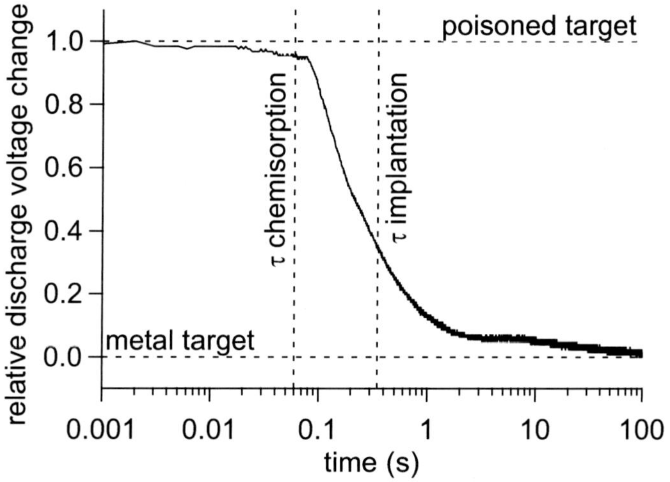

Figure 8: Sputter cleaning of a poisoned Ta target. Prior to the sputter cleaning process the target was sputtered in pure oxygen (0.3 Pa, discharge current 0.3 A) until a stable discharge voltage was reached. After the poisoning process the oxygen gas was replaced by pure argon (0.3 A) and the discharge voltage was recorded during the sputter cleaning process (discharge current 0.3 A). The calculated time constants (τ) for the sputter cleaning of an oxide layer formed by chemisorption (1 monolayer) and an oxide layer formed by direct ion implantation (thickness 2 nm) are indicated.

One can of course "fix" the time constant by increasing the number of monolayers in equation (14) by changing the value of the target atomic surface density, or including more monolayers. However, there is a fundamental problem with this approach. Generally speaking, chemisorption is a fast process for the oxidation of the first monolayer, but the formation of a thicker compound layer is too slow to compete with the sputtering process. This simple reasoning shows that there is a need for extra oxidation processes. One, suggested by D. Depla *et al.* is reactive ion implantation [4]. The formed reactive ions in the plasma are accelerated towards the negative target, and can be implanted in the target. For the typically discharge voltage used during magnetron sputtering, the implantation depth is in the order of 2 nm. The thicker layer gives rise to a time constant, for the same example, of 0.35 seconds, which corresponds with the sputter cleaning experiments (see Fig. **8**). A second oxidation process for the target was suggested by Berg *et al.*: knock-on implantation of chemisorbed species into target. TRIDYN simulations show that chemisorbed oxygen atoms can be knock-on implanted into the target by the bombardment with inert gas ions [5]. The current status of modelling includes therefore three target processes: i) chemisorption, ii) knock-on implantation and direct reactive ion implantation. A model describing the influence of these processes on the reactive behaviour has been published [4] using one reactive gas, and the text below takes the model a step further, *i.e.,* the implementation of these processes for two different reactive gases as in the case of oxynitride deposition.

Although the focus of reactive sputter modelling has been for a long time on the target processes, the importance of the substrate processes may not be neglected. The shape of the hysteresis behaviour is influenced by the deposition profile, and hence to reach a better fit between model and experiment, the deposition profile should be included. Simulation of deposition profiles can be performed in an analytical way, but using Monte Carlo (MC) based particle trajectory codes, the same goal can be reached. Analytical methods are typically valid for a given set-up, and these methods do not include the details of the deposition set-up. This kind of problems can be tackled using the MC method. An example of such calculation will be discussed in the following paragraphs. Including the deposition profile does not influence the fundamental description of the substrate processes. Most authors describe the compound formation on the substrate by a chemisorption process. The large discrepancy between published sticking coefficient for chemisorption of oxygen on metals, and measured values of the sticking coefficient during deposition shows that the compound formation is better described as the incorporation of the reactive gas in the growing layer [6, 7]. The impact of this result is not clear yet. Nevertheless, the last results just show that the modelling of reactive magnetron sputtering is still a vivid research topic.

The Target Processes

The different target processes are schematically shown in Fig. **9**.

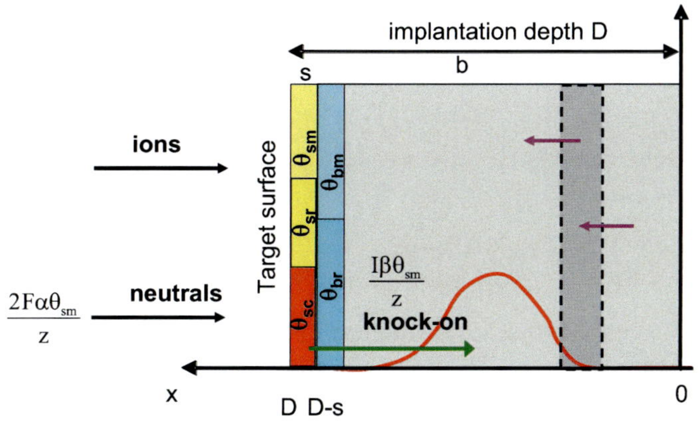

Figure 9: The different target processes during reactive magnetron processes. The target is characterized by three fractions, *i.e.,* θ_m, θ_r and θ_c.To distinguish with the first bulk layer these surface fractions are indicated in the figure by θ_{sm}, θ_{sr} and θ_{sc}. The latter fraction is related to chemisorption and is only present at the surface. Reactive ions and knock on implanted chemisorbed species are implanted into the target according an implantation profile. The transfer due to the target erosion is depicted schematically be the right to left arrows.

For one target/one gas combination the three target processes have been described before [4], and the derived equations, in a somewhat different form, are given below.

$$
\begin{aligned}
&n_{o,s}\frac{d\theta_m}{dt} = v_s n_m^{i=2} - I_d Y_m \theta_m - 2F\alpha_t\theta_m + I_d\beta\theta_c && (a)\\
&n_{o,s}\frac{d\theta_r}{dt} = v_s\left(n_o - n_m^{i=2}\right) - I_d Y_r \theta_r && (b)\\
&n_{o,s}\frac{d\theta_c}{dt} = 2F\alpha_t\theta_m - I_d\beta\theta_c - I_d Y_c\theta_c && (c)\\
&\frac{dn_m^i}{dt} = v_s\nabla \bullet n_m^i - kn_m^i n^i && (d)\\
&\frac{dn_f^i}{dt} = v_s\nabla \bullet n_f^i - kn_m^i n_f^i + \left(2f + \frac{\beta}{z}\theta_{cs}\right) I_d p(x) && (e)
\end{aligned}
\qquad (15)
$$

The first three equations (a), (b), and (c) describe the target surface condition. The target surface condition is described by three fractions: i) the metal fraction θ_m, ii) the compound formed by implanted reactive ions θ_r, and iii) the compound formed by chemisorption of reactive gas molecules θ_c. Equation (a) describes the time dependence of the metal surface fraction θ_m. The surface metal fraction is defined by the transfer of metal from the bulk to the surface. This transfer process is defined by the erosion rate v_s, and the concentration of metal in the subsurface layer (n_m for layer I = 2, in Fig. **9** indicated as θ_{bm}). The sputter process removes metal atoms from the surface towards the vacuum. The flux of sputter metal atoms is calculated from the ion current density I_d, the metal sputter yield Y_m, and the metal surface fraction θ_m. Another process which reduces the surface metal fraction is the chemisorption of reactive gas molecules. The latter process is defined by the flux of reactive gas molecules F, the sticking coefficient α_t, and the surface metal fraction. This process is balanced by the knock-on ion implantation of the chemisorbed species. By the knock-on process the compound formed by chemisorption is converted back into metal. The last term of equation (a) describes this process. Besides the ion current density and the compound fraction formed by chemisorption θ_c, the knock-on yield β defines this process. The second surface fraction, *i.e.,* the compound fraction formed by ion implantation θ_r, is described by equation (b). The equation contains only two terms: the transfer of compound from the bulk towards the surface (see Fig. **9**, θ_{br}), and the sputter removal of the compound with Y_r the compound sputter yield. The last surface fraction, *i.e.,* the compound fraction formed by chemisorption, is described by equation (c). This equation is given by the balance between chemisorption (see description of equation (a)), the knock-on implantation (see description of equation (a)), and the sputter removal of the compound with Y_c the compound sputter yield.

Equations (d) and (e) describe the bulk processes. To describe these processes the target is subdivided in i layers. Material (compound and metal) is transferred from one cell to another by the erosion process, which is described by the first term in both equation (d) and equation (e). This transfer process is defined by the erosion rate v_s. The erosion rate is defined by the target surface condition,

$$v_s = \frac{\left(Y_m\theta_m + Y_c\theta_c + Y_r\theta_r\right)I_d}{n_o} \quad \textbf{(16)}$$

i.e., the target sputter yield is weighted according to the fraction of the different surface species.

The chemical reaction between the implanted reactive atoms is given by the second term in both equation (d) and equation (e). The reaction rate constant k defines the rate of the chemical reaction. The last term in equation (e) describes the direct ion implantation and the knock-on implantation. It is assumed in a first approximation that the implantation profile p(x) is identical for both mechanisms and that the implantation profile does not change as a function of the target oxidation. The number of reactive gas atoms implanted in the target depends on the reactive gas fraction of the ion current. It is assumed that this fraction is equal to the reactive gas fraction in the plasma. The knock-on implantation process has been described above.

In reference [4] this model was used to explain several aspects of the reactive gas process for one target/one gas combination. When a second reactive gas is introduced in the vacuum chamber, one can wonder about the target processes. Of course more equations will be needed to describe the presence of the second reactive gas, but in essence little will change.

$$
\begin{aligned}
& n_{o,s}\frac{d\theta_m}{dt} = v_s n_m^{i=2} - I_d Y_m \theta_m - 2F_O\alpha_{tN}\theta_m - 2F_O\alpha_{tO}\theta_m + I_d\beta_O\theta_{cO} + I_d\beta_N\theta_{cN} && (a)\\
& n_{o,s}\frac{d\theta_{rN}}{dt} = v_s n_{rN}^{i=2} - I_d Y_{rN}\theta_{rN} && (b)\\
& n_{o,s}\frac{d\theta_{cN}}{dt} = 2F_N\alpha_{tN}\theta_m - I_d\beta_N\theta_{cN} - I_d Y_{cN}\theta_{cN} && (c) \qquad \mathbf{(17)}\\
& \frac{dn_m^i}{dt} = v_s\nabla\bullet n_m^i - k_O n_m^i n_{fO}^i - k_N n_m^i n_{fN}^i && (d)\\
& \frac{dn_{fN}^i}{dt} = v_s\nabla\bullet n_{fN}^i - k n_m^i n_{fN}^i + \left(2f_N + \frac{\beta}{z}\theta_{cN}\right) I_d p_N(x) && (e)
\end{aligned}
$$

Similar equations as in equation (15) can be written down. Equations (17) show the changes needed to describe the interaction of the target with the second reactive gas. As shown in the previous section, the less stable compound (generally the nitride) can be converted into the more stable compound (generally the oxide). The description of this process brings two additional terms in these equations. The first is the chemical reaction between the formed compound n_{rN} with the implanted species n_{fO}, described as $k_c n_{rN} n_{fO}$ with k_c the reaction rate constant for this process. The second term is similar as described in the previous section, *i.e.*, the chemisorption of the first reactive gas (oxygen) on the formed compound, *i.e.*, $2F_O\alpha_{ON}\theta_{cN}$ with α_{ON} the conversion coefficient from oxide to nitride.

Before studying the time dependence of the poisoning process (see section 3.2), the model was tested by comparing it with published experimental measurements [8]. It is not the goal to completely fit the experimental results, but just to see if the simulated results follow the same trend as the experiment. In the paper of Martin *et al.* [8] a titanium target is sputtered in two reactive gasses, *i.e.*, oxygen and nitrogen, to deposit an oxynitride coating. Beside the trapping effect discussed in the previous section, they also show some 2D plots indicating three regions as a function of the two reactive gas flows. These three regions correspond with the metal mode, the instability region, and the poisoned mode. A synergy between oxygen and nitrogen has been observed for these 2D diagrams since the boundary delimiting the metallic mode to the complete poisoning of the target is linear and only depends on the oxygen and nitrogen flow rates. This is an interesting experimental result to compare with the simulations. To simulate such a result experimental parameters should be given. This would mean a detailed analysis and dedicated experiments to retrieve these parameters. Instead of this approach, we have reduced to problem to its essence, *i.e.*, the reactive sputtering of a target material in a mixture of two reactive gasses with one gas more reactive than the other. Hence, the reaction rate constant for the first gas is set at $5\text{x}10^{-23}$ $cm^3.s^{-1}$ while for the second gas its value was five times lower, *i.e.*, $1\text{x}10^{-23}$ $cm^3.s^{-1}$. A same reasoning was made for the sticking coefficients. For gas 1 the sticking coefficient on the target and substrate was set at 0.15 while 0.05 was used for the second gas. As it is known that the sputter yield of nitrides is higher than oxides, compound 1 has a lower sputter yield (0.05) than compound 2 (0.1). The implantation of oxygen and nitrogen in a metal results in a very similar implantation profile and therefore the parameters defining the Gaussian implantation profile were equal for both gases (2 nm for the implantation depth, and 1 nm for the ion straggle). The same value for the knock on yield was used based on the same reasoning (β = 0.3). The other experimental parameters are given in the caption of Fig. **10**.

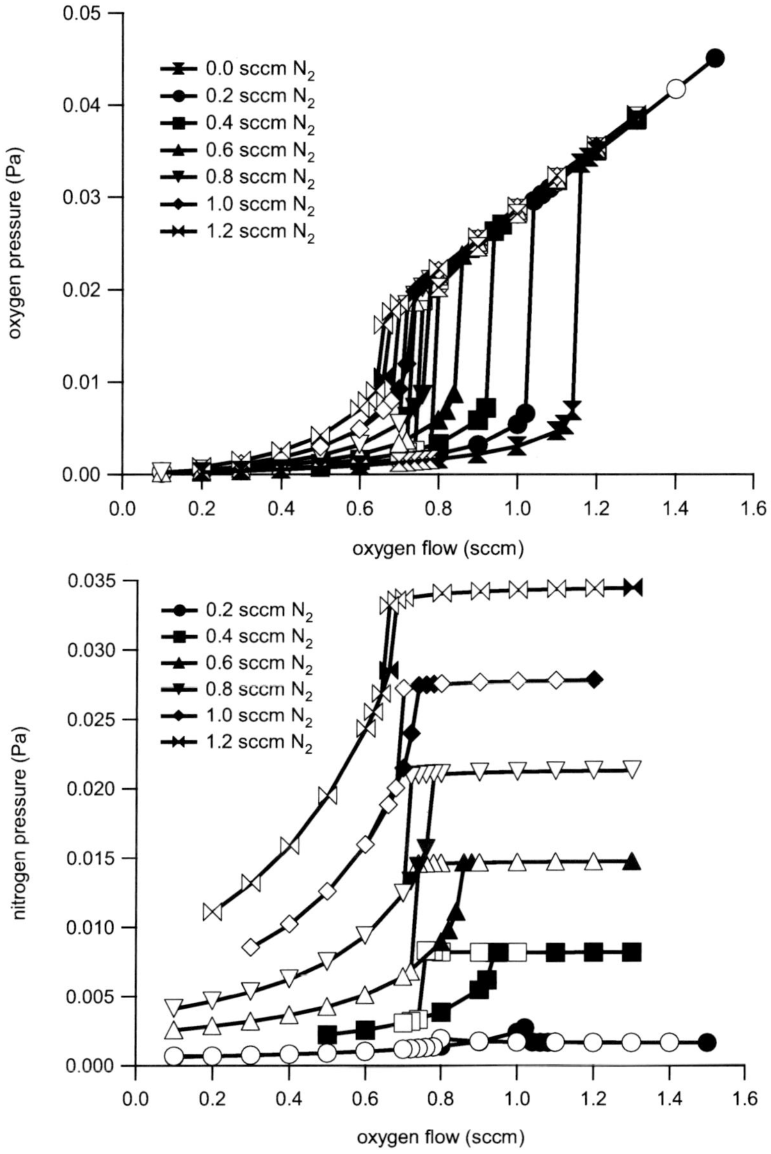

Figure 10: Hysteresis behaviour of the oxygen and nitrogen pressure as a function of the oxygen flow. Simulation parameters: I = 0.4 A, target size =10 cm^2, S (pumping speed) = 50 L/s, Substrate area = 1000 cm^2.

Fig. **10** shows the hysteresis behaviour of the reactive gases as a function of the oxygen flow rate for different fixed values of the nitrogen flow rate. The addition of nitrogen to the plasma results in a shift of the critical point towards lower oxygen gas flows, and a narrowing of the hysteresis. A better way to describe this behaviour using a 2D plot indicating the critical point as function of the two reactive gas flows. This is shown in Fig. **11**.

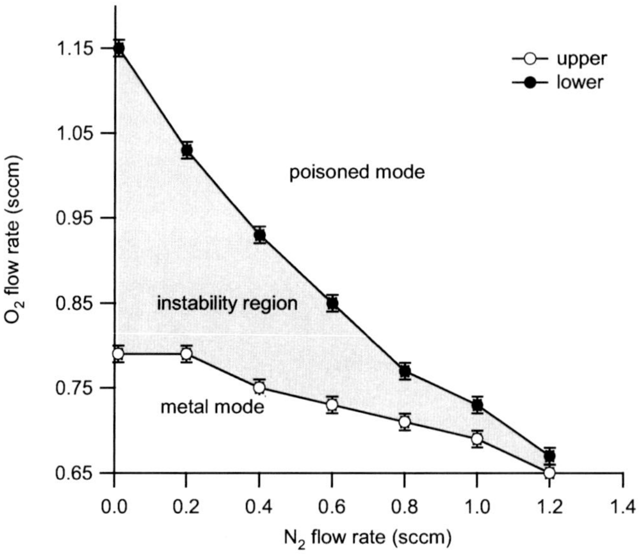

Figure 11: Critical point as a function of the oxygen and nitrogen flow for the hysteresis shown in Fig. **10**.

An almost linear behaviour is found similar to the experiments by Martin *et al.* [8]. Also the narrowing of the hysteresis by increasing the N_2 flow rate was shown by Martin *et al.* [8]. In summary, the experimental tendency can be easily mimicked by the experiment. In the following section this linear behaviour can be understood when studying the time dependence of the poisoning process.

The Time Dependence of the Poisoning Process

In Fig. **12** the simulated time evolution of the poisoning process is shown for the same experimental parameters as in Fig. **11**. First the target is sputtered in a mixture of argon and nitrogen. In Fig. **12** this is indicated by the first grey zone. The target condition is hardly affected by the presence of nitrogen. Indeed, θ_m the metal fraction remains almost equal to one, *i.e.,* a pure metal target. The reason for this behaviour can be explained by the getter action of the deposited metal. Indeed, the partial pressure remains quite low, and the nitride substrate fraction (right figure) increases up to approximately 0.4. After 30 seconds the oxygen flow rate is increased. Target poisoning sets immediately in, and after 100 seconds the target is completely poisoned.

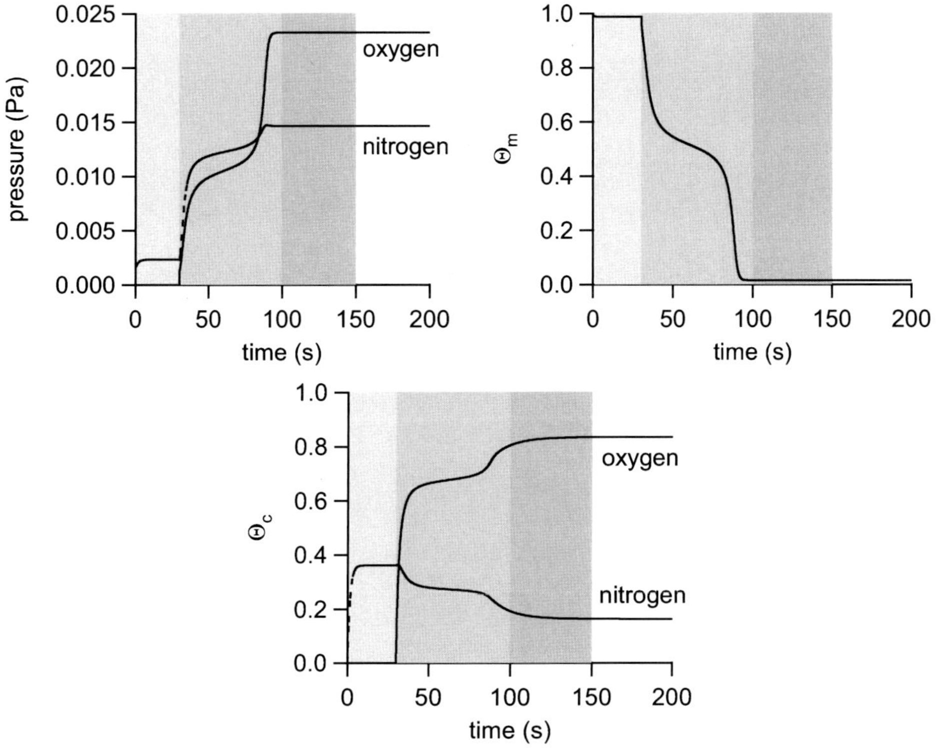

Figure 12: Time evolution of the partial pressure of nitrogen and oxygen (figure left). Time evolution of the target metal fraction (figure middle). Time evolution of the fraction nitride and oxide on the substrate (figure right).

From this behaviour one can understand the almost linear behaviour of the critical points as a function of the nitrogen flow (see Fig. **11**). Although the target is not affected by the nitrogen gas, a substantial fraction of the deposited metal is converted into nitride. This limits the getter ability for oxygen. Hence, as less oxygen can be consumed by the deposited metal, poisoning starts at a lower oxygen flow. After complete poisoning of the target (θ_m = 0, 100 seconds), one notices a further change of the substrate condition as indicated by the third grey zone in Fig. **12**. The target condition changes faster than the substrate condition, and therefore the substrate condition changes in a minor way after complete target poisoning.

The Influence of the Substrate

The equations describing the substrate condition (equations (9) and (10)) assume a uniform deposition profile. The model can be improved by using a more realistic deposition profile. Hence equations (9) and (10) should be modified to

$$\begin{aligned} &\psi_i R^N\left(1-\theta_{cN}^i\right)A_t + 2\alpha_N F_N\left(1-\theta_{cO}^i-\theta_{cN}^i\right) = \psi_i\left(R^m+R^O\right)A_t\theta_{cN}^i + 2\alpha_{ON}F_O\theta_{cN}^i \\ &\psi_i R^O\left(1-\theta_{cO}^i\right)A_t + 2\alpha_O F_O\left(1-\theta_{cO}^i-\theta_{cN}^i\right) + 2\alpha_{ON}F_O\theta_{cN}^i = \psi_i\left(R^m+R^N\right)A_t\theta_{cO}^i \end{aligned} \quad \textbf{(18)}$$

with ψ_i the deposition profile, given by

$$\sum_i \psi_i dA_{c,i} = 1 \quad \textbf{(19)}$$

With $dA_{c,i}$ the area of the substrate element i. In agreement with equations (9) and (10), $\psi_i = 1/A_c$ for each substrate element when a uniform deposition profile is assumed.

Measuring the deposition profile can be quite complicated when the vacuum chamber has a complex design. Hence, it is easier to simulate the deposition profile. This can be achieved using the particle trajectory code SIMTRA, developed by Van Aeken *et al.*; the details of this Monte Carlo code are described in reference [9]. At present the code does not account for the presence of the reactive gas. Hence, the code is ideal to simulate the deposition profile in metallic mode, but should be handled with a bit more care when used in poisoned mode. Indeed, the code does not account for the scattering of the sputtered material by the present oxygen/nitrogen gas and the possible changes of the emission profile by target poisoning. Even with these two limitations, it is instructive to see the impact of including the deposition profile on the simulated hysteresis behaviour. To simplify the simulation, only one gas is included in the model. Fig. **13** shows a calculated deposition profile.

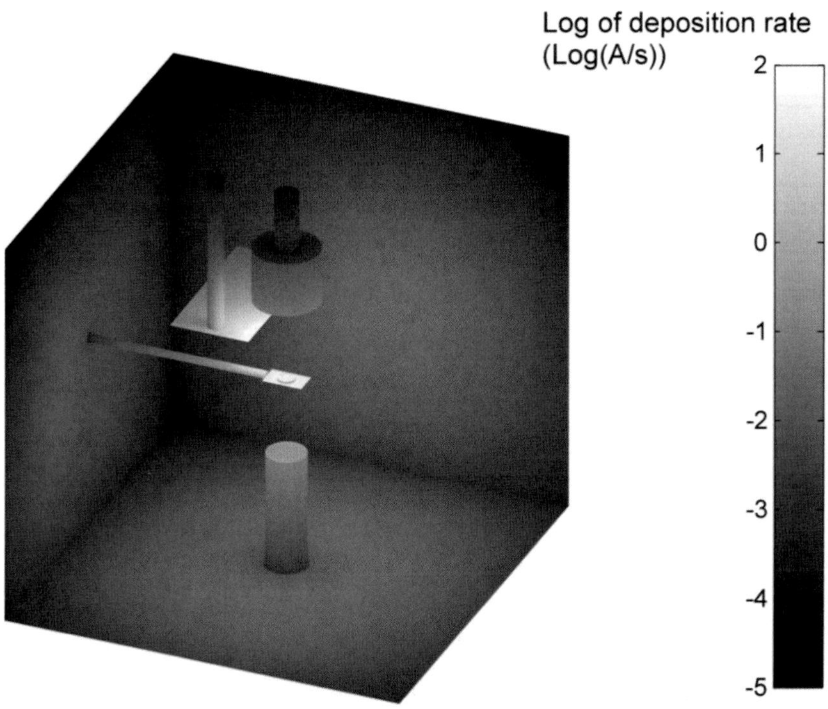

Figure 13: Deposition profile for a cube vacuum chamber (size 0.5 m x 0.5 m x 0.5 m). In the top, one notices the magnetron pointing downwards. At the moment a part of a mass spectrometer is shown. The substrate holder is mounted on a rod pointing from the left to the middle of the vacuum chamber. Finally, a shutter, located between magnetron and substrate, is in open position.

In Fig. **14** the hysteresis behaviour of the oxygen partial pressure is shown for different substrate definitions. It can be clearly seen that for an adequate description of the overall process the deposition profile must be included. The critical oxygen flow rate can be mimicked by using an average and uniform deposition flux. The surfaces with lowest deposition flux (walls, backside of the magnetron) will hardly contribute that the oxygen flow rate as these surfaces will be completely oxidized (θ_c=1) before the critical point is reached. If these surfaces are excluded from the calculation, an average deposition flux of 1/485 cm^{-2} can be calculated. This more or less corresponds with the total calculation (see Fig. **13**). However, it is clear that using a uniform deposition profile it will be hard to fit an experimental curve.

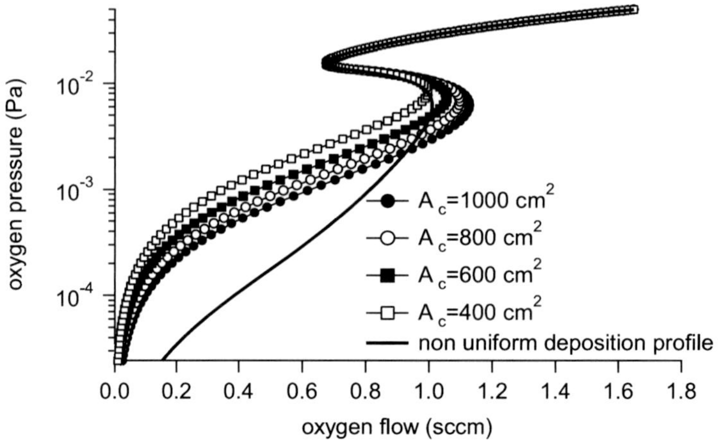

Figure 14: Simulated hysteresis behaviour of the oxygen partial pressure. The influence of the substrate description is shown. The full line corresponds with a calculation based on a non uniform deposition profile based on SIMTRA calculations (see Fig. **13**). The other, marked lines are the result of a calculation based on a uniform deposition profile with a substrate A_c as indicated. The simulation parameters were identical to the parameters related with oxygen as discussed in the previous section.

The properties of oxynitrides depend strongly on their chemical composition. Hence, one can wonder if this chemical composition depends on the deposition flux, or stated differently on the position in the vacuum chamber even when the gas is uniformly distributed. Solving equation (17), assuming no conversion from nitride to oxide, the ratio can be calculated as,

$$\frac{\theta^i_{cO}}{\theta^i_{cN}} = \frac{R^O\left(\psi_i\left[R^N + R^O + R^M\right] + 2\alpha_N F_N + 2\alpha_O F_O\right) + 2R^M \alpha_O F_O}{R^N\left(\psi_i\left[R^N + R^O + R^M\right] + 2\alpha_N F_N + 2\alpha_O F_O\right) + 2R^M \alpha_N F_N} \quad \mathbf{(20)}$$

and depends on the deposition profile. From this equation two extreme situations can be retrieved. In full poisoning, the metal deposition rate R^M becomes zero and equation (20) reduces to

$$\frac{\theta^i_{cO}}{\theta^i_{cN}} = \frac{R^O}{R^N} \quad \mathbf{(21)}$$

When working in metal mode, *i.e.,* when the target is not poisoned, R^O and R^N are zero, and the composition does not depend on the deposition profile,

$$\frac{\theta_{cO}^{i}}{\theta_{cN}^{i}} = \frac{\alpha_O F_O}{\alpha_N F_N} \tag{22}$$

However, at high pumping speed, the target is generally already partially poisoned before the critical point, and also when working in the transition region by feedback control, one must realize that the composition will depend on the deposition profile. The influence of the deposition flux on the composition is simulated with the same parameters as for Fig. **11**. The result is shown in Fig. **15**. This figure shows the calculated composition as a function of the oxygen pressure.

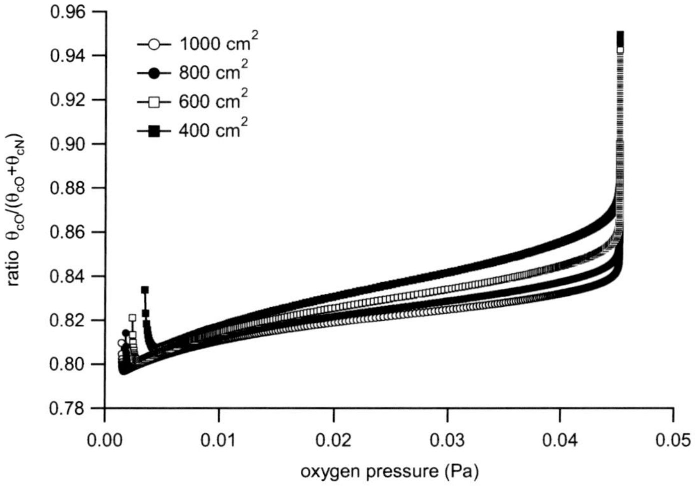

Figure 15: The simulated composition of an oxynitride thin film as function of the oxygen partial pressure. The different traces correspond with a different size of the substrate, which defines the uniform deposition flux. At high pressure, *i.e.,* for a fully poisoned target, the composition does not depend on deposition flux.

CONCLUSION

It is possible to predict trends and general behavior of the reactive sputtering process from calculations that are based on rather basic and simple assumptions. This type of simulations predicts surprisingly well the general processing behaviour observed in experiments. However, results from such basic modelling are of qualitative nature. To be able to quantitatively predict effects in a more precise and accurate way, one has to make use of more complex and advanced modelling. Although several target processes and a more accurate substrate description are included, there is still a further need for a more complete model for reactive magnetron sputtering. At first sight one can interpret the research activities related to this topic as hairsplitting. However, there are still some fundamental questions which are more than interesting.

ACKNOWLEDGEMENT

Declared none.

CONFLICT OF INTEREST

The author(s) confirm that this chapter content has no conflict of interest.

REFERENCES

[1] S. Berg, T. Nyberg. Thin Solid Films, vol. 476, pp. 215-230, 2005.
[2] W. D. Sproul, D. J. Christie, D. C. Carter, *et al.* in *46th Annual SVC Technical Conference* (Society of Vacuum Coaters, San Francisco, CA, pp. 98-103, 2003.
[3] D. Severin, O. Kappertz, T. Kubart, *et al.* Appl. Phys. Lett., vol. 88, pp. 161504-3, 2006.
[4] D. Depla, S. Heirwegh, S. Mahieu, R. De Gryse. J. Phys. D: Appl. Phys., vol. 40, pp. 1957-1965, 2007.
[5] D. Rosen, I. Katardjlev, S. Berg, W. Moller. Nucl. Instr. & Meth. B, vol. 228, pp. 193-197, 2005.
[6] W.P. Leroy, S. Mahieu, R. Persoons, D. Depla. Thin Solid Films, vol. 518, pp. 1527-1531, 2009.
[7] W. P. Leroy, S. Mahieu, R. Persoons, D. Depla. Plasma Process. Polym., vol. 6, pp. 342-346, 2009.
[8] C. Rousselot, N. Martin. Surf. Coat. Technol., vol. 142, pp. 206-210, 2001.
[9] K. Van Aeken, S. Mahieu, D. Depla. J. Phys. D: Appl. Phys., vol. 41, pp. 205307-6, 2008.

Metallic Oxynitride Thin Films by Reactive Sputtering and Related Deposition Methods, **2013, 27-50**

CHAPTER 2

Reactive Gas Pulsing Process for Oxynitride Thin Films

Nicolas Martin[1,*], Jan Lintymer[2], Aurélien Besnard[3] and Fabrice Sthal[1]

[1]Institut FEMTO-ST, 32, Avenue de l'observatoire, 25044 Besançon, France; [2]Comadur SA, Le Col 33, CH-2400 Le Locle, Switzerland and [3]LaBoMaP, Centre Arts et Métiers ParisTech de Cluny, Rue porte de Paris, 71250 Cluny, France

Abstract: An original reactive sputtering method, namely the reactive gas pulsing process (RGPP) was developed for the synthesis of titanium oxynitride thin films. Such a method implements a metallic titanium target dc sputtered, a constant supply of argon and nitrogen gases and a pulsing oxygen mass flow rate, which is periodically controlled *vs.* time. Various period times and different patterns can be generated: rectangle, sine, isosceles triangle, mounting or descending triangle and exponential. Real-time measurements of the target potential as well as total sputtering pressure are recorded in order to study the instability phenomena of the process. They are also pertinent diagnostic tools to select the most suitable pulsing patterns required to alternate the process between the nitrided and the oxidized sputtering modes. As a result, alternation is produced for exponential and rectangular patterns. For the latter, the influence of the duty cycle α, defined as the ratio of the injection time of oxygen by the pulsing period, on the behaviour of the reactive sputtering process and optical properties of deposited films, is systematically investigated. Finally, the added value brought by the exponential patterns is examined. It is shown that the exponential pulse leads to significant improvements of the oxygen injection. The purpose is to introduce the right amount of oxygen so as to poison the titanium target surface without saturating the sputtering atmosphere by oxygen. Thus, the speed of pollution of the target surface appears as an appropriate parameter to better understand the beneficial effect of the exponential shape on the control of the RGPP method.

Keywords: Reactive sputtering, reactive gas pulsing process (RGPP), titanium oxynitride, pulse shape, rectangular pulses, exponential pulses, duty cycle, process stability, target poisoning, hysteresis, optical transmittance, sputtering yield, deposition rate, pumping speed, target potential, nitrided sputtering mode, oxidized sputtering modes, pulsing period, multilayer structure, target pollution speed.

INTRODUCTION

Reactive magnetron sputtering process is a widely used technique to deposit many different kinds of compounds. Basically, oxides, nitrides, carbides or sulphides of

***Address correspondence to Nicolas Martin:** Institut FEMTO-ST, 32, avenue de l'observatoire 25044 Besançon, Cedex, France; Tel: +33 (0)3 81 85 39 69; Fax: +33 (0)3 81 85 39 98; E-mail: nicolas.martin@femto-st.fr

Filipe Vaz, Nicolas Martin and Martin Fenker (Eds)

single or several metallic compounds can be produced by this method [1-4]. However, a typical phenomenon of the reactive sputtering process is a hysteresis loop of some experimental parameters as the reactive gas is injected into the chamber. This non-linear evolution of the deposition parameters has been thoroughly investigated and modelled by many researchers [5-12]. Such attractive field of studies is mainly due to the reactive mode as an exciting way to synthesize thin solid films with variable compositions. Nevertheless, because of the non-linear effects *vs.* the reactive gas flow rate, some compositions are quite difficult, or indeed impossible to achieve (*e.g.* metalloid deficient concentrations) without using improved devices [13-15] or feedback control systems [16-18]. Thus, the strong majority of these studies have been focused on the avalanche-like transition between the metallic and the compound sputtering modes involving a single metallic target and only one reactive gas like oxygen, nitrogen, acetylene, *etc.* [19-23]. For the deposition of ternary compounds like metallic oxynitrides, a single metallic target and two reactive gases – oxygen and nitrogen – can be used. Hence, the reactive sputtering process becomes more complex and it restrains the final properties of the films. In order to reach some desired compositions, the conventional way employs a very high nitrogen mass flow rate in comparison to that of the oxygen. Since oxygen exhibits a stronger reactivity than nitrogen with regards to the metal, this classical method limits the range of metalloid concentrations and does not solve the drawbacks of the reactive sputtering mode for the synthesis of oxynitride films.

An original approach was recently proposed by Sproul *et al.* [24, 25] using a feedback control device of oxygen and nitrogen gases by mass spectrometry. It successfully led to the preparation of silicon and titanium oxynitride thin films with adjustable chemical concentrations and a judicious process control. Aronson *et al.* [26] and recently others [27-34] proposed an innovative approach for the deposition of metallic oxides or nitrides in which the reactive gas was pulsed on and off at regular times. Finally, Martin *et al.* [35-39] and afterwards other authors [40-45] developed the reactive gas pulsing process (RGPP) for the deposition of metallic oxynitride thin films.

The motivation of this chapter is a reporting on this RGPP method, especially applied for the synthesis of tuneable titanium oxynitride coatings. A metallic titanium target was sputtered using oxygen and nitrogen as reactive gases. The nitrogen mass flow rate was kept constant whereas that of oxygen was periodically introduced with various period times and different patterns. A strong emphasis was put in the rectangular pulse. A systematic change of the duty cycle α from 0 to 100 % of a constant pulsing period was carried out. Last but not least, the exponential patterns were particularly examined and led to an enhanced control of the oxygen injection, *i.e.,* a better control to introduce the right amount of oxygen gas so as to poison the surface of the target without saturating the sputtering atmosphere by the most reactive gas. From real time measurements of the target potential and sputtering pressure, poisoning of the target surface was discussed. It was correlated with the deposition rates and optical transmittance of the films in order to better understand the reason why RGPP can improve the control of the reactive sputtering for oxynitride thin films.

MATERIALS AND METHODS

Titanium oxynitride thin films were sputter deposited by dc reactive magnetron sputtering. A stainless-steel chamber (60 L) was evacuated with pumping units (turbomolecular pump backed by a mechanical pump) down to a residual vacuum close to 10^{-5} Pa. The metallic titanium target (purity 99.6 at. % and 51 mm diameter) was sputtered with a constant current density J_{Ti} = 100 A m^{-2}. The substrate holder was grounded and maintained at room temperature during the deposition. The target-to-substrate distance was 50 mm. The pumping speed was kept constant at S = 10 L s^{-1}. Argon and nitrogen mass flow rates were maintained at 2.30 and 0.80 sccm, respectively. These operating conditions correspond to argon and nitrogen partial pressures of 0.24 and 0.09 Pa, respectively. Thus, these partial pressures lead to the full nitridation of the titanium target surface. Oxygen was introduced using a home-made computer controlled system, namely the reactive gas pulsing process (RGPP). Such a system allows a well-controlled injection *vs.* time of the oxygen gas according to various patterns and characteristics (cf. next section). Deposition rates were calculated from thickness measurements and the corresponding deposition time. Optical transmittance was measured at 633 nm for coatings deposited on glass substrates with a constant thickness close to 400 nm.

STRATEGY FOR PULSING

Shape of the Pulse

The RGPP system allows an accurate control of the oxygen injection *vs.* time. Various periodic mass flow rates *vs.* time can be generated: rectangle, sine, triangle (isosceles, mounting, descending) and exponential patterns (Fig. **1**).

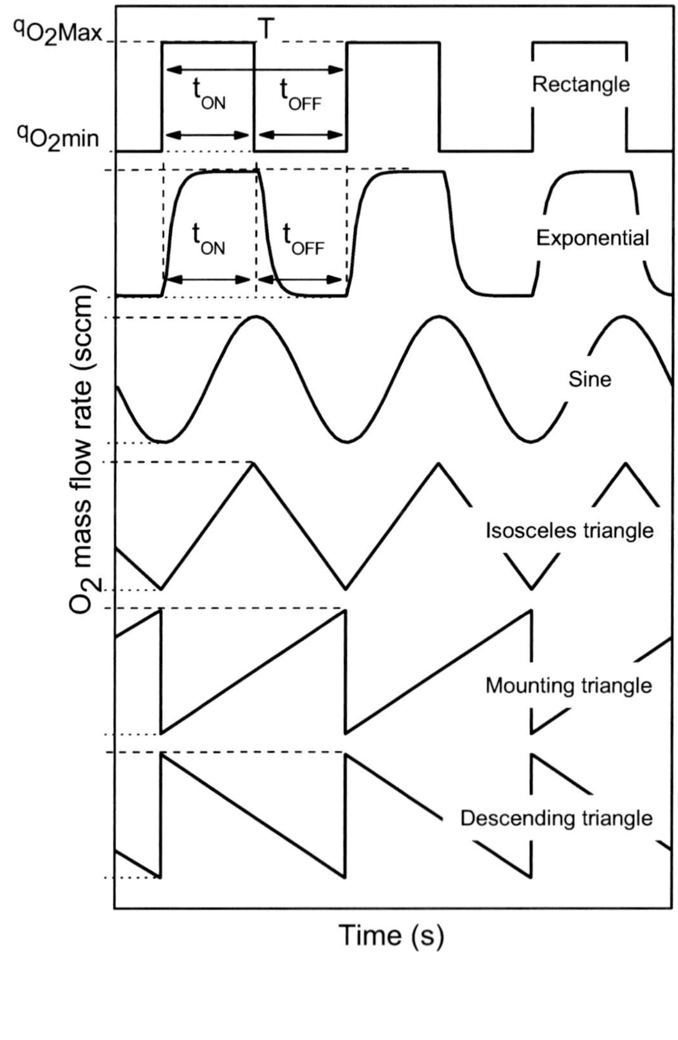

Figure 1: Different periodic shapes of pulsing patterns of the oxygen mass flow rate *vs.* time. Pulsing period T, injection times t_{ON} and t_{OFF} as well as maximal q_{O2Max} and minimal q_{O2min} oxygen mass flow rates can be controlled by RGPP.

Period of pulses T, times of injection t_{ON} and t_{OFF}, as well as maximal and minimal oxygen mass flow rates q_{O2Max} and q_{O2min}, respectively, can accurately be modified. For every deposition, the titanium target was pre-sputtered in argon atmosphere at 0.24 Pa for five minutes. Afterwards, nitrogen was supplied up to a partial pressure corresponding to the nitrided sputtering mode, *i.e.*, N_2 pressure close to 0.09 Pa. The process was stabilised for five minutes in the nitride mode. At last, oxygen gas was introduced according to the different pulsing patterns as illustrated in Fig. **1**. The period was first changed from 10 to 400 s. In addition, for rectangular and exponential pulses, equal t_{ON} and t_{OFF} times were used. For exponential pulses, it must be noted that both τ times were kept constant at τ_{mou} = τ_{des} = 1 s (cf. next section for the description of exponential pulses). The minimal oxygen mass flow rate was fixed at q_{O2min} = 0 sccm in order to completely stop the oxygen injection during the t_{OFF} time. On the other hand, the maximum oxygen flow rate was set at q_{O2Max} = 2.0 sccm. Such a value corresponds to the minimum amount of oxygen required to avalanche the basic Ti – O_2 process in the compound (oxidised) sputtering mode for the same sputtering chamber and operating conditions.

First measurements were focused on the deposition rate as a function of the type of pulsing pattern (Fig. **2**).

It is worth to notice that the rate does not only depend on the pulsing period T, but it is also influenced by the shape of the pulse. One can suggest that the pulses can be divided into two groups. The first one is composed of all triangular pulses whereas the second one puts together rectangular, exponential and sine pulses. For all triangular pulses, the lowest rates are measured from 115 to 145 nm h^{-1} as the pulsing period rises from 10 to 400 s. It is mainly assigned to the trapping of the process in the oxidised sputtering mode. In spite of q_{O2min} = 0 sccm and increasing the triangular pulsing period, the time of nitridation of the target surface associated with the minimal amount of oxygen required to restore the process in the nitrided mode trap the process in the oxidised mode, even for the longest period T of 400 s. As a result, the process remains in the oxidised sputtering mode and transparent titanium oxynitride thin films are always produced for any triangular pattern.

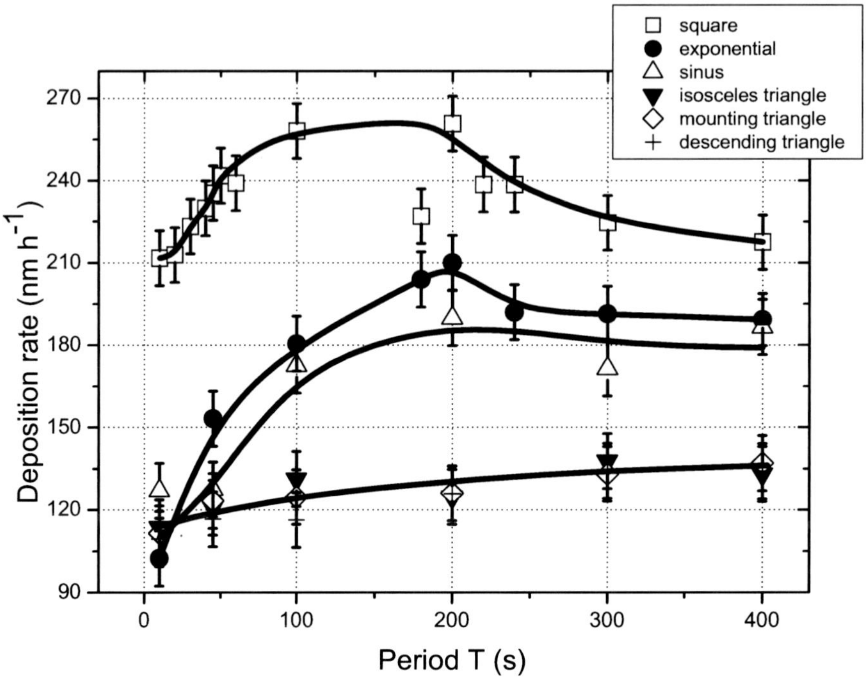

Figure 2: Deposition rate of titanium oxynitride thin films *vs.* the pulsing period T for various pulsing patterns.

For the second group, deposition rates show a very similar evolution *vs.* the pulsing period. A maximum rate is measured for a pulsing period T in-between 150 to 200 s. A significant drop of the deposition rate is noticed when the period decreases and is shorter than 200 s. It is mainly attributed to the alternation of the process between nitrided and oxidised sputtering mode. Such alternation occurs more and more frequently as the pulsing period is reduced. On the other hand, for pulsing periods longer than 300 s, deposition rates of the three types of pulses tend to become constant close to 180 nm h^{-1}. An increasing pulsing period leads to longer time sequences in nitrided or oxidised sputtering modes. Thus, during the t_{ON} time, titanium oxide is mainly deposited whereas during the t_{OFF} time, titanium nitride is produced. Long enough pulsing periods lead to the synthesis of periodic TiN/TiO_2 multilayers and a constant deposition rate.

The highest deposition rate is obtained for rectangular patterns. For this type of pulse, oxygen injection is abruptly changed from 2.0 to 0 sccm at the end of the

t_{ON} time. It means that the process rapidly comes back to the nitrided mode compared to the exponential pulse. For this latter pulse, the oxygen mass flow rate does not change so rapidly *vs.* time when t_{OFF} time starts since τ_{des} = 1 s. Thus, during a single pulsing period, oxidised sputtering mode predominates for the exponential pulse in spite of an alternation of the process between each mode for both pulses. In addition, assuming that the sputtering yield Y of titanium dioxide is lower than that of titanium nitride (Y_{TiO2} = 0.015 and Y_{TiN} = 0.08 for an argon ion energy of 400 eV [46, 47]), rectangular pulses lead to the highest deposition rate.

From optical transmittance at 633 nm of the films deposited on glass substrates (film thickness is 400 nm), one can easily assess if titanium oxynitrides behave as a nitride or an oxide compound (Fig. **3**).

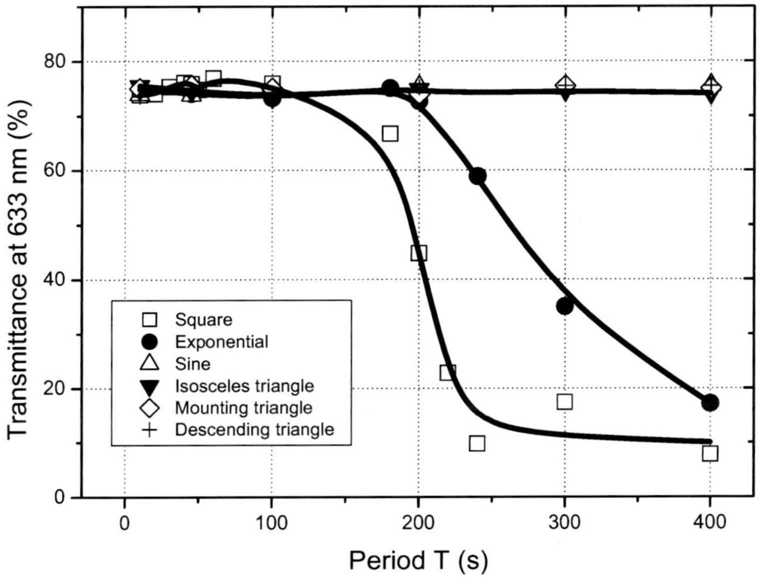

Figure 3: Optical transmittance measured at 633 nm *vs.* pulsing period T for titanium oxynitride thin films (400 nm thick) deposited on glass substrates for various pulsing patterns.

For transparent thin films typical of titanium dioxide material deposited on glass substrates, the optical transmittance at 633 nm is higher than 70 % whereas a transmittance lower than few % or null correlates with absorbent compounds, commonly met in titanium nitride. At first, it is worth to note that optical transmittance remains constant close to 75 % for sine and all triangular patterns and for all pulsing periods. It corresponds to the deposition of transparent TiO_2 films with nitrogen as dopant. These highest optical transmittances correlate well with the lowest deposition rates measured in Fig. **2**. Titanium dioxide films are also obtained with the sine pulse in spite of enhanced deposition rates observed for pulsing periods higher than 200 s. Thus, a fully oxidized state of the target surface is not systematically required to produce transparent titanium dioxide films.

An important drop from 75 to 10 % of the optical transmittance can be noticed with rectangular and exponential pulses when the pulsing period is higher than 200 s. These results illustrate well that rectangular and exponential pulses are the most suitable patterns to produce tuneable titanium oxynitride thin films with behaviours included between those of metallic titanium nitride and dielectric titanium oxide. Such a drop is even more marked for rectangular pulses since this kind of patterns abruptly switches the process between nitrided and oxidized sputtering mode.

In order to understand the pertinence of rectangular and exponential pulses, real-time measurements of the sputtering pressure (Fig. **4**) and titanium target potential (Fig. **5**) *vs.* time were systematically recorded.

The type of patterns can be easily distinguished from the sputtering pressure *vs.* time measurements. It is especially significant for sine and all triangular pulses. Because of the process is completely trapped in the oxidized sputtering mode (there is no t_{OFF} time for these patterns), the target surface is mainly poisoned by the oxygen gas and the sputtering pressure follows well the changes of the oxygen mass flow meter. For ascending and descending triangular patterns, pressure does not accurately show triangular shape at the beginning of the pulsing period. For ascending triangles, the oxygen flow rate is abruptly stopped and the drop of the sputtering pressure depends on the pumping speed and on the capacity of the process to be restored to the nitrided mode. For descending triangles, peaks of the sputtering pressure are not sharp because of the low pumping speed ($S = 10\ L\ s^{-1}$), but also due to the kinetics of the target oxidation.

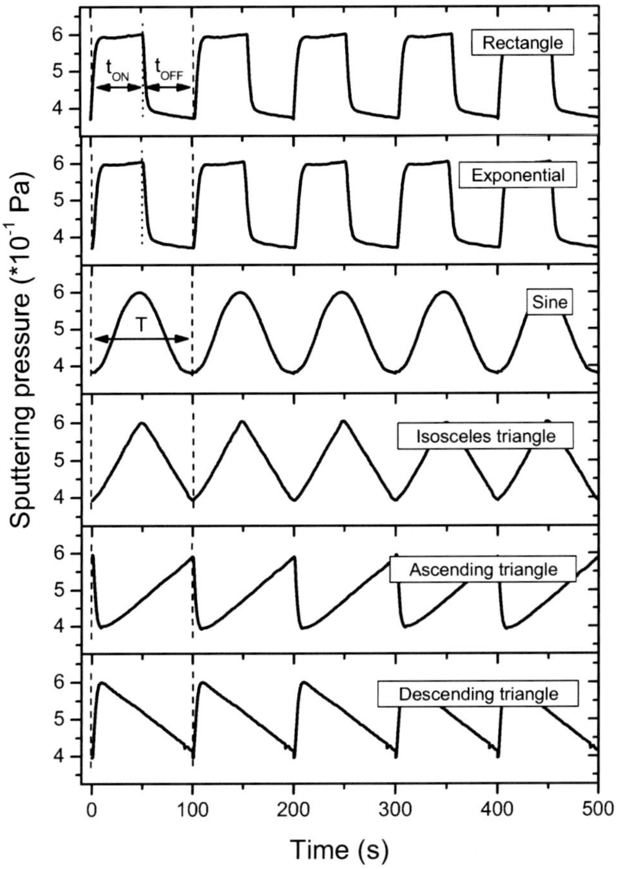

Figure 4: Sputtering pressure *vs.* time for used pulsing patterns and for a constant pulsing period T = 100 s. For rectangular patterns, $t_{ON} = t_{OFF} = 50$ % of the pulsing period T. The same operating conditions were used for exponential pulses with $\tau_{mou} = \tau_{des} = 1$ s.

Sputtering pressure measured for rectangular and exponential patterns exhibit very similar features because the τ times involved in these experiments, lead to oxygen mass flow rates very close to the rectangle ones. At the beginning of the t_{ON} time, pressure abruptly increases from $P_{sput} = 0.33$ to 0.63 Pa and becomes nearly constant after few tens seconds. It correlates with the transition of the process from the nitrided to the oxidized sputtering mode, which depends on the operating conditions such as current density applied to the target, maximum oxygen mass flow rate q_{O2Max}, pumping speed S, *etc.* When oxygen injection is stopped (beginning of the t_{OFF} time), a reverse evolution can be noticed. The sputtering pressure rapidly drops and tends to its original value corresponding to the nitrided mode, *i.e.,* $P_{sput} = 0.33$ Pa. Here again, the abruptness of the pressure decrease depends on the sputtering parameters, especially the pumping speed of the process. Afterwards, the sputtering pressure slightly reduces and tends to the nitrided steady state value ($P_{sput} = 0.33$ Pa). These pressure variations can be correlated with the non-linear evolutions of the target potential (Fig. **5**).

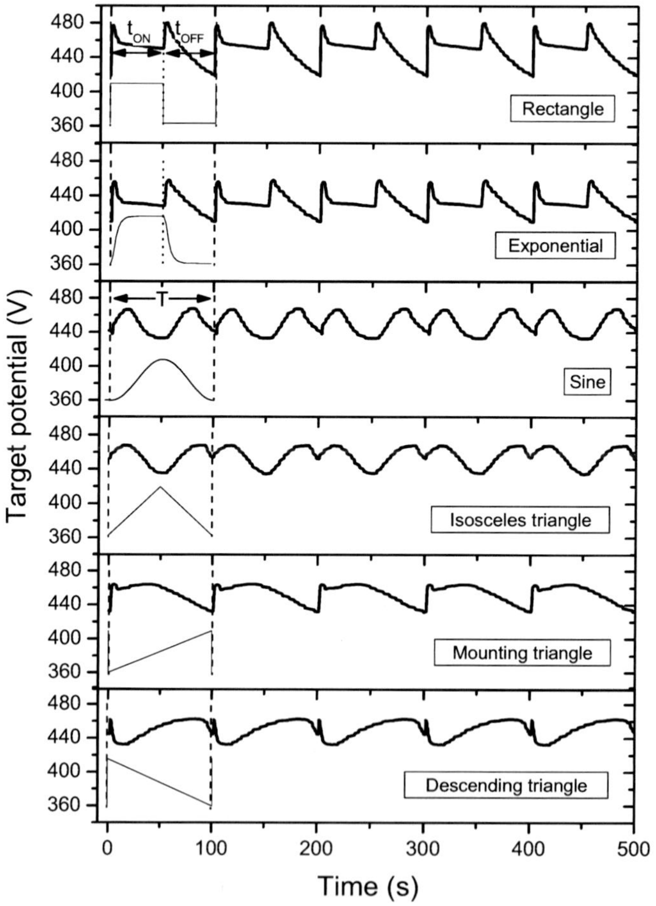

Figure 5: Titanium target potential as a function of time for used pulsing patterns and for a constant pulsing period T = 100 s. For rectangular patterns, $t_{ON} = t_{OFF} = 50$ % of the pulsing period T. The same operating conditions were used for exponential pulses with $\tau_{mou} = \tau_{des} = 1$ s.

Such a potential is a fundamental parameter, which gives important knowledge on kinetics and phenomena occurring on the target surface. For sine and isosceles triangular patterns, the evolution of the target potential *vs.* time is very similar. The process is trapped in the oxidized sputtering mode. For the maximum oxygen flow rate (time corresponding to the half period T), the target potential is minimum ($U_{Ti} < 440$ V) because the process tends to the steady state in the oxidized sputtering mode. As the oxygen mass flow rate drops, the process trends to be restored to the nitrided mode since the potential exhibits a maximum. However, this time spent at a very low or null oxygen introduction is too short. The next period starts and the oxidized state prevails. It is also the case for mounting and descending triangular patterns. The process is mainly in the oxidized sputtering mode in spite of some possible times to alternate with the nitrided one at the beginning and at the end of the pulses for mounting and descending triangular pulses, respectively.

As previously noticed for the sputtering pressure measurements, the target potential *vs.* time curves are very similar for rectangular and exponential patterns. Abrupt changes and peaks of the potential are observed at the beginning of each t_{ON} and t_{OFF} times. It well correlates with the start or stop of oxygen injection. During the t_{ON} time, the target potential tends to be constant and corresponds to the oxidized sputtering mode. The potential exhibit sharp peaks, which correspond to a rapid oxidation of the target surface compared to the broader peaks of potential noticed when t_{OFF} starts. The decrease of potential is slower without reaching a constant value. It means that the process alternates between oxidized and nitrided mode, but nitridation is not completely reached.

Thus, taking into account sputtering pressure in Fig. **4** and target potential measurements in Fig. **5**, the process is not necessarily at the steady state conditions just because the sputtering pressure has reached its steady state values [48, 49]. In addition, these results show that rectangular and exponential pulses appear as the most suitable patterns to get an alternation of the process between the nitrided and oxidized sputtering modes, and consequently to produce titanium oxynitride thin films with tuneable properties. Pulsing parameters have to be adjusted, especially the t_{ON} and t_{OFF} times, *i.e.,* the duty cycle parameter.

Duty Cycle

Exponential and rectangular pulses are the only patterns involving t_{ON} and t_{OFF} times. The previous part of this chapter particularly showed that the periodic injection and stop of oxygen allowed an alternation of the reactive sputtering process between the nitrided and oxidized sputtering modes. Thus, this second part aims at investigating the role of some temporal parameters of RGPP, especially the influence of the duty cycle α for the rectangular pulses, on the behaviours of the reactive sputtering (Fig. **6**).

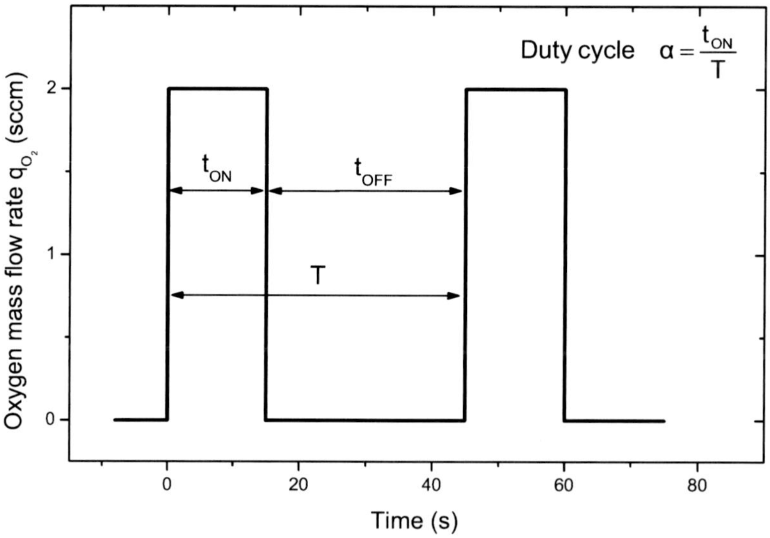

Figure 6: Typical rectangular pulses showing a schematic view of pulsed oxygen mass flow rate *vs.* time with a pulsing period T = 45 s and a duty cycle α = 0.33, *i.e.*, t_{ON} = 15 s and t_{OFF} = 30 s.

To this aim, a constant pulsing period T = 45 s was used. The t_{ON} time is related to the time when the oxygen flows into the sputtering chamber at its maximum value, *i.e.*, q_{O2Max}. It was set to q_{O2Max} = 2.0 sccm since this value is the minimum amount required to stabilize the process in the oxidized sputtering mode for single Ti – O_2 system in our experimental set-up described in the previous section. The t_{OFF} time corresponds to the minimum oxygen mass flow rate q_{O2min}. In this study and during this time, oxygen injection is completely stopped (*i.e.*, q_{O2min} = 0 sccm), in order to avoid the trapping phenomenon of the process in the oxidized sputtering mode [50, 51]. So, the duty cycle α is defined as:

$$\alpha = \frac{t_{ON}}{t_{ON} + t_{OFF}} = \frac{t_{ON}}{T} \quad \textbf{(1)}$$

Such a parameter was systematically changed from 0 to 100 % of the pulsing period T so as to show that this is one of the most relevant parameters, which can be adjusted to tune the properties of titanium oxynitride thin films.

The influence of the duty cycle α on the deposition rate is firstly investigated (Fig. 7).

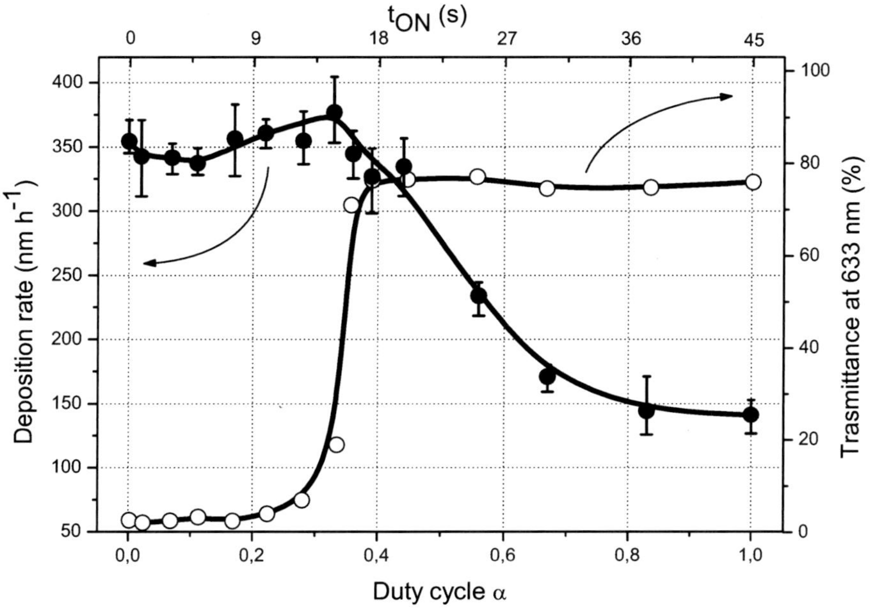

Figure 7: Deposition rate and average optical transmittance at 633 nm of titanium oxynitride thin films sputter deposited on glass substrate as a function of the duty cycle α and the corresponding t_{ON} oxygen injection time.

A continuous evolution of the rate can be noticed as α rises. This behaviour can be considered as a strong advantage compared to the conventional reactive sputtering, where a typical feature is the sudden drop of the deposition rate as a function of the reactive gas flow rate. In addition, non-linear phenomena of the process are not observed by RGPP, whereas a metallic target sputtered with argon and oxygen or nitrogen or both, commonly exhibits hysteresis effects of the deposition parameters. In Fig. **7**, the deposition rate is slightly enhanced for α = 0.33 and is even higher than that of titanium nitride (*i.e.,* 355 nm h^{-1}). A maximum rate of 375 nm h^{-1} is reached at α = 0.33. Afterwards, the deposition rate regularly reduces and tends to that of titanium dioxide (*i.e.,* 140 nm h^{-1}) measured for continuous introduction of oxygen gas.

Similarly, the average optical transmittance measured at 633 nm for titanium oxynitride thin films deposited on glass exhibits a strong increase for duty cycles corresponding to the maximum deposition rates. These optical measurements are representative of the transparency of the film-substrate system in the visible region. As a result, up to duty cycles α = 0.22, films are strongly absorbent with an optical transmittance lower than few percents. Inversely, transparent and dielectric thin films are produced with duty cycles higher than α = 0.39, where transmittance is close to 75 %. Some typical interference fringes are observed (not shown here). A transition zone occurs for duty cycles in-between 0.22 and 0.39. Transmittance abruptly increases from few to 75 % and semi-transparent titanium oxynitride thin films are synthesized for this range of duty cycles. Such a range appears as the most relevant conditions to reach tuneable oxynitride compounds. As previously claimed, real-time measurements of the target potential and sputtering pressure *vs.* time can also be taken into account to understand the smooth transition of the deposition rate *vs.* duty cycle and the gradual change of synthesis from nitride to dielectric compounds. Up to α = 0.28, the process predominantly remains in the nitrided sputtering mode. During a single pulsing period, the average target potential is enhanced leading to an increase of the deposition rate. Absorbent coatings are obtained since the nitrided mode prevails. For duty cycles higher than α = 0.28, the poisoning time of the target surface by the oxygen becomes longer and the time available to recover the nitrided mode is reduced. A titanium oxide layer is formed on the target surface for a longer time. Thus, it decreases the particles ejection from the target since the sputtering yield of titanium dioxide is lower than that of titanium nitride. Consequently, the deposition rate decreases and transparent titanium dioxide thin films are produced. So, the transition from nitride to oxide compounds can be obtained with rectangular pulses playing with the duty cycle. This transition is located for α

included between 0.3 and 0.4 assuming the most significant change of the optical transmittance from few to 75 %. It means that about (0.4 – 0.3) × 45 = 4.5 s can be spent to adjust the process so as to get various titanium oxynitride compounds. It will be shown in the next paragraph that this narrow temporal window can be extended using exponential patterns, particularly optimising τ_{mou} and τ_{des} times.

Exponential Pulse

Typical exponential equations were used in order to control oxygen mass flow rate *vs.* time according to exponential patterns. The following equation (2) describes well the flow rate during the t_{ON} time.

$$q_{O_2}(t) = \left(q_{O_2\,Max} - q_{O_2\,min}\right) \times \left(\frac{1 - e^{\frac{-t}{\tau_{mou}}}}{1 - e^{\frac{-t_{ON}}{\tau_{mou}}}} \right) + q_{O_2\,min} \qquad \textbf{(2)}$$

where $q_{O2}(t)$ is the oxygen mass flow rate *vs.* time (sccm), t the time (s), q_{O2Max} the maximum oxygen flow rate (sccm), q_{O2min} the minimum oxygen mass flow rate (sccm), τ_{mou} the tau mounting time (s). Similarly, equation (3) gives the oxygen mass flow rate *vs.* time during the t_{OFF} time.

$$q_{O_2}(t) = \left(q_{O_2\,Max} - q_{O_2\,min}\right) \times \left(\frac{1 - e^{\frac{-t}{\tau_{des}}}}{1 - e^{\frac{-t_{OFF}}{\tau_{des}}}} \right) + q_{O_2\,min} \qquad \textbf{(3)}$$

where τ_{des} is the tau descending time (s). These equations well show that the exponential pulses can provide an extended range of pulsing patterns as illustrated in Fig. **8**. Taking into account previous results claim in previous paragraphs especially about the relevance of using rectangular and exponential patterns, as well as the requirement to implement a t_{OFF} time long enough to allow an alternation of the process between the nitrided and the oxidized sputtering mode, the tau descending time τ_{des} was kept constant at 10^{-1} s and the minimum oxygen flow rate q_{O2min} was set to 0 sccm. It corresponds to a complete stop of the oxygen injection during the t_{OFF} time (Fig. **8**). On the other hand, during the t_{ON} time, maximum oxygen mass flow rate q_{O2Max} was maintained at 2 sccm as before and the tau mounting time τ_{mou} was systematically varied according to the following values: -10, -1, 10^{-1}, 1, 10 and 10^2 s (Fig. **8**). Such operating conditions allow tuning more accurately the amount of oxygen introduced into the sputtering process.

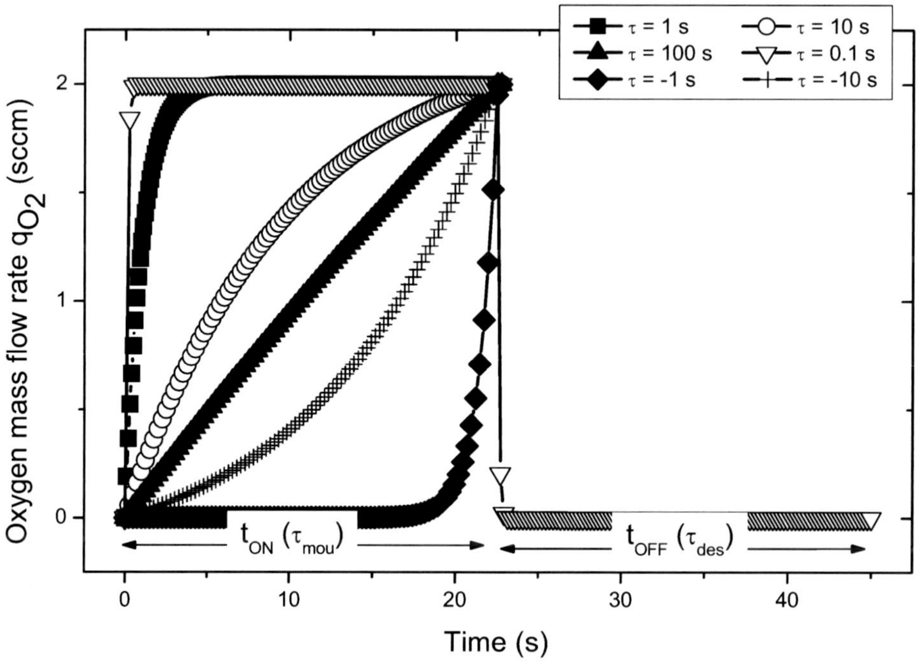

Figure 8: Various shapes of pulses can be generated with exponential pulses adjusting tau mounting time τ_{mou} and tau descending time τ_{des}. Tau descending time τ_{des} is set to 0.1 s so as to stop the oxygen injection as similarly employed for rectangular pulses.

As previously performed with rectangular pulses, the duty cycle was systematically changed from $\alpha = 0$ to 100 % of a constant pulsing period T = 45 s. Achieved deposition rates as a function of the duty cycle and for various τ_{mou} are nearly constant, close to 350 nm h^{-1}. The maximum of deposition rate *vs.* α previously observed for rectangular pulses (Fig. **7**) vanishes with exponential ones. Moreover, the optical transmittance at 633 nm of the films deposited in glass substrates exhibit the gradual transition from absorbent titanium nitride to transparent titanium dioxide compound as shown previously (Fig. **9**).

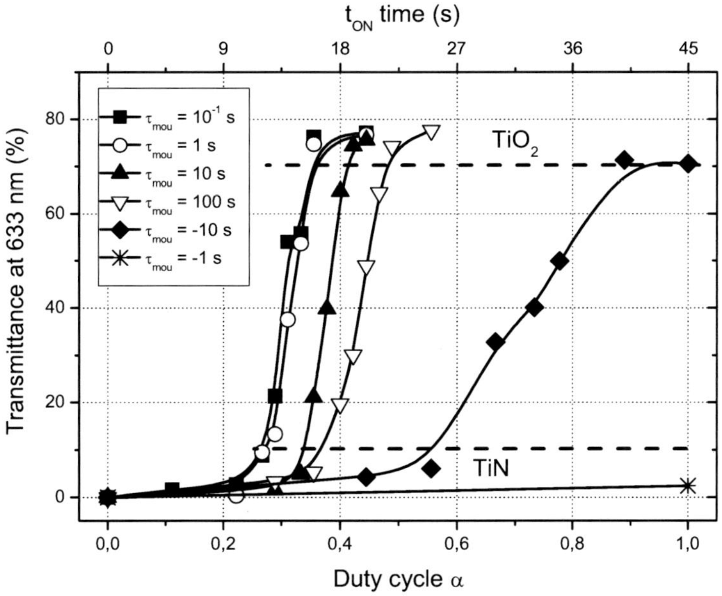

Figure 9: Optical transmittance at 633 nm of titanium oxynitride thin films deposited on glass *vs.* duty cycle α and for τ_{mou} = -10, -1, 10^{-1}, 1, 10 and 10^2 s. The gradual transition zone from absorbent TiN to transparent TiO_2 compound depends on duty cycle α and tau mounting time τ_{mou}.

The transition is always produced for all positive τ_{mou} times and a shift to high duty cycles is clearly evidenced as τ_{mou} rises from 10^{-1} to 10^2 s. In addition, it is also worth noting that the enhancement of transmittance is not so abrupt as τ_{mou} time increases. As a result, a widening of the duty cycle window suitable to gradually deposit titanium oxynitrides appears. These shift and widening effects are correlated with the amount of oxygen introduced in the process during the t_{ON} time. Changing the τ_{mou} time allows controlling the shape of the pulse and thus the injected oxygen. A rectangular-like pulse is generated if $\tau_{mou} \rightarrow 0$ s whereas a mounting triangular-like pulse is produced for $\tau_{mou} \rightarrow +\infty$ s.

For negative τ_{mou} times, duty cycles lower than 0.44 always lead to absorbent coatings. The gradual and shifted transition of the optical behaviours from TiN to TiO_2 is even more extended and moved to higher duty cycles for τ_{mou} = -10 s. This duty cycle window corresponds to more than 20 s of the t_{ON} time against 4.5 s previously determined with rectangular pulses and with deposition rates similar to that of titanium nitride (about 350 nm h^{-1}). This enhancement is again due the possibility to reduce the amount of oxygen introduced into the process during the t_{ON} time. Besides for negative τ_{mou} times in-between - ∞ to 0 s, the pulsing signal can be varied from triangle to Dirac-like shape. A finer injection of oxygen is then achieved. For $\tau_{mou} \rightarrow 0$ s, the shape of the pulse looks like an artefact and injected oxygen becomes negligible. Therefore, negative τ_{mou} times must be lower than -10 s, but higher than -100 s to obtain the maximum size of the transition window from titanium nitride to titanium dioxide compounds.

The widening of this transition zone can be correlated with poisoning kinetics of the target surface *via* the real time measurements of the target potential (Fig. **10**).

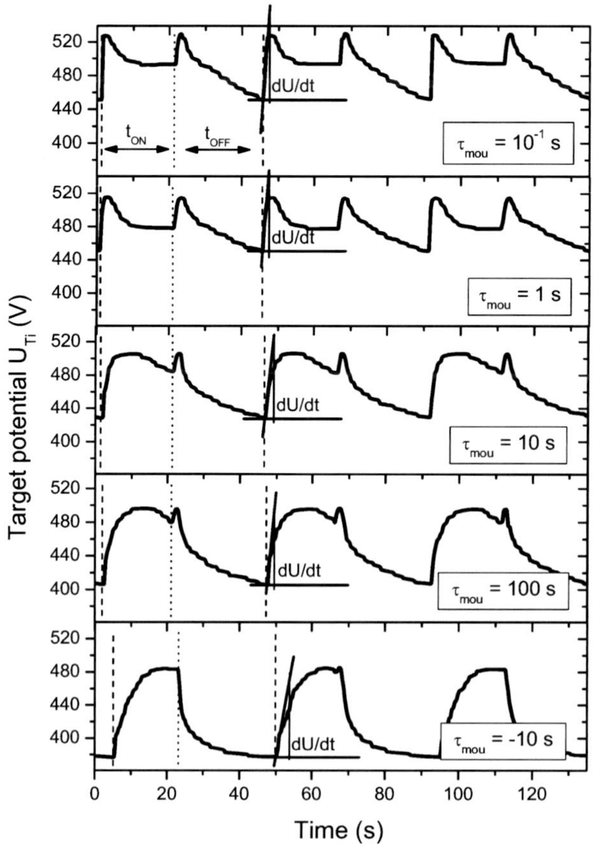

Figure 10: Titanium target potential *vs.* time for exponential patterns with various tau mounting times τ_{mou}. The pulsing period was kept at T = 45 s, the duty cycle α = 0.44 and the tau descending time τ_{des} = 10^{-1} s. Derivative dU/dt was calculated at the beginning of each t_{ON} time in order to determine the speed of pollution of the target surface.

For $\tau_{mou} = 10^{-1}$ and 1 s, the potential evolution *vs.* time exhibits a very similar feature to that previously measured for rectangular pulses. At the beginning of the t_{ON} and t_{OFF} times, potential peaks are systematically observed. They are mainly assigned to the oxidation of the target surface when oxygen is injected as the t_{ON} time starts, and when the oxygen is stopped (beginning of the t_{OFF} time), the process tends to be restored to the nitrided mode. Thus, an alternation of the process between both modes occurs for these operating conditions.

Increasing the τ_{mou} time, the abrupt rise of the target potential observed for low τ_{mou} times at 10^{-1} or 1 s, becomes less significant. In addition, the potential peak is not so sharp. It is shifted at the end of the t_{ON} time and the maximum target potential decreases from 528 to 495 V as τ_{mou} changes from 10^{-1} to 10^{2} s. A lower amount of oxygen is injected into the process during the t_{ON} time and the poisoning of the target surface is less efficient. Consequently, the time required to restore the process in the nitrided mode is reduced. The maximum of the target potential recorded at the beginning of the t_{OFF} time, is strongly diminished and vanishes for $\tau_{mou} = -10$ s.

Target potential does not exceed 485 V for a single period, and exponential growths and decays are periodically measured during the t_{ON} and t_{OFF} times, respectively. These behaviours are closely related to the kinetics of pollution of the target surface. Formation and thickness of the poisoned layer depends on the amount of reactive gas introduced into the chamber. A given time is required to reach the steady state equilibrium and to build up the oxide (or nitride) layer on the target surface. Taking into account the target potential evolution *vs.* time measured for $\tau_{mou} = -10$ s, one can claim that a thinner oxide layer is formed on the target surface than at positive τ_{mou} times. Moreover, the potential is down to 385 V for $\tau_{mou} = -10$ s at the end of the t_{OFF} time. This value corresponds to a predominance of the nitrided state of the target.

The abrupt increase of the target potential noticed at the beginning of thc t_{ON} time, supplies valuable information about poisoning kinetics of the target surface. The speed of pollution can be estimated from the time derivative of the potential dU/dt [52]. Such a speed of pollution was determined *vs.* the tau mounting time τ_{mou} and for a constant duty cycle $\alpha = 50$ % of the pulsing period T, *i.e.*, $t_{ON} = t_{OFF}$ (Fig. **11**).

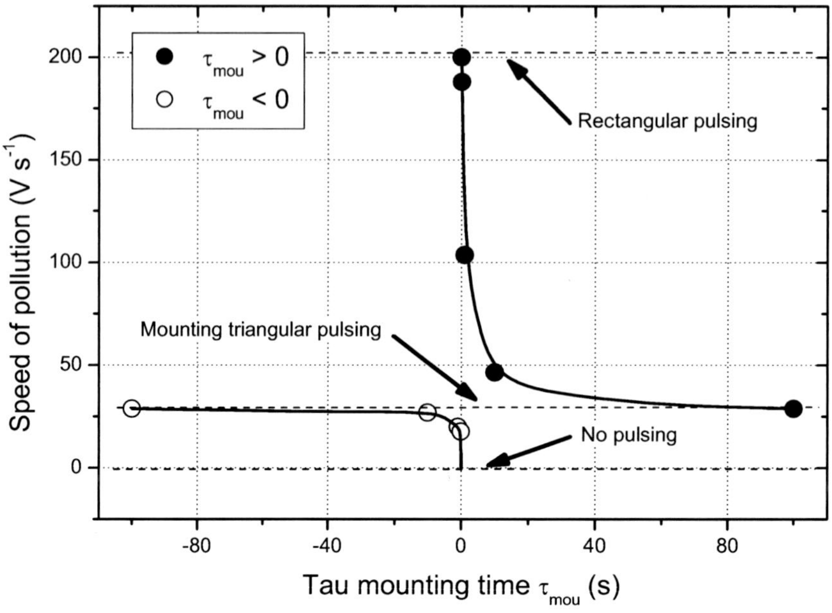

Figure 11: Speed of pollution of the target surface *vs.* τ_{mou} time calculated from the time derivative of the target potential. Speed values obtained with rectangular (τ_{des} = 0 s) and mounting triangular pulses ($\tau_{mou} \rightarrow + \infty$ or $-\infty$) are given. No pulsing means no injection of the oxygen gas.

The hyperbolic-like graph produced can be divided into two parts: positive and negative $\tau_{mou} = 10^{-1}$ s times. The maximum speed of pollution reaches 200 V s^{-1} and is obtained for positive τ_{mou} times as $\tau_{mou} \rightarrow 0$ s. It corresponds to rectangular patterns and thus, a maximum amount of oxygen is injected during the t_{ON} time. A slight increase of the τ_{mou} time leads to a sudden drop of the speed and finally, to an asymptotic value close to 29 V s^{-1} as $\tau_{mou} \rightarrow + \infty$ s. This increasing τ_{mou} time also gives rise to a longer time to reach a critical oxygen flow in order to avalanche the process in the oxidized sputtering mode. It is well correlated with the maximum of the peak potential, which is shifted and attenuated at the end of the t_{ON} time as the τ_{mou} time increases. In addition, the asymptotic evolution of the speed does not only depend on the kinetics of the process, it is also due to the pulsing signal itself since a mounting triangular pattern is generated for τ_{mou} higher than 10^2 s. So, it is worth to note that implementing positive τ_{mou} times only, the range of pollution speeds is restricted.

The negative τ_{mou} times also exhibit the same asymptotic value of the speed, *i.e.*, dU/dt = 29 V s^{-1}. Mounting triangular pulses are similarly produced for τ_{mou} lower than 10^{-2} s as previously generated for $\tau_{mou} \rightarrow +\infty$ s. On the other hand, the speed can be chiefly reduced and becomes null when τ_{mou} times are in-between 0 and -10 s. For this negative range of τ_{mou} times, the amount of oxygen introduced into the process becomes negligible since a peak-like shape is generated (Fig. **8**). Technical limitations of RGPP appear, *i.e.*, instead of pulsing a sharp and narrow peak (Dirac-like shape), artefact is rather obtained or no pulsing occurs for τ_{mou} between -10^{-2} and 0 s. However, very low speeds of pollution and thus an improved control of the RGPP can be achieved for τ_{mou} times lower than -10^{-2} s.

CONCLUSION

The reactive gas pulsing process (RGPP) is an original and new method successfully implemented for the synthesis of titanium oxynitride thin films by reactive sputtering. Nitrogen mass flow rate is kept constant whereas that of the oxygen gas is periodically introduced *vs.* time. Various shapes of pulses can be generated including rectangle, exponential, sine, isosceles, mounting and descending triangles. Many temporal and flow parameters can be adjusted in order to improve the control of the reactive gas pulsing process.

It is clearly established that exponential and rectangular patterns are the most relevant pulsing shapes required to deposit tuneable titanium oxynitride thin films. A large panel of physical and chemical properties from the metallic titanium nitride to the dielectric titanium dioxide can be prepared using rectangular or exponential patterns by RGPP. For both pulses, an alternation of the reactive process between the nitrided and oxidized sputtering modes is achieved without trapping the system in the fully oxidized state, as produced with sine and triangular patterns. For such patterns, transparent TiO_2-like compounds are systematically deposited for any operating conditions.

The requirement of a complete stop of oxygen injection (*i.e.*, t_{OFF} time) is demonstrated thanks the most significant results produced with rectangular and exponential pulses. To this aim, the role of the duty cycle α on the behaviours of the reactive sputtering process and the nature of as-deposited films is of fundamental interest. A systematic change of the duty cycle α from 0 to 100 % of a constant pulsing period T clearly demonstrates a gradual transition from nitride to oxide compounds. Deposition rates vary monotonously, and with them the physico-chemical properties such as optical transmittance in the visible region. The full oxidation or nitridation of the target surface can be avoided and a wide range of titanium oxynitrides is achieved by only changing the duty cycle parameter.

Last but not least is the particular case of exponential pulses. They dot not only allow some technical improvements of the RGPP method but they mainly bring a strong upgrading of the oxygen injection thanks to some adequate adjustments of the temporal parameters peculiar to the exponential pulses. Speed of pollution of the target surface is again reduced and alternation of the process between nitrided and oxidized sputtering modes is even more controlled. As a result, implementation of exponential pulses widens the pertinent duty cycles window compared to rectangular ones.

Finally, RGPP is not only an exciting method, which was developed to deposit tuneable and homogeneous titanium oxynitride thin films, it can also be involved to reach periodic multilayered structures of oxide/nitride, oxynitride/oxide, *etc.* at the micro- and nanoscale. In addition, the method can be extended to more complex systems implementing the pulsing of several reactive gases in a multitarget sputtering process.

ACKNOWLEDGEMENTS

RGPP was developed in the framework of the European project NMP3-CT-2003-505948 “HARDECOAT”. European Union is acknowledged for its financial support.

CONFLICT OF INTEREST

The author(s) confirm that this chapter content has no conflict of interest.

REFERENCES

[1] I. Safi, Surf. Coat. Technol., vol. 127, pp. 203-219, 2000.
[2] W.D. Sproul, Science, vol. 273, pp. 889-892, 1996.

[3] R.P. Howson, Pure Appl. Chem., vol. 66, pp. 1311-1318, 1994.
[4] S. Schiller, K. Goedicke, J. Reschke, V. Kirchhoff, S. Schneider, F. Milde, Surf. Coat. Technol., vol. 61, pp. 331-337, 1993.
[5] S. Berg, H.O. Blom, T. Larsson, C. Nender, J. Vac. Sci. Technol., vol. A5, pp. 202-207, 1987.
[6] D. Depla, R. De Gryse, Surf. Coat. Technol., vol. 183, pp. 184-189, 2004.
[7] E. Kusano, J. Appl. Phys., vol. 70, pp. 7089-7096, 1991.
[8] L. Pekker, Thin Solid Films, vol. 312, pp. 341-347, 1998.
[9] A. Billard, C. Frantz, Surf. Coat. Technol., vol. 59, pp. 41-47, 1993.
[10] A. Kinbara, E. Kusano, S. Baba, J. Vac. Sci. Technol., vol. A10, pp. 1483-1487, 1992.
[11] M. Hecq, A. Hecq, M. Fontignes, Thin Solid Films, vol. 115, pp. 45-48, 1984.
[12] S. Kadlec, J. Musil, J. Vyskocil, J. Vac. Sci. Technol., vol. A8, pp. 1560-1565, 1990.
[13] G. Este, W.D. Westwood, J. Vac. Sci. Technol., vol. A2, pp. 1238-1247, 1984.
[14] T. Larsson, H.O. Blom, C. Nender, S. Berg, J. Vac. Sci. Technol., vol. A6, pp. 1832-1836, 1988.
[15] J. Danroc, A. Aubert, R. Gillet, Surf. Coat. Technol., vol. 33, pp. 83-90, 1987.
[16] W.D. Sproul, P.J. Rudnik, C.A. Gogol, R.A. Mueller, Surf. Coat. Technol., vol. 39-40, pp. 499-506, 1989.
[17] W.D. Sproul, Surf. Coat. Technol., vol. 33, pp. 73-81, 1987.
[18] A.F. Hmiel, J. Vac. Sci. Technol., vol. A3, pp. 592-595, 1985.
[19] P. Carlsson, C. Nender, H. Barankova, S. Berg, J. Vac. Sci. Technol., vol. A11, pp. 1534-1539, 1993.
[20] N. Martin, C. Rousselot, J. Vac. Sci. Technol., vol. A17, pp. 2869-2878, 1999.
[21] H. Barankova, S. Berg, P. Carlsson, C. Nender, Thin Solid Films, vol. 260, pp. 181-186, 1995.
[22] T. Kubart, D.H. Trinh, L. Liljeholm, L. Hultman, H. Hogberg, T. Nyberg, S. Berg, J. Vac. Sci. Technol., vol. A26, pp. 565-570, 2008.
[23] F. Vaz, P. Cerqueira, L. Rebouta, S.M.C. Nascimento, E. Alves, P. Goudeau, J.P. Riviere, K. Pischow, J. De Rijk, Thin Solid Films, vol. 447, pp. 449-454, 2004.
[24] W.D. Sproul, D.J. Christie, D.C. Carter, Thin Solid Films, vol. 491, pp. 1-17, 2005.
[25] W.D. Sproul, D.J. Christie, D.C. Carter, S. Berg, T. Nyberg, 46th Annual Technical Conference Proceedings, pp. 98-103, 2003.
[26] A.J. Aronson, D. Chen, W.H. Class, Thin Solid Films, vol. 72, pp. 535-540, 1980.
[27] H. Sekiguchi, A. Kanzawa, T. Imai, T. Honda, J. Vac. Sci. Technol., vol. A12, pp. 3176-3179, 1994.
[28] H. Sekiguchi, T. Murakami, A. Kanzawa, T. Imai, T. Honda, J. Vac. Sci. Technol., vol. A14, pp. 2231-2234, 1996.
[29] D. Gall, R. Gampp, H.P. Lang, P. Oelhafen, J. Vac. Sci. Technol., vol. A14, pp. 374-379, 1996.
[30] R.P. Howson, N. Danson, I. Safi, Thin Solid Films, vol. 351, pp. 32-36, 1999.
[31] N. Martin, J. Lintymer, J. Gavoille, J. Takadoum, J. Mater. Sci., vol. 37, pp. 4327-4332, 2002.
[32] N.M.G. Parreira, T. Polcar, A. Cavaleiro, Surf. Coat. Technol., vol. 201, pp. 5481-5486, 2007.
[33] N.M.G. Parreira, T. Polcar, N.J.M. Carvalho, A. Cavaleiro, European Phys. J.: Appl. Phys., vol. 43, pp. 331-337, 2008.

[34] N.M.G. Parreira, T. Polcar, N. Martin, O. Banakh, A. Cavaleiro, Plasma Process. Polym., vol. 4, pp. 69-75, 2007.

[35] N. Martin, O. Banakh, A.M.E. Santo, S. Springer, R. Sanjinès, F. Lévy, Appl. Surf. Sci., vol. 185, pp. 123-133, 2001.

[36] N. Martin, R. Sanjinès, J. Takadoum, F. Lévy, Surf. Coat. Technol., vol. 142-144, pp. 615-620, 2001.

[37] N. Martin, J. Lintymer, J. Gavoille, J.M. Chappé, F. Sthal, J. Takadoum, F. Vaz, L. Rebouta, Surf. Coat. Technol., vol. 201, pp. 7720-7726, 2007.

[38] N. Martin, J. Lintymer, J. Gavoille, J.M. Chappé, F. Sthal, J. Takadoum, F. Vaz, L. Rebouta, Surf. Coat. Technol., vol. 201, pp. 7727-7732, 2007.

[39] N. Martin, J. Lintymer, J. Gavoille, J.M. Chappé, F. Sthal, J. Takadoum, F. Vaz, L. Rebouta, Surf. Coat. Technol., vol. 201, pp. 7733-7738, 2007.

[40] P. Carvalho, L. Cunha, E. Alves, N. Martin, E. Le Bourhis, F. Vaz, J. Phys. D: Appl. Phys., vol. 42, pp. 195501-7, 2009.

[41] C. Petitjean, M. Grafouté, C. Rousselot, J.F. Pierson, Surf. Coat. Technol., vol. 202, pp. 4825-4829, 2008.

[42] H. Le Dréo, O. Banakh, H. Keppner, P.A. Steinmann, D. Brian, N.F. De Rooij, Thin Solid Films, vol. 515, pp. 952-956, 2006.

[43] J.M. Chappé, P. Carvalho, S. Lanceros-Mendez, M.I. Vasileskiy, F. Vaz, A.V. Machado, M. Fenker, H. Kappl, N.M.G. Parreira, A. Cavaleiro, E. Alves, Surf. Coat. Technol., vol. 202, pp. 2363-2367, 2008.

[44] M. Fenker, H. Kappl, C.S. Sandu, Surf. Coat. Technol., vol. 202, pp. 2358-2362, 2008.

[45] E. Aubry, E. Weber, A. Billard, N. Martin, Appl. Surf. Sci., vol. 257, pp. 10065-10071, 2011.

[46] T. Kubart, D. Depla, D.M. Martin, T. Nyberg, S. Berg, Appl. Phys. Lett., vol. 92, pp. 221501-3, 2008.

[47] N. Martin, C. Rousselot, Surf. Coat. Technol., vol. 110, pp. 158-167, 1998.

[48] L.B. Jonsson, T. Nyberg, S. Berg, J. Vac. Sci. Technol., vol. A18, pp. 503-508, 2000.

[49] L.B. Jonsson, T. Nyberg, I. Katardjiev, S. Berg, Thin Solid Films, vol. 365, pp. 43-48, 2000.

[50] N. Martin, A.R. Bally, P. Hones, R. Sanjinès, F. Lévy, Thin Solid Films, vol. 377-378, pp. 550-556, 2000.

[51] J.M. Chappé, N. Martin, J. Lintymer, F. Sthal, G. Terwagne, J. Takadoum, Appl. Surf. Sci., vol. 253, pp. 5312-5316, 2007.

[52] C. Rousselot, N. Martin, Surf. Coat. Technol., vol. 142-144, pp. 215-219, 2001.

CHAPTER 3

Exploring the Potential of High Power Impulse Magnetron Sputtering for Tailoring the Chemical Composition and the Properties in Metal Oxynitride Films

Kostas Sarakinos*

Departement of Physics, Chemistry and Biology (IFM), Linköping University, 58183 Linköping, Sweden

Abstract: Metal oxynitrides (MONs) are a new class of functional materials in which tuning of the O-to-N ratio in the non-metal sublattice enables to tailor their properties between those of the reference metal oxide and the metal nitride systems. In thin film form MONs are frequently deposited by reactive magnetron sputtering techniques under the presence of two reactive gases (O_2 and N_2). In this case the ability to control the O-to-N ratio in the growing MON films is largely determined by the stability of the reactive sputtering process in the transition zone between metallic and compound target. Recently it has been shown that the implementation of a newly developed highly ionized magnetron sputtering technique, *i.e.,* the high power impulse magnetron sputtering (HiPIMS), can also allow for a stable operation during reactive deposition of metal oxides. In this chapter the state of the art regarding the process characteristics in reactive HiPIMS processes are reviewed and the potential of HiPIMS for tailoring the composition and the properties of MON films is discussed.

Keywords: Material science, thin films, coatings, PVD, ionized-PVD, magnetron sputtering, HiPIMS, HPPMS, reactive sputtering, process stability, transition zone, metal oxides, metal nitrides, metal oxynitrides, chemical composition, functional properties.

INTRODUCTION

Metal oxynitrides (MONs) are a new class of functional materials which in film form are employed in a variety of applications in surface protection, in optics and in microelectronics [1-5]. The research and technological interest on MON films emerges from the fact that control of the O-to-N ratio in the non-metal sublattice enables to design materials that combine properties of the reference metal oxide (MOs) and the metal nitride (MNs) systems [1-5]. A technique that is widely employed for the synthesis of MO, MN and MON films is the magnetron

***Address correspondence to Kostas Sarakinos:** Department of Physics, Chemistry and Biology (IFM), Linköping University, 58183 Linköping, Sweden; Tel: +46-13-281241; Fax: +46-13-287568; Email: kostas@ifm.liu.se

Filipe Vaz, Nicolas Martin and Martin Fenker (Eds)

sputtering in direct current (dc), radio frequency (rf) and pulsed configurations under the presence of a reactive gas (O_2 and/or N_2). The latter is known to result in chemisorption [6] and/or implantation [7] of reactive gas species on the target surface and the surface adjacent layers, respectively. This process is referred to as target coverage [6] and depending on the relative amount of the reactive gas in the sputtering atmosphere it can lead to target surface compositions ranging from a metallic (*i.e.,* non-covered) to a compound (*i.e.,* fully covered) target [6]. The target coverage, in turn, determines largely the chemical composition of the growing film. Commonly in reactive sputtering processes of MOs, but also of MNs in several cases, the relationship between the amount of the reactive gas and the target coverage is non-linear [6]. This leads to process instabilities in the target coverage range between the metallic and the compound sputtering mode (referred to as the transition zone) [6] and to hysteresis in the process parameters [6]. The process instabilities and the presence of the hysteresis have as a consequence that the stable target coverage conditions and therefore film compositions which can be achieved are limited and lie outside the transition zone [6]. It is therefore evident that the stabilization of the transition zone and the elimination of the hysteresis are prerequisites for extending the composition and thus the properties range of deposited films. In order for this to be achieved, several solutions have been suggested in the literature, including the use of feedback control systems [8], the pulsing of the reactive gas flow [9-11] and the implementation of multichamber arrangements for a gradual variation of the incorporation of reactive gas in the film [12]. Other studies [13, 14] have reported that in some MON systems and under certain conditions the simultaneous presence of O_2 and N_2 in the sputtering atmosphere can facilitate the stabilization of the transition zone and the elimination of the hysteresis. Recently it has been shown that the implementation of a newly developed highly ionized magnetron sputtering technique, *i.e.,* the high power impulse magnetron sputtering (HiPIMS) [15] can also allow for a stable and hysteresis free reactive deposition of MOs [16, 17]. In this chapter the state of the art regarding the process characteristics in reactive HiPIMS processes are reviewed and the potential of HiPIMS for tailoring the composition and the properties of MON films is discussed.

HIGH POWER IMPULSE MAGNETRON SPUTTERING: A NOVEL IONIZED PHYSICAL VAPOR DEPOSITION TECHNIQUE

Ionized physical vapor deposition (IPVD) describes a category of PVD techniques in which ions are a significant fraction of the film-forming species [18]. In IPVD the trajectories as well as the kinetic energy of ions are manipulated employing electric and/or magnetic fields in order to facilitate control over the film deposition process and tailor the properties of the growing films [18]. High power impulse magnetron sputtering (HiPIMS) (also referred to as high power pulsed magnetron sputtering-HPPMS) is a novel IPVD technique which was developed in the late 1990s by Kouznetsov [19] at the University of Linköping, Sweden. The seed of HiPIMS probably dates back to investigations performed in the former USSR in the 1980s [20-22]. In this section a brief overview of the fundamentals and applications of HiPIMS is provided. For additional information the reader is referred to the review articles available in the literature [23-27]. It should also be mentioned here that part sections "High Power Impulse Magnetron Sputtering: A Novel Ionized Physical Vapour Deposition Technique" and "Process Characteristics in Reactive HIPIMS of Oxides" in the present chapter are based on a previous review article by the author [24].

In HiPIMS the power is applied to a conventional magnetron source in short unipolar pulses (5 – 500 μs in length) of a relatively low duty cycle (<10%) and frequency (<10 kHz) as illustrated in Fig. **1**.

This mode of operation gives rise in maximum values of target power density during the pulse on-time (referred to as peak target power density) in the order of 1 – 5 $kWcm^{-2}$ [24]. At the same time the average target power density is maintained at a level of 5 – 10 Wcm^{-2} which is similar to that in conventional magnetron sputtering techniques (*e.g.* dc magnetron sputtering – dcMS) [23-27]. These extremely high values of peak target power density generate plasmas with electron densities in the range 10^{17} – 10^{19} m^{-3} [28] which are significantly larger than the values of 10^{14} – 10^{16} m^{-3} commonly obtained in dcMS discharges [29]. The dense plasmas, in turn, allow for a relatively high degree of ionization for both Ar gas atoms and sputtered species [30], *e.g.* up to 90% for Ti [30]. It has to be pointed out here that the ionization of the sputtered material in a conventional dcMS discharge is typically below 10% [29]. Apart from their relatively large

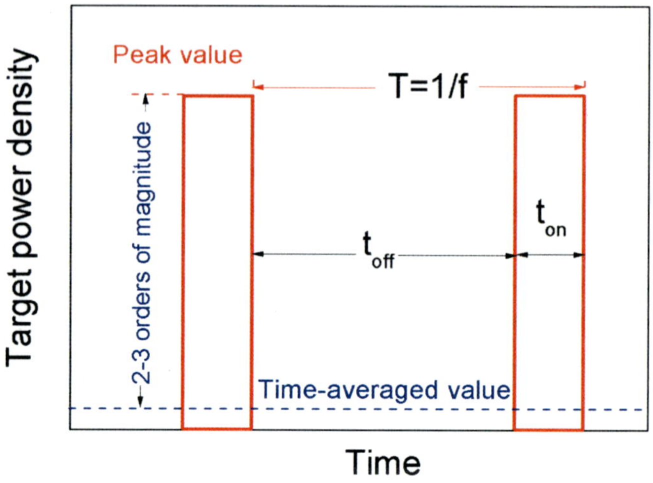

Figure 1: Schematic illustration of the HiPIMS operation principle. The power is applied to the target in short unipolar pulses (5 – 500 μs in length) of a relatively low duty cycle (<10%) and frequency (<10 kHz) resulting in peak target power densities in the order of 1 – 5 kWcm^{-2} At the same time the average target power density is maintained at a level of 5 – 10 Wcm^{-2}, which is similar to that in conventional magnetron sputtering techniques.

number, the ionized sputtered species in a HiPIMS discharge exhibit broader energy distribution functions and larger mean energies in comparison to those in a dcMS discharge [30]. Another feature of the HiPIMS discharge is the transport of ionized sputtered material to off-normal with respect to the target directions [31]. This phenomenon is caused by plasma instabilities in the HiPIMS discharge [32-34], which are in turn triggered by the time-dependent character of the HiPIMS process [35]. It is therefore evident that in HiPIMS processes relatively large fluxes of energetic ions are made available at the growing film. These growth conditions have been shown to enable the deposition of ultra-smooth and dense films, even without the utilization of additional bias voltage potential for increasing the energy of the impinging ions [23, 24, 26]. The intense energetic bombardment during HiPIMS has also been used to control the morphology and the microstructure of the growing films, allowing for instance for the deposition of nanocrystalline metal nitride films at room temperature [36, 37]. Other areas where the beneficial nature of HiPIMS has manifested itself is for tailoring the phase formation in metal [38], metal oxide [39-42], and metal carbide films [43] facilitating deposition of certain phases at temperatures significantly lower than those required by conventional magnetron sputtering techniques. The relatively large ionized fluxes in combination with the off-normal transport in a HiPIMS discharge has been utilized to deposit homogeneous films on complex shaped substrates, such as cutting tools [44] and patterned microelectronic devices [45]. Moreover, the ionized sputtered species available in the HiPIMS discharge have been utilized for the engineering of the film/substrate interface and enhancing the film adhesion [46].

PROCESS CHARACTERISTICS IN REACTIVE HIPIMS OF OXIDES

The incidence of energetic plasma ions on the target surface during the sputtering process results in the ejection of target atoms [47], as well as the emission of electrons (referred to as secondary electrons) [47]. The number of atoms ejected per incident plasma ion (referred to as sputtering yield) determines the arrival rate of condensing particles onto the target surface and thus the film growth rate. The number of secondary electrons emitted per incident ion (referred to as secondary electron emission yield) determines the plasma impedance and through these electrical characteristics of the glow discharge, such as the discharge (or target) voltage. The ejection of target atoms and the emission of secondary electrons depend, for a given ion type and energy, on the composition of the target surface [6]. As mentioned in section "Introduction" in reactive sputtering processes the composition of the target surface is largely determined by the relative amount of reactive gas in the sputtering atmosphere [6]. Therefore, monitoring of the film growth rate and the target voltage as a function of the composition of the gas atmosphere enables to determine the sputtering regimes (metallic, transition and oxidic zone) and evaluate the stability of the process. The target voltage (V_T) and growth rate (R_g) as a function of the reactive gas flow (q_{O2}) during reactive dc magnetron sputtering of Zr in an Ar-O_2 atmosphere are presented in Fig. **2** (a) and (b), respectively.

The process is operated at an average target current of 0.8 A [17]. The value of q_{O2} is varied between 0 and 3 sccm while the Ar flow is adjusted accordingly in order to maintain a constant working pressure of 0.8 Pa [17]. It can be seen that V_T remains nearly constant at a value of ~215 V in the q_{O2} range 0 – 2.2 sccm.

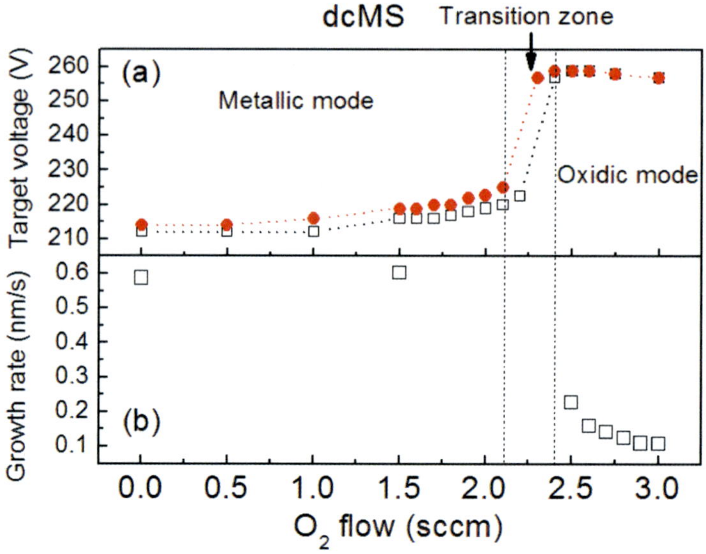

Figure 2: (a) Target voltage and (b) film growth rate as a function of O_2 flow during reactive operation of a Zr target in an Ar-O_2 atmosphere employing dcMS. In (a) the hollow squares and the filled circles are values for recorded for increasing and decreasing O_2 flow, respectively (data taken from [17]).

This behavior implies no significant change in the secondary electron emission yield, *i.e.,* the process is in the metallic mode. The latter is also justified by the nearly constant R_g values obtained in the same q_{O2} range. For O_2 flows larger than 2.4 sccm a larger V_T of ~260 is recorded signifying that the process is in the oxidic mode. The transition from the metallic to the oxidic mode takes place *via* an abrupt increase of the target voltage in the q_{O2} range 2.2 – 2.4 sccm. In this transition zone no stable working points can be obtained, *i.e.,* the process ends up either in the metallic or in the transition zone after a certain time. Moreover, upon decreasing the O_2 flow, a hysteresis in the V_T-q_{O2} curve is observed. Due to the unstable nature of the transition zone, films can only be deposited in the metallic or in the oxidic mode. The latter results in R_g values lower by ~50% as compared to those achieved in the metallic mode which is a result of the lower sputtering yield of the oxide in comparison to the corresponding metal. The V_T-q_{O2} and R_g-q_{O2} curves of the reactive HiPIMS process of a Zr in an Ar-O_2 ambient for a pulse on-time of 50 μs and off-time of 450 μs at otherwise identical to dcMS conditions are presented in Fig. **3** (a) and (b), respectively.

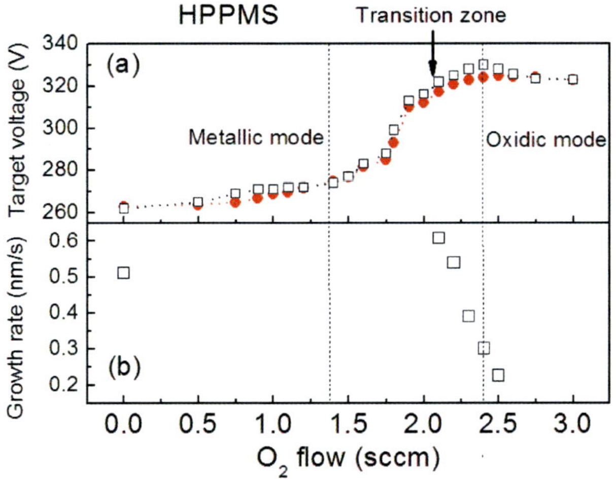

Figure 3: (a) Target voltage and (b) film growth rate as a function of O_2 flow during reactive operation of a Zr target in an Ar-O_2 atmosphere employing HiPIMS. In (a) the hollow squares and the filled circles are values for recorded for increasing and decreasing O_2 flow, respectively (data taken from [17]).

It can be seen that the transition zone is stable and hysteresis-free and it is extended over a larger q_{O2} range as compared to the dcMS process. The stabilization of the transition results in a gradual decrease of the R_g values when O_2 flow is increased. Similar behavior has also been reported for reactive HiPIMS of Al in Ar-O_2 atmosphere [16].

In general, the stability of the transition zone in reactive sputtering processes is determined by the competition between the formation and the removal (sputtering) of the compound from the target surface [6]. According to the formalism developed by Berg and Nyberg [6], the steady-state target coverage, ϑ_t, is equal to

$$\frac{J_i Y_c \vartheta_t}{e} - a2F(1-\vartheta_t) = 0 \tag{1}$$

where J_i is the ion target current density, Y_c the compound sputtering yield, e the elementary charge, a the sticking coefficient of the reactive gas, and F the flux of the reactive gas molecules towards the target [6]. Stabilization of the transition mode can be achieved, if for a certain nominal partial pressure/flow of the reactive gas the term "$\frac{J_i Y_c \vartheta_t}{e}$" (removal of the compound) increases and/or the term

"$a2F(1-\vartheta_t)$" (formation of the compound) decreases. The pulsed character of the HiPIMS discharge affects both terms in Eq. (1). The high peak currents (up to three orders of magnitude higher than the average values) result in a significantly higher instantaneous erosion rate during the pulse on-time, which may effectively clean the target surface and displace the oxidation onset at higher reactive gas flows [16]. In addition, the absence of plasma during the pulse off-time results in a limited activation of the reactive species [16]. Under these conditions relatively high levels of reactive gas exposure are necessary for the formation of the compound on the target surface, *i.e.,* this mechanism leads to lower effective values of the sticking coefficient a in Eq. (1). Furthermore, as a result of the high target voltage in HiPIMS [17], higher compound sputtering yields Y_c and, as a consequence, higher values of the compound erosion rate (normalized to the ion target current density) can be achieved [17]. However, a necessary condition for this to occur is that Ar^+ is a significant fraction of the ion target current [17], *i.e.,* the effect of self-sputtering on the target erosion rate is low [17]. The latter is the case when relatively moderate values of peak target current are applied [48]. The high peak target current in HiPIMS results in rarefaction of the neutral species in the target's vicinity [48, 49] which affects not only the Ar but also the reactive gas [50]. The rarefaction implies that the reactive species flux F is lower than the value which corresponds to the nominal partial pressure of the reactive gas should be expected [50].

THE POTENTIAL OF HIPIMS FOR TAILORING THE CHEMICAL COMPOSITION AND THE PROPERTIES OF METAL OXYNITRIDE FILMS

As mentioned in section "Introduction" the properties of MON's are to a large degree determined by the O-to-N ratio in the non-metal sublattice. Thus, in the case of MON films grown by reactive sputtering methods, the stabilization of the transition zone and the elimination of the hysteresis are prerequisites for extending the composition and thus the properties range of deposited films. This concept has been demonstrated for a number of MON systems, *e.g.* FeON [51], NiON [52], ZrON [13] and HfON [14]. In the case of FeON [51], stabilization of the transition zone was achieved by supplying the O_2 into the chamber using well-defined pulses. This approach enabled to extend the composition range of the deposited FeON films and through this to form the so-called intermediate iron oxides, *i.e.,* oxides with O incorporation smaller than the one that corresponds to the composition Fe_2O_3 [51]. Moreover, the wider process window achieved by the stabilization of the transition zone allowed for tuning of the optical properties of the films between those of the Fe-N and Fe-O films [45]. In the cases of NiON [52], ZrON [13] and HfON [14] the stabilization was facilitated by the coexistence of N_2 and O_2 in the sputtering atmosphere. The mechanisms that lead to the stabilization in this case have been discussed by Severin *et al.* [53]. Deposition of NiON films in the transition zone allowed for the growth of films with the crystal structure of the metastable NiN_2 phase [20]. In the case of ZrON [13] and HfON [14] the stabilization of the transition zone facilitated the formation of films with the crystal structure of the high temperature tetragonal and cubic ZrO_2 and HfO_2 phases, respectively. These phases exhibit high potential for applications as ionic conductors [54] and high-k dielectrics [55].

HiPIMS has not been so far employed for the growth of MON films. However, it can be anticipated that this technique exhibits a great potential with respect to the control over the chemical composition and the properties of MONs. One reason for this is the ability of HiPIMS to stabilize the transition zone in reactive sputtering processes. This feature has only been demonstrated for MO systems, as discussed in section "Process Characteristics in Reactive HIPIMS of Oxides". Based on this fact one can expect that reactive HiPIMS processes where N_2 is employed as a reactive gas will also exhibit an enhanced stability. This is because, for a given target material, N_2 is commonly less reactive than O_2 which is known to promote the process stability [53]. Thus HiPIMS might turn out to be an elegant approach which allows stabilization of the process during reactive deposition of a vast range of MON systems even without the need for complicated feedback control arrangements. Moreover, it was mentioned in the previous paragraph that in many MON reactive deposition experiments, the process can be intrinsically stable when N_2 and O_2 coexist in the reactive gas atmosphere. However, in order for this to occur a minimum amount of N_2 is necessary. HiPIMS may enable to stabilize reactive deposition processes with a smaller amount of N_2 as compared to those required in conventional techniqucs. This in turn means that a larger process window would be available to tune the composition and thus the properties of the resulting MONs. This, in combination with the large fluxes of the energetic species available during HiPIMS manifests a potential for the deposition of high quality functional MON films.

CONCLUSION

High power impulse magnetron sputtering (HiPIMS) is an ionized PVD technique in which ultra-dense plasmas and highly ionized deposition fluxes are generated by applying power to a sputtering cathode in short unipolar pulses of large amplitude. These plasmas have been shown to facilitate growth of ultra smooth and dense films with controlled microstructure and phase composition and enhanced adhesion on underlying substrates. As far as it concerns reactive deposition, the use of HiPIMS has been reported to lead to stabilization of the transition zone between the metallic and the compound zone, in particular during the deposition of metal oxide films. It can be anticipated that this feature of the HiPIMS process can be utilized for stable reactive deposition of transition metal oxynitrides (MONs). Thus, MON films with O-to-N ratios over a relatively wide range can be synthesized. The latter in combination with the large fluxes of energetic species available in HiPIMS can facilitate enhanced control over the properties of MON films.

ACKNOWLEDGEMENT

Declared none.

CONFLICT OF INTEREST

The author(s) confirm that this chapter content has no conflict of interest.

REFERENCES

[1] C.A. Jong, and T.S. Chin, "Optical characteristics of sputtered tantalum oxynitride - TaO_xN_y films", Mater. Chem. Phys., vol. 74, pp. 201, 2002; P. Wilhartitz, S. Dreer, and P. Ramminger, "Can oxygen stabilize chromium nitride?—Characterization of high temperature cycled chromium oxynitride" Thin Solid Films, vol. 447/448, pp. 289, 2004.

[2] J.M. Chappé, N. Martin, J.F. Pierson, G. Terwagne, J. Lintymer, J. Gavoille, and J.Takadoum, "Influence of substrate temperature on titanium oxynitride thin films prepared by reactive sputtering", Appl. Surf. Sci., vol. 225, pp. 29, 2004.

[3] M. Fenker,H. Kappl, K. Petrikowski, and R. Bretzler, "Pulsed power magnetron sputtering of a niobium target in reactive oxygen and/or nitrogen atmosphere", Surf. Coat. Technol., vol. 200, pp. 1356, 2005.

[4] C. Petitjean, M. Grafouté, J.F. Pierson, C. Rousselot, and O. Banakh, "Structural, optical and electrical properties of reactively sputtered iron oxynitride films", J. Phys. D: Appl. Phys., vol. 39, pp. 1894, 2006.

[5] S. Venkataraj, O. Kappertz, R. Jayavel, and M. Wuttig, "Growth and characterization of zirconium oxynitride films prepared by reactive direct current magnetron sputtering", J. Appl. Phys., vol. 92, pp. 2461, 2002.

[6] S. Berg, and T. Nyberg, "Fundamental understanding and modelling of reactive sputtering processes", Thin Solid Films, vol. 476, pp. 215, 2005.

[7] D. Depla, and R. De Gryse, "Target poisoning during reactive magnetron sputtering: Part I: The influence of ion implantation", Surf. Coat. Technol., vol. 183, pp. 184, 2004.

[8] W.D. Sproul, M.E. Graham, M.S. Wong, S. Lopez, D. Li, and R.A. Scholl, "Reactive direct current magnetron sputtering of aluminum oxide coatings", J. Vac. Sci. Technol. A, vol. 13, no. 3, pp. 1188-1191, 1995.

[9] A.J. Aronson, D. Chen, and W.H. Class, "Preparation of titanium nitride by a pulsed D.C. magnetron reactive deposition technique using the moving mode of deposition", Thin Solid Films, vol. 72, pp. 535, 1980.

[10] N. Martin, J. Lintymer, J. Gavoille, J.M. Chappé, F. Sthal, J. Takadoum, F. Vaz, and L. Rebouta, "Reactive sputtering of TiO_xN_y coatings by the reactive gas pulsing process. Part I: Pattern and period of pulses", Surf. Coat. Technol., vol. 201, pp. 7720, 2007.

[11] C. Petitjean, M. Grafouté, C. Rousselot, and J.F. Pierson, "Reactive gas pulsing process: A method to extend the composition range in sputtered iron oxynitride films", Surf. Coat. Technol., vol. 202, pp. 4825, 2008.

[12] M. Scherer, "Magnetron sputter-deposition on atom layer scale", Vakuum in Forschung und Praxis, vol. 4, pp. 24, 2009.

[13] D. Severin, O. Kappertz, T. Kubart, T. Nyberg, S. Berg, A. Pflug, M. Siemers, and M. Wuttig, "Process stabilization and increase of the deposition rate in reactive sputtering of metal oxides and oxynitrides", Appl. Phys. Lett., vol. 88, pp. 161504, 2006.

[14] K. Sarakinos, D. Music, S. Mraz, M. to Baben, K. Jiang, F. Nahif, A. Braun, C. Zilkens, S. Konstantinidis, F. Renaux, D. Cossement, F. Munnik, and J. M. Schneider, "On the phase formation of sputtered hafnium oxide and oxynitride films", J. Appl. Phys., vol. 108, pp. 014904, 2010.

[15] V. Kouznetsov, K. Macak, J.M. Schneider, U. Helmersson, and I. Petrov, "A novel pulsed magnetron sputter technique utilizing very high target power densities", Surf. Coat. Technol., vol. 122, pp. 290, 1999.

[16] E. Wallin, and U. Helmersson, "Hysteresis-free reactive high power impulse magnetron sputtering", Thin Solid Films, vol. 516, pp. 6398, 2008.

[17] K. Sarakinos, J. Alami, C. Klever, and M. Wuttig, "Process stabilization and enhancement of deposition rate during reactive high power pulsed magnetron sputtering of zirconium oxide", Surf. Coat. Technol., vol 202, pp. 5033, 2008.

[18] J. Hopwood, Ionized Physical Vapor Deposition, San Diego: Academic Press, 2000.

[19] V. Kouznetsov, K. Macak, J.M. Schneider, U. Helmersson, and I. Petrov, "A novel pulsed magnetron sputter technique utilizing very high target power densities", Surf. Coat. Technol., vol. 122, pp. 290, 1999.

[20] D.V. Mozgrin, I.K. Fetisov, and G.V. Khodachenko, "High-current low-pressure quasi-stationary discharge in a magnetic field: experimental research", Plasma Phys. Rep., vol. 21, pp. 400, 1995.

[21] S.P. Bugaev, N.N. Koval, N.S. Sochugov, and A.N. Zakharov, "Investigation of a high-current pulsed magnetron discharge initiated in the low-pressure diffuse arc plasma", XVIIth International on Discharges and Electrical Insulation in Vacuum, vol. 2, pp. 1074-1076, 1996.

[22] I.K. Fetisov, A.A. Filippov, G.V. Khodachenko, D.V. Mozgrin, and A.A. Pisarev, "Impulse irradiation plasma technology for film deposition", Vacuum, vol. 53, pp. 133, 1999.
[23] U. Helmersson, M. Lattemann, J. Bohlmark, A.P. Ehiasarian, and J.T. Gudmundsson, "Ionized physical vapor deposition (IPVD): A review of technology and applications", Thin Solid Films, vol. 513, pp. 1, 2006.
[24] K. Sarakinos, J. Alami, and S. Konstantinidis, "High power pulsed magnetron sputtering: A review on scientific and engineering state of the art", Surf. Coat. Technol., vol. 204, pp. 1661, 2010.
[25] A. Anders, "Discharge physics of high power impulse magnetron sputtering", Surf. Coat. Technol., vol. 205, pp. S1, 2011.
[26] D. Lundin and K. Sarakinos, "An introduction to thin film processing using high-power impulse magnetron sputtering", J. Mat. Res., vol. 27, pp. 780, 2012.
[27] J. T. Gudmnundsson, N. Brenning, D. Lundin, and U. Helmersson, "High power impulse magnetron sputtering discharge", J. Vac. Sci. Technol. A, vol. 30, pp. 030801, 2012.
[28] J.T. Gudmundsson, J. Alami, and U. Helmersson, "Evolution of the electron energy distribution and plasma parameters in a pulsed magnetron discharge", Appl. Phys. Lett., vol. 78, pp. 3427, 2001.
[29] C. Christou, and Z.H. Barber, "Ionization of sputtered material in a planar magnetron discharge", J. Vac. Sci. Technol. A, vol. 18, pp. 2897, 2000.
[30] J. Bohlmark, J. Alami, C. Christou, A.P. Ehiasarian, and U. Helmersson, "Ionization of sputtered metals in high power pulsed magnetron sputtering", J. Vac. Sci. Technol. A, vol. 23, pp. 18, 2005.
[31] D. Lundin, P. Larsson, E. Wallin, M. Lattemann, N. Brenning, and U. Helmersson, "Cross-field ion transport during high power impulse magnetron sputtering", Plasma Sources Sci. Technol., vol. 17, pp. 035021, 2008.
[32] D. Lundin, U. Helmersson, S. Kirkpatrick, S. Rohde, and N. Brenning, "Anomalous electron transport in high power impulse magnetron sputtering", Plasma Sources Sci. Technol., vol. 17, pp. 025007, 2008.
[33] A. P. Ehiasarian, A. Hecimovic, T. de los Arcos, R. New, V. Schulz-von der Gathen, M. Böke, and J. Winter, "High power impulse magnetron sputtering discharges: Instabilities and plasma self-organization", Appl. Phys. Lett., vol. 100, pp. 114101, 2012.
[34] A. Anders, "Self-organization and self-limitation in high power impulse magnetron sputtering", Appl. Phys. Lett., vol. 100, pp. 224104, 2012.
[35] N. Brenning, I. Axnäs, M.A. Raadu, D. Lundin, and U. Helmerson, "A bulk plasma model for dc and HiPIMS magnetrons", Plasma Sources Sci. Technol., vol. 17, pp. 045009, 2008.
[36] J. Alami, K. Sarakinos, F. Uslu, and M. Wuttig, "On the relationship between the peak target current and the morphology of chromium nitride thin films deposited by reactive high power pulsed magnetron sputtering", J. Phys. D: Appl. Phys., vol. 42, pp. 015304, 2009.
[37] A.P. Ehiasarian, P.E. Hovsepian, L. Hultman, and U. Helmersson, "Comparison of microstructure and mechanical properties of chromium nitride-based coatings deposited by high power impulse magnetron sputtering and by the combined steered cathodic arc/unbalanced magnetron technique", Thin Solid Films, vol. 457, pp. 270, 2004.
[38] J. Alami, P. Eklund, J.M. Andersson, M. Lattemann, E. Wallin, J. Bohlmark, P. Persson, and U. Helmersson, "Phase tailoring of Ta thin films by highly ionized pulsed magnetron sputtering", Thin Solid Films, vol. 515, pp. 3434, 2007.
[39] J. Alami, K. Sarakinos, F. Uslu, C. Klever, J. Dukwen, and M. Wuttig, "On the phase formation of titanium oxide films grown by reactive high power pulsed magnetron sputtering", J. Phys. D: Appl. Phys., vol. 42, pp. 115204, 2009.

[40] E. Wallin, T.I. Selinder, M. Elfwing, and U. Helmersson, "Synthesis of α-Al_2O_3 thin films using reactive high-power impulse magnetron sputtering", Europhys. Lett., vol. 82, pp. 36002, 2008.
[41] R. Bandorf, M. Vergöhl, P. Giesel, T. Wallendorf, and G. Mark, "Investigation of HPPMS Titania Thin Films Prepared by Unipolar, DC-Superimposed and Bipolar Sputtering", Society of Vacuum Coaters 50th Annual Technical Conference Procedings, Louisville, KY, pp. 160, 2007.
[42] S. Konstantinidis, A. Hemberg, J.P. Dauchot, and M. Hecq, "Deposition of zinc oxide layers by high-power impulse magnetron sputtering", J. Vac. Sci. Technol. B, vol. 25, pp. L19, 2007.
[43] J. Alami, P. Eklund, J. Emmerlich, O. Wilhelmsson, U. Jansson, H. Högberg, L. Hultman, and U. Helmersson, "High-power impulse magnetron sputtering of Ti-Si-C thin films from a Ti_3SiC_2 compound target", Thin Solid Films, vol. 515, pp. 1731, 2006.
[44] J. Alami, S. Bolz, and K. Sarakinos, "High power pulsed magnetron sputtering: Fundamentals and applications", J. All. Compd., vol. 483, pp. 530, 2009.
[45] R. Chistyakov, B. Abraham, W.D. Sproul, J. Moore, and J. Lin, "Modulated Pulse Power Technology and Deposition for Protective and Tribological Coatings", Society of Vacuum Coaters 50th Annual Technical Conference Proceedings, Louisville, KY, pp. 139, 2007.
[46] A.P. Ehiasarian, J.G. Wen, and I. Petrov, "Interface microstructure engineering by high power impulse magnetron sputtering for the enhancement of adhesion", J. Appl. Phys., vol. 101, pp. 054301, 2007.
[47] B. Chapman, Glow Discharge Processes, New York: John Wiley & Sons, 1981.
[48] J. Alami, K. Sarakinos, G. Mark, and M. Wuttig, "On the deposition rate in a high power pulsed magnetron sputtering discharge", Appl. Phys. Lett., vol. 89, pp. 154104, 2006.
[49] D. Horwat, and A. Anders, "Compression and strong rarefaction in high power impulse magnetron sputtering discharges", J. Appl. Phys., vol. 108, pp. 123306, 2010.
[50] K. Sarakinos, J. Alami, and M. Wuttig, "Process characteristics and film properties upon growth of TiO_x films by high power pulsed magnetron sputtering", J. Phys. D: Appl. Phys., vol. 40, pp. 2108, 2007.
[51] C. Petitjean, M. Grafoute, C. Rousselot, and J. F. Pierson, "Reactive gas pulsing process: A method to extend the composition range in sputtered iron oxynitride films", Surf. Coat. Technol., vol. 202, pp. 4825, 2008.
[52] F. Uslu, "Structural and optical properties of copper- and nickel-oxynitride films", Ph.D. thesis, RWTH Aachen University, Aachen, 2008.
[53] D. Severin, O. Kappertz, T. Kubart, T. Nyberg, S. Berg, A. Pflug, M. Simers, and M. Wuttig, "Process stabilization and increase of the deposition rate in reactive sputtering of metal oxides and oxynitrides", Appl. Phys. Lett., vol. 88, pp. 161504, 2006.
[54] A.S. Kao and G. L. Gorman, "Modification of zirconia film properties by low-energy ion bombardment during reactive ion-beam deposition", J. Appl. Phys., vol. 67, pp. 3826, 1990.
[55] X. Zhao and D. Vanderbilt, "First-principles study of structural, vibrational, and lattice dielectric properties of hafnium oxide", Phys. Rev. B., vol. 65, pp. 233106, 2002.

CHAPTER 4

Tuneable Properties of Zirconium Oxynitride Thin Films

Pedro Carvalho[1], Luis Cunha[1], Nuno Pessoa Barradas[2], Eduardo Alves[2], Juan Pedro Espinós[3] and Filipe Vaz[1,*]

[1]Centro de Física, Universidade do Minho, 4710-057 Braga, Portugal; [2]Instituto Superior Técnico/ITN, Universidade Técnica de Lisboa, E.N. 10, 2686-953 Sacavém, Portugal and [3]Instituto de Ciencia de Materiales de Sevilla (CSIC-Univ. Sevilla), Avda. Américo Vespucio s/n, 41092 Sevilla, Spain

Abstract: This chapter is related to the investigation of decorative zirconium oxynitride, ZrO_xN_y, films prepared by dc reactive magnetron sputtering, using a 17:3 nitrogen-to-oxygen-ratio gas mixture. The films changed appearance from metallic-like, very bright yellow-pale and golden yellow, for low gas mixture flows to red-brownish for intermediate gas flows. A significant decrease of brightness was associated with this colour change. With further increase of the reactive gas flow, the colour of the samples changed from red-brownish to dark blue. The films deposited with relatively high gas flows showed only apparent colorations due to interference effects. The variations in composition disclosed the existence of four different zones, which were correlated to different crystalline structures. X-ray diffraction analysis revealed significant changes in the films crystalline structure, which changed at relatively low gas flows from B1 NaCl face-centred cubic zirconium nitride-type phase to a poorly crystallized over-stoichiometric nitride phase, which may be similar to that of Zr_3N_4. For higher flows, the texture changed from an oxynitride-type phase, similar to that of γ-Zr_2ON_2, to a ZrO_2 monoclinic-type structure. The composition/structure variations were consistent with the chemical bonding analysis carried out by X-ray Photoelectron Spectroscopy (XPS), which showed oxygen doping in both Zr_3N_4 and ZrN-type grown films. The electronic, optical and mechanical properties of the films exhibited significant changes from zone to zone, consistent with the different compositions, as well as wit the variations in the crystalline phases and bonding states.

Keywords: Reactive sputtering, zirconium oxynitride, decorative applications, thin films.

INTRODUCTION

PVD (*P*hysical *V*apour *D*eposition) thin film technology has been applied with

***Address correspondence to Filipe Vaz:** Centro de Física, Universidade do Minho, 4710-057 Braga, Portugal; Tel: 253 601530/510471; Fax: 253 604398; E-mail: fvaz@fisica.uminho.pt enormous success in several technological fields due to a wide range of possible material combinations and, therefore, a significant number of possible property gradients [1]. Metallic oxynitrides represent a group of modern ceramic materials of increasing technological importance. They have been extensively investigated over the past few years mainly due to their remarkable properties in the domain of protective applications, such as wear, diffusion and corrosion-resistance [2-4]. Furthermore, their relatively high hardness, chemical stability, durability and electronic properties lead to a wide variation of uses in other fields, ranging from tribological to electronic applications [5]. Their relevance arises from the fact that the presence of oxygen allows the tailoring of film properties between those of the base metal (Me) nitride, MeN_y, and those of the correspondent oxide, MeO_x. Tuning the oxide/nitride ratio allows one to tune the band-gap, and crystallographic order between oxide and nitride and hence the electronic and the overall set of properties that maybe required for a particular application. The particular system that will be discussed in this chapter is zirconium oxynitride, ZrO_xN_y.

In order to better define the physical limits of this family of ceramic compounds, let us start by recalling the properties of zirconium nitride, ZrN, and of zirconium oxide, ZrO_2. Zirconium nitride, ZrN, has a melting point temperature of about 3250 K, a typical electrical resistivity of the order of few hundreds of $\mu\Omega\cdot cm$, a thermal conductivity of the order of few $W\cdot m\cdot K^{-1}$ and a superconducting transition temperature of 9 K [6]. These properties are closely related to its electronic structure [7-9]. In fact, it is known that the non-vanishing DOS (*D*ensity *O*f *S*tates) at the Fermi energy is responsible for the metallic conductivity of ZrN. K. Schwarz *et al.* [10] reported that the particular electronic structure of ZrN consists of a low-lying N 2*s* band, followed by the *p- band*, which origins mainly from N 2*p* states, but with significant contribution from Zr 4*d* orbitals. The covalent interactions between Zr *4d* and N *2p* orbitals are responsible for the zirconium nitride's high hardness and high melting point. The following band is primarily derived from Zr 4*d* states and, thus, is denoted as *d band*, which, in the case of ZrN, overlaps the p-band. Between the "*p band*" and "*d band*" there is a minimum in the DOS at eight valence electrons. From this minimum the DOS steadily increases until all nine valence electrons of ZrN are accommodated, and so the

Fermi energy, E_F, lies in the "*d band*" with a relatively high DOS. Due to these properties, ZrN is widely used in several technological domains, such as mechanical rotation tools [11], cryogenic thermometers [12], Josephson junctions [13], gates in MESFET transistor technology [14], diffusion barriers in p/n junctions [15] and in Cu/Si contacts [16,17]. It is also worth to notice their inert matrix, used in fast reactor fuels for incineration of plutonium and minor actinides [18]. Besides these applications and properties of the ZrN phase, zirconium nitride shows also an interesting transformation from the conducting to the insulating phase, when the composition changes from ZrN to Zr_3N_4 [19-21].

On the other hand, transition metal oxide films, MeO_x, are frequently applied in several optical devices [22,23], such as anti-reflection and dense wavelength multiplexing devices and switchable windows. The uses of such systems are driven by the excellent mechanical properties of oxide-based films and by their good chemical stability, high refractive index and wide optical band gap. Zirconium oxide, ZrO_2, is one of the most refractory of all transition metal oxide compounds and exhibits both high thermal shock resistance, corrosion resistance, high refractive index and broad region of low absorption from the near-UV (above 240 nm) to the mid-IR (below 8 μm) [24-26]. Due to these properties, it is widely used for the thermal, oxidation and hot corrosion protection of high temperature components, providing insulation for metallic structures [27-30].

Taking into account that the properties of the materials significantly depend on the interdependence of composition and type of bonds, structure and morphology, the study of these correlations will be a major concern of this chapter, in order to better understand the overall behaviour of zirconium oxynitride thin films. For this purpose, a set of Zr-O-N films will be presented and analyzed in terms of all the above items. The study of the properties of this set of films basically allows the establishment of limits for practical applicability of devices coated with these materials, providing new areas of application and supplying criteria for new materials design for specific applications. Moreover, this knowledge will be of major importance to understand the electrical, optical, mechanical and tribological behaviour of the prepared set of samples and the nature of the metallic-to-insulating transition with increasing oxygen concentration. The main focus of this work is given to the detailed study of the different phases that are formed in

ZrO_xN_y ternary systems in comparison with those that are characteristic of ZrN_y and ZrO_x binary systems.

MATERIAL AND METHODS

In order to investigate the behaviour of zirconium oxynitrides, a set of ZrO_x, ZrN_y and ZrO_xN_y films were prepared on glass and silicon (100) substrates by reactive DC magnetron sputtering in a laboratory-sized deposition system. The deposition system is formed by two vertically opposed rectangular magnetrons (unbalanced of type 2), in a closed field configuration. The films were prepared with the substrate holder positioned at 70 mm from the target in all runs, using a DC current density of 100 A m^{-2} on the zirconium target (99.6 at.%). The working atmosphere was composed by Ar+N_2 (for the preparation of the ZrN_y films), and by Ar+O_2 (for the preparation of the ZrO_x films). For the preparation of the oxynitride films, ZrO_xN_y, the atmosphere was composed by Ar and a reactive N_2+O_2 gas mixture (17:3 ratio). The Ar flow was kept constant at 60 sccm for all depositions. The set of films of the ZrO_x system were prepared with oxygen flows varying from 0.5 to 15 sccm (corresponding to a partial pressure variation between 0.9×10^{-2} and 12×10^{-2} Pa). Regarding the preparation of the set of ZrN_y thin films, the nitrogen flow was varied from 3 to 40 sccm (corresponding to a partial pressure variation between 9.9×10^{-3} and 2.9×10^{-1} Pa). In the case of the oxynitrides, the reactive gas flow of the gas mixture (N_2+O_2) varied from 6 to 40 sccm (corresponding to partial pressure variation between 7.8×10^{-3} and 3.2×10^{-1} Pa). The working pressure was approximately constant during the depositions (varying slightly between ~0.4 and 0.5 Pa). The substrates were grounded and no external heating was used. The temperature evolution of the coated substrates (resulting from the deposition process itself) was monitored with a thermocouple placed close to the surface of the substrate holder. A delay time of five minutes was used before positioning the surface of the samples in front of the Zr target. This delay time was used to avoid target poisoning resulting from previous depositions and also to assure a constant deposition temperature of the substrates during film growth.

The atomic composition of the as-deposited samples was measured by Rutherford Backscattering Spectroscopy (RBS) using (1.4, 1.75) MeV and 2 MeV for the proton and He^+ beams, respectively. The scattering angles were 140º (IBM geometry) and 180º (annular detector). The tilt angles were 0º and 30º. Composition profiles for the as-deposited samples were generated using the software code IBA NDF [31]. For the determination of ^{14}N, ^{16}O and ^{28}Si concentrations, the correspondent cross-sections data given by Gurbich [32] were used. The analyzed area was about 0.5×0.5 mm^2. In addition, for some samples, Particle Induced X-ray Emission (PIXE) measurements were performed to check for impurities.

Ball cratering tests were used to measure the thickness of the films. The results were used for the determination of the deposition rate. The characterization of the bond-types was carried out in a XPS spectrometer (12 kV, 20 mA) from VG (ESCALAB 210). The spectra were obtained by using non-monochromalized Mg K_α (1253.6 eV) radiation. XPS spectra were recorded after successive sputtering-clean processes with Ar^+ ions (3.5 kV, 30º incidence), until achieving a compositional stationary regime. The spectra were taken in the ΔE constant mode with a pass energy of 50 eV. The Shirley algorithm was applied for the background subtraction [33]. In order to eliminate the "charging effect" of the analysis, all spectrums where the N 1s peak was clearly observed, were calibrated to the N 1s position - 396.2 eV. This peak position is referred to in the literature as the energy of the N 1s peak in ZrO_xN_y systems [34,35]. However, this approach does not correspond to the "ideal" calibration, since the N 1s peak positions should be different in ZrN, Zr_3N_4 or ZrO_xN_y systems. The selected calibration energy value for the N 1s peak results from two factors: i) the inexistence of Ar peaks in some of the sputter-cleaned samples and, ii) the surface oxide degradation of the samples. The peak located at a binding energy (E_B) value of 396.2 eV is referred in the literature as representative of the Zr-N bond in Zr_3N_4 phase [36,37] and of the Zr-O-N bond type in ZrO_xN_y compounds. Since zirconium and nitrogen can present similar ionic valences in zirconium oxynitride and in the Zr_3N_4 form [35], the selected calibration should be the most adequate choice. The oxide coatings were calibrated to the Ar $2p_{3/2}$ peak - 242.6 eV and to the N 1s peak - 396.2 eV.

The structure and phase distribution of the coatings were accessed by X-ray diffraction (XRD), using a conventional Philips PW 1710 diffractometer, operating with Cu Kα radiation, in a Bragg–Brentano configuration. The electrical resistivity at room temperature and as a function of temperature was obtained using Van der Pauw geometry for resistivity values lower than 1.82×10^6 μΩ·cm and by I-V curves, between Al parallel contacts, for higher resistivity values.

TUNEABLE PROPERTIES

Composition

The ternary system Zr-N-O presents composition variations lying in a ternary diagram that includes the well-known Zr_3N_4; ZrO_2 and ZrN, where the first two are actually occupying two of the triangle's vertices [38], see Fig. **1**. The third vertex of the triangle is located in a position that corresponds to sub-stoichiometric ZrN_y compounds.

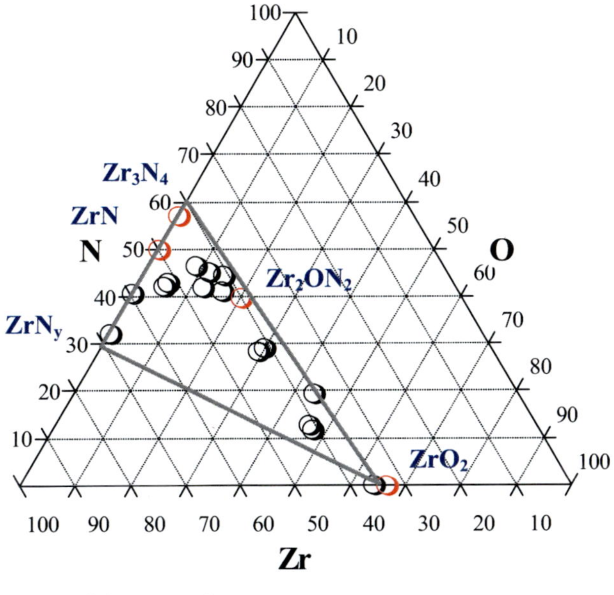

Figure 1: Ternary diagram of the Zr-N-O system.

The different compositions of zirconium oxynitride coatings deposited by PVD can be controlled by careful adjustment of the kinetic and thermodynamic parameters, *i.e.,* the rate of metallic atoms arriving at the substrate *versus* the thermodynamic affinity of the metalloids with zirconium. Taking into account the available thermochemical data for zirconium oxides and nitrides, a complete oxidation of the film would be expected as long as enough oxygen is available, since the formation of a metal–oxygen bond is energetically more favourable ($\Delta H_f^{\circ}{}_{ZrO2}$: -1100.6 kJ/mol [39]) than that one of a metal–nitrogen bond ($\Delta H_f^{\circ}{}_{ZrN}$: -365.5 kJ/mol [40]).

Several parameters could be chosen to change the kinetic *versus* the thermodynamics parameters in a PVD process; target current density, partial pressure of the reactive gases flows, ion bombardment, vacuum pumping speed, temperature, *etc.* Among this group of possibilities, we have selected the change of the partial pressure of the gas mixtures containing oxygen and nitrogen [41,42]. Due to high affinity of zirconium towards to oxygen rather than to nitrogen [43,44], the possibility of tuning the kinetics *versus* the thermodynamics parameters implies limitations on the composition of the gas mixture, which should always contain an excess of nitrogen. Among the different possibilities, the most promising results arose from the use of reactive gas mixtures varying from 95% N_2 + 5% O_2 (19:1 ratio) to 85% N_2 + 15% O_2 (17:3 ratio). As an example, Fig. **2** shows the variation of the concentration of the three elements of the ternary system as a function of the reactive gas mixture flow, for the films produced using the mixture with N_2:O_2 ratio 17:3 [42,45].

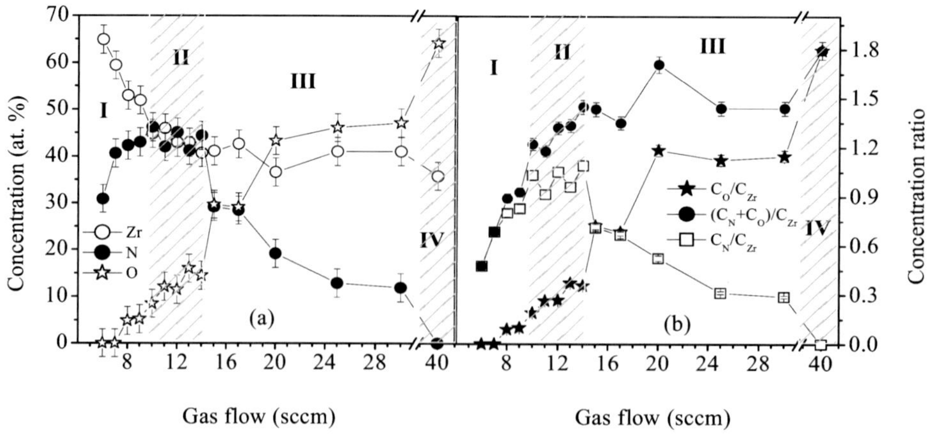

Figure 2: (a) Variation of the atomic concentration and; (b) Variation non-metals over the zirconium atomic ratios, C_O/C_{Zr}, C_N/C_{Zr} and $(C_N+C_O)/C_{Zr}$ as a function of the reactive gas mixture flow. An average uncertainty of 5 at.% was estimated for the RBS concentration. The uncertainty of the atomic ratios was determined by error propagation.

Fig. **2**. (a) shows that the composition of the prepared set of zirconium oxynitride films can be divided into 4 distinct regions, according to the particular features revealed by the plot. This figure shows that the films prepared with reactive gas flows below 10 sccm, evidence a continuous increase of the oxygen content up to approximately 6 at.% with the increase of the reactive gas flow, associated with an increase of the nitrogen content from ~31 to ~43 at.% and a systematic decrease of the zirconium content from about 69 down to about 52 at.%, respectively. This first group of samples will be noted hereafter as belonging to a first zone, named zone I. As a consequence, (Fig: **2b**) shows that, for the films prepared within this zone I, all the non-metals to zirconium atomic ratios, C_O/C_{Zr}, C_N/C_{Zr} and $(C_N+C_O)/C_{Zr}$ are also increasing when the reactive gas mixture flow increases. Particularly noticeable is the increase of the C_N/C_{Zr} ratio, which approaches 0.8, and thus providing evidence for some tendency to induce the formation of sub- and stoichiometric Zr-N-type compounds.

For gas flows increasing from 10 to 14 sccm, some important changes in composition seem to occur. These changes are characterized by an approximately constant concentration of both zirconium and nitrogen, associated to a slight increase of oxygen from about 9 to 15-16 at.%. This zone will be indexed in the text as zone II. Within this second zone, the C_N/C_{Zr} atomic ratio is roughly constant and close to one, which might be a good indication for the formation of close-stoichiometric ZrN compounds. The C_O/C_{Zr} atomic ratio increases from approximately 0.20 to 0.36-0.37, which gives indications about some possible oxygen inclusion in the film compounds. One possibility is the increase of oxygen incorporation/substitution in the zirconium nitride matrix, resulting in an over-stoichiometric oxygen-doped zirconium nitride matrix [42]. For the samples prepared within this second zone, the ratio $(C_N+C_O)/C_{Zr}$ increases from 1.22 to 1.45 due to the increased oxygen content, indicating the possibility of having other types of compounds such as thermodynamic non-stable phases (Zr_3N_4), which may also result from the oxygen doping of growing nitride films.

For the samples prepared with reactive gas flows between 15 and 30 sccm, Fig. **2**. (a) shows that the zirconium concentration is still approximately constant, but a clear decrease in the nitrogen concentration from about 44 at.%, at 14 sccm, to ~12 at.%, at 30 sccm, is observed. This change is followed by an increase of the oxygen concentration from about 15 at.%, at 14 sccm, to about 30 at.%, at 15 and 17 sccm, and from about 43 at.% to ~47 at.% from 20 to 30 sccm. In terms of atomic ratios, the value of C_O/C_{Zr} is about 0.7 for the first samples of this zone (15 and 17 sccm), evolving to approximately 1.2 for the remaining samples. The values of the C_N/C_{Zr} ratio decrease significantly from 0.7 to ~0.3, while the values of $(C_N+C_O)/C_{Zr}$ are roughly constant at a value close to 1.5. This evolution shows a significantly different behaviour in comparison to that one of zone II. This suggests a possible change on the atomic arrangements and thus possible changes in both compound formation and crystalline structures.

The samples deposited with the reactive gas flow higher than 30 sccm (indexed to zone IV) show a stoichiometry close to that of zirconium oxide, ZrO_2, with no detectable nitrogen by RBS. These results indicate a progressive replacement of nitrogen by oxygen within the matrix as a result of the dominant role of the thermodynamic factors over the kinetic ones.

Crystalline Structure

Due to the nature and characteristics of PVD processes, the structural properties of thin films are, in general, strongly dependent of the deposition parameters, such as deposition temperature, gas (working and reactive ones) pressures, impurities, ion bombardment *etc.* These deposition parameters are directly correlated with the final composition of the coatings and both factors (depositions conditions *per se* and composition of the coatings) influence the crystalline structure.

Thin films prepared by sputtering (such as those studied in this work) can range, in structural terms, from polycrystalline towards quasi-amorphous, depending on the composition of deposited materials and on the specific preparation conditions such as those mentioned above. Nevertheless, the structurally and/or compositionally designed thin films are often in a non-equilibrium state *e.g.*, characterized by compressive stresses, small grain sizes in the nanometre range, or metastable phases. In the polycrystalline case, the different composition, crystalline phase and even the specific orientation of the grains (with or without texture) may strongly influence the properties of the films and their final functional behaviour.

The structural characterization of the produced films revealed that four distinct behaviours are also present. These behaviours may be directly correlated with the previously defined composition zones, the set of samples prepared presents also four distinct behaviours. Fig. **3** shows X-ray diffraction patterns representative of each one of the four different compositional zones.

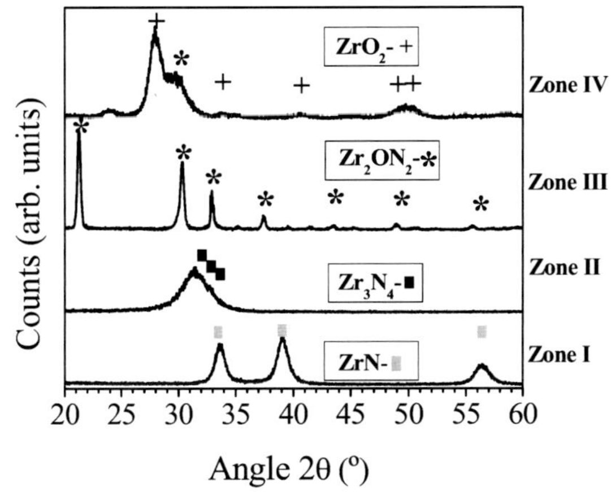

Figure 3: XRD patterns for samples from the four compositional zones.

This figure shows that in fact, the changes in composition induced by the variation of the deposition parameters (the mixed gas flow rate), result in different structural arrangements of the films, and thus, as it will be verified, have a major importance for all the properties of the coatings.

For films within compositional zone I, which revealed atomic ratios C_N/C_{Zr} up to ~0.8 and $(C_N+C_O)/C_{Zr}$ ~ 0.9, the XRD patterns revealed that the films crystallize in a B1-NaCl-type crystal structure (face centred cubic, fcc), typical of ZrN [46,47]. Analyzing the patterns of Fig. **4**, it is noticed that the increase of the reactive gas flow promotes a progressive change on the preferential growth from direction (111) to direction (200).

In general, all the DC sputtered ZrN_y thin films deposited under these conditions show strong <111> preferred orientations compared to ZrN powders, where the <200> intensity is the highest. For the films of zone I, when the C_N/C_{Zr} ratio approaches stoichiometry, the intensity of the (200) peak increases. In the literature it is found that the (200) preferential orientation is also associated to high nitrogen partial pressures [48].

Regarding the X-ray diffractograms of the films prepared with reactive gas flows between 10 and 14 sccm – zone II (Fig. **4**), the results suggest the development of a crystalline structure that does not match with that of *fcc* ZrN, revealing a smooth transition from the *fcc*-NaCl type structure of zone I towards a new one.

The reduced number of diffraction peaks with a significant broadness cannot allow a fully accurate indexing of this structure. The broadening of the peak at $2\theta \approx 32.2°$, which would correspond to the (320) plane of the oxygen doped orthorhombic, *orth*-Zr_3N_4 structure, could result from the small grain size and/or some reduction of the crystalline volume fraction of the coating, or even to a superposition of several peaks predicted in this region for this structure. The possibility of having a mixture of both *fcc*-ZrN(O) and a different type of structure (*orth*-Zr_3N_4) must be assumed, particularly for the first two samples in this zone, as suggested by the large tail observed in the diffraction peak between 30 and 33°. Moreover, for the sample prepared with the highest flow within this zone (14 sccm), the indexing of the peak located at $2\theta \approx 32.2°$ to a (111) peak of a ZrN-type structure is very unlikely, since its shift relatively to the ICDD standard position is too large (about 2°). The diffraction peak at $2\theta \approx 32.2°$ would then correspond to the (320) plane of the oxygen doped *orth*-Zr_3N_4 type structure or even to a mixture of several peaks.

Other authors have already indicated the shift of the (111) peak of the ZrN (NaCl type phase) towards smaller angles for films prepared with high nitrogen flows [49,50], which resulted in the development of a poorly crystallized *orth*-Zr_3N_4 phase. Lévy and co-authors [51] have noticed similar behaviour on magnetron sputtered ZrN_y thin films, claiming that for $y > 1.27$ the XRD diffraction patterns suggest the development of a particular structural arrangement consisting on a mixture of defect-rich *fcc*-ZrN and poorly crystallized *orth*-Zr_3N_4.

The development of over-stoichiometric structures in metal nitrides has being discussed for some time in the literature and it is known, and widely accepted, that in the case of heavier transition metals, there is a tendency towards an increased stability when chemical bonds are established, corresponding to highest oxidation states. For example, in the case of Zr and Hf nitrides, this enhanced stability of the +4 oxidation state in comparison to the common +3, allows the coexistence of the metallic nitride, MN (oxidation state +3) and insulating M_3N_4 (oxidation state +4) phases [52]. For zirconium oxynitrides prepared by PVD techniques or by nitridation of ZrO_2, the higher oxidation state (+4) is stabilized [52]. As a result, the stoichiometry of the oxynitrides tends to a mixture of ZrO_2 with Zr_3N_4, rather than the constant development of a *fcc*-ZrN-type phase (doped or pure) [52].

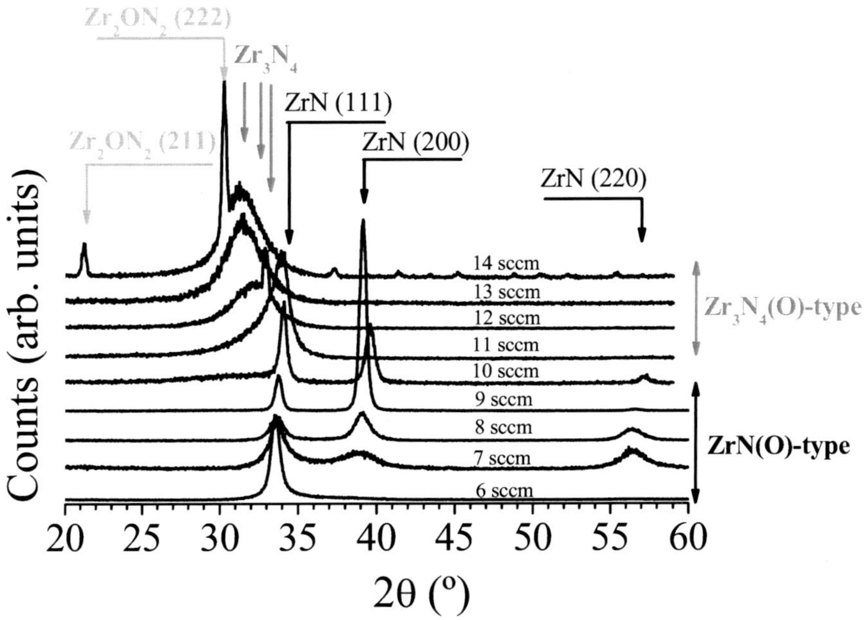

Figure 4: X-ray diffraction patterns of the ZrO_xN_y samples (deposited on Si substrates) prepared within the first two composition zones.

With further increase of the reactive gas flow from 15 to 30 sccm, the composition shows an abrupt increase of the oxygen content (Fig. **2**). This change is well correlated with the formation of another structure, which is observed by the change in the diffraction patterns (Fig. **5**), if compared to those of films from zone I and II (Fig. **4**). With the deposition conditions of zone III (gas flows from 15 up to 30 sccm), the diffraction peaks, including the very small ones, correspond to a cubic structure, which can be indexed to that of γ-Zr_2ON_2 (ICDD card Nº 48-1635) [53,35].

An analysis of the films composition ratios shows that the ratio C_N/C_{Zr}, in the films from zone III, decreases from about 0.8 down to approximately 0.3, and that C_O/C_{Zr} increases from about 0.7, in the first two samples, to about 1.2, for the last ones. Taking into account that in the stoichiometric Zr_2ON_2 compound, the ratios C_N/C_{Zr} and C_O/C_{Zr} are 1 and 0.5, respectively, it is concluded that the films in zone III have different atomic arrangements within the cubic lattice, where again oxygen is playing a fundamental role. In fact, oxygen is always in excess, in contrast to the deficiency of nitrogen, providing some evidence that the films may have significant lattice substitution/replacement in the nitrogen positions. Clarke *et al.* [54] claimed that this structure (cubic Bixbyite-type crystal structure [55]) is similar to that of fluorite- type from cubic ZrO_2, in which one-fourth of anion positions are unoccupied and two-thirds of the remaining oxygen atoms are replaced by nitrogen atoms. This arrangement means that in most oxynitrides, N and O atoms are statistically disordered, and M_2N_2O can be formally considered as a M_2X_3-type compound [54], where X corresponds to the non-metallic elements (O and N, in this particular case).

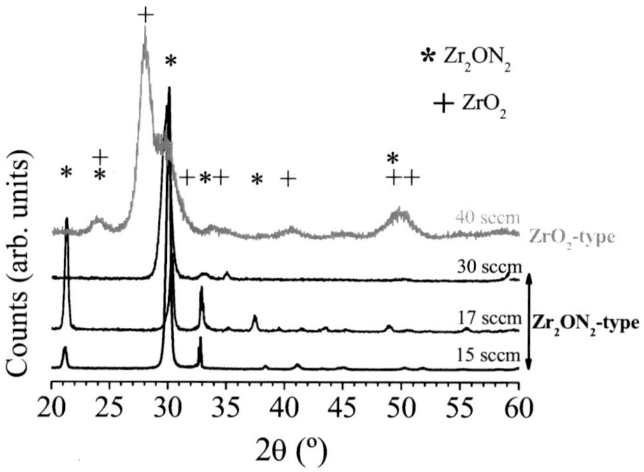

Figure 5: X-ray diffraction patterns of the ZrO_xN_y samples (deposited on Si substrates) prepared within zones III and IV.

Regarding the samples deposited with the highest reactive gas mixture flows (zone IV), the stoichiometry is close to zirconium dioxide. C_O/C_{Zr} is about 1.8 and

X-ray diffractograms confirmed the oxide phase formation, revealing the existence of a monoclinic-zirconium oxide phase, *mon*-ZrO_2 (ICDD card Nº37-1484).

Electronic Structure

Fig. **6** shows experimental XPS valence band spectra for representative samples lying in the different compositional and structural zones previously identified.

Assuming that there is no shift in the position of the Fermi energy, the valence band spectra characteristics of the previously identified four zones can be derived. The spectra are dominated by the Zr 4*p* levels, but the O 2*s,* N 2*s* and molecular orbitals associated with O2*p*/N2*p* and Zr 4*d* valence levels can also be observed. Bazhanov *et al.* [52] already reported that the valence band top in zirconium oxynitride is formed mainly by N 2*p* orbitals, while the valence band bottom is composed of O 2*p* orbitals, and that the conduction band in zirconium oxynitride structures consists predominantly of Zr 4*d* orbitals, with also a significant contribution of Zr 4*d* states to the valence band, indicating that the Zr–(O)N bonds in this compound have a significantly covalent character.

The first important observation is that the Zr 4*p* levels in the oxide zone (zone IV) appear chemically shifted by ~ 3.5 eV to higher binding energy with respect to the nitride (zone I). This shift is most likely due to the increase of the ionic character of the bond. Moreover, some authors have claimed that the zirconium oxide Zr *4p* peak presents a shift of ~ 4 eV to higher binding energies regarding to the metal [56,57]. In the nitride form, the Zr 4*p* core level exhibits an energy shift towards higher binding energy, being of about 1.5 eV in the case of Zr_3N_4-type films, if compared to the value of metallic zirconium [58]. Associated with this shift of the Zr levels, there is also a decrease of the band at the Fermi level, until it disappears for samples lying in zone III (oxynitride-type films zone), and a parallel growth of a broad band at energies between 3 and 8 eV. This could indicate a progressive transfer of zirconium 4*d* valence electrons to the O 2*p* and N 2*p* bands, until the complete depopulation of the *d-band.* The films become, therefore, insulator-type. From films within zone I to those in zone IV, the decrease of the *d band* at the Fermi level is associated to the decrease of the intensity of the N 2*s* peak and of the increase of the O 2*s* peak. This variation is resultant of the change from N dominant films to O dominant ones. The difference in energy of the *2s* peaks (N *2s* and O *2s*) reflects the increase of the nuclear charge, which results in contracted orbitals and lower-lying energy levels [59].

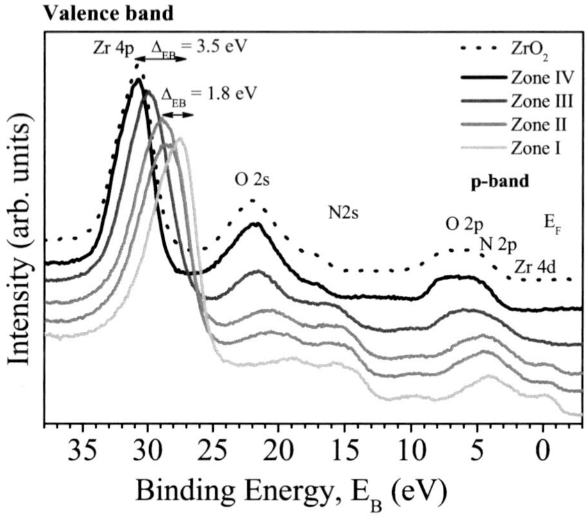

Figure 6: XPS valence band spectra for representative samples lying in the different zones.

Also from Fig. **6** it is possible to determine that the energy gap between the top of the valence band and the Fermi level is ~5eV (energy distance between the *p* band and Fermi level) for the films corresponding to zones III and IV, reinforcing the idea of a progressive tendency to form insulating-type films. This could be interpreted as coming from the large energy difference between the O 2*p* and N 2*p* bands and the *nd* orbitals that promotes an intrinsic gap between the *p* and *d* bands in the material. In contrast to fcc-zirconium nitride (films within zone I), which is a metallic conductor, monoclinic zirconium oxide (and corresponding films from zone IV) is an insulator. The electronic configuration of $4d^0$ valence electrons located in the O 2*p*-derived band, at energies between 4 and 11 eV, with a large energy difference between the 2*p* and the *nd* orbitals originate an intrinsic band gap [59].

Morphology

Fig. **7** presents SEM micrographs of the cross-section of the films that were previously indexed to zone I, showing the correspondent morphologies. For comparison purposes, Fig. **7** (d) shows the cross section of a "pure" ZrN film prepared in similar conditions as those of the ZrO_xN_y ones.

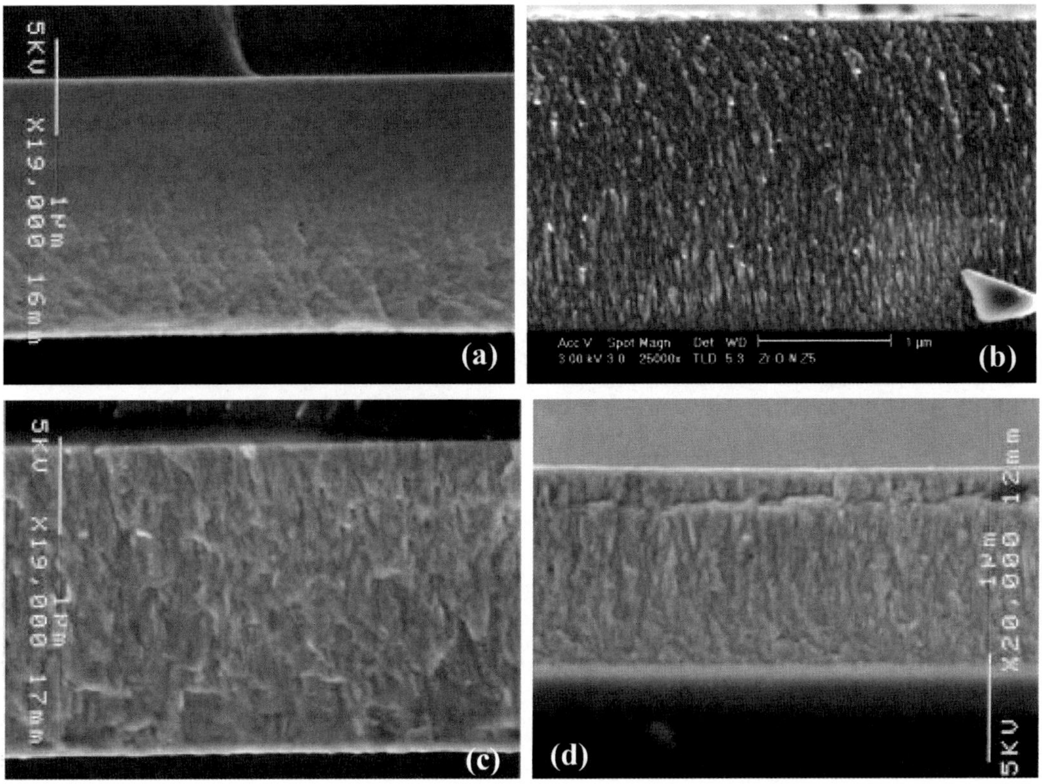

Figure 7: SEM cross-sections of the films lying in the first zone. Films prepared with a gas mixture flows of: a) 6 sccm; b) 8 sccm and c) 9 sccm. The micrograph presented in d) corresponds to a "pure" ZrN film.

From this figure it is possible to observe that the films grow with a compact fibrous microstructure, with no clear development of the very common PVD columnar-type morphology. The increase of gas flow (towards the stoichiometric condition) seems to result in less compact microstructures and more fibrous-type films, or even to what may be called a "quasi" columnar-type growth. If related to the Thornton's zone model, the samples in this zone could be indexed to zone I, which is consistent with the ratio of the film melting and deposition temperature since these coatings present a T_s/T_m ratio lower than 0.14. Fig. **8** presents SEM cross-section micrographs of the samples that were indexed to zone II. From the observed micrographs, a first point seems quite clear.

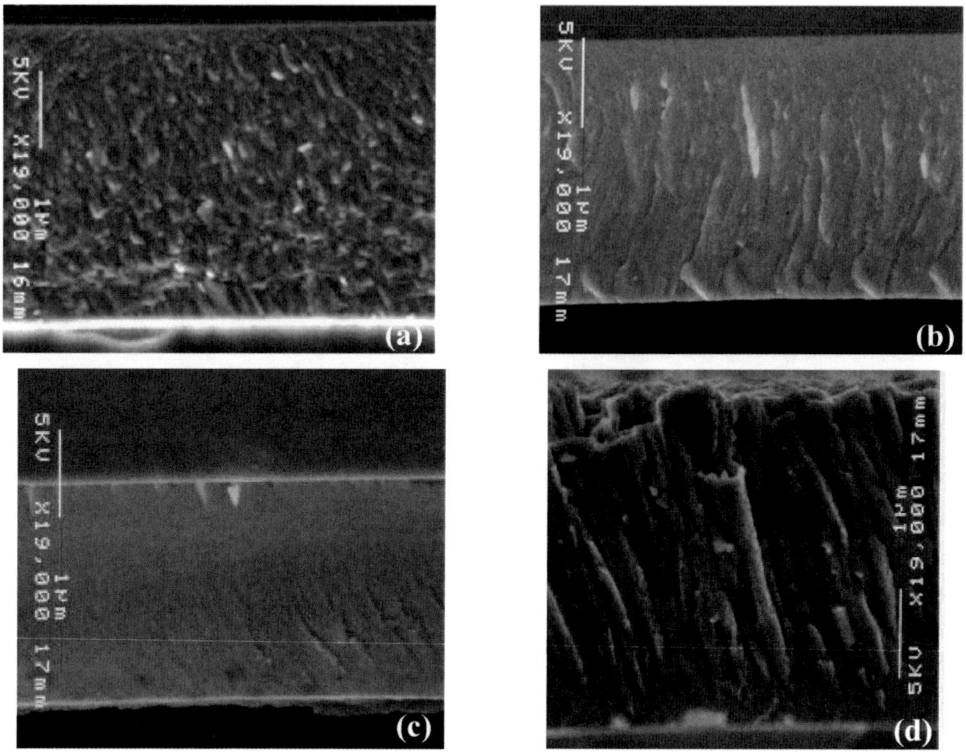

Figure 8: Cross-sections observed by SEM of the films lying in the second zone; film prepared with a gas mixture flow of: (a) 10 sccm;(b) 11 sccm; (c) 13 sccm; and (d) 14 sccm.

There is a notorious difference between the edge samples (samples prepared with gas flows of 10 sccm, Fig. **8** a) and 14 sccm, Fig. **8** d)) and those within this second zone. In fact, the sample deposited with a reactive gas flow of 10 sccm still presents clear traces of the fibrous microstructure characteristic of the films prepared within zone 1, while the sample prepared with 14 sccm is already revealing some differences from the previous samples, similar to what will be shown later for samples of zone III.

Excluding these edge samples, SEM observation indicates that there is a clear change in the type of growth, which changed from the fibrous microstructure (zone I) to quasi-featureless morphologies, (Fig. **8** (b) trough (c)). In spite of these differences, the characteristic morphologies of these films should also be indexed to zone I of Thornton's zone model. Regarding the upper limit of this second zone (sample prepared with a gas flow of 14 sccm represented in Fig. **8** (d)), the sample starts to develop a typical columnar microstructure with a characteristic V-shaped morphology, where the columns present a dome-rounded shape top. This sample already shows some similarities to the morphological features of Thornton's transition zone, zone T.

With further increase of the reactive gas flow, zones III and IV, the micrographs of samples cross-sections (Fig. **9**) show that there are no significant changes in the films morphology. Anyway, the samples from zone III (Fig. **10** (a)) present a less pronounced columnar morphology when compared to the sample deposited with a reactive gas flow of 14 sccm (upper limit of zone II), but the columns present also a dome-rounded shape top.

The sample from the last zone (zone IV) presents a columnar morphology, as well, but the top has less pronounced dome-rounded shape. In terms of an indexation to the Thornton's zone model, both zones seem to develop thin films with a morphology that could be located between zone I and zone T. The overall set of morphological features shows that the different compositional zones are also present in the microstructural arrangements, which reveal some changes according to the transition between the previous identified zones.

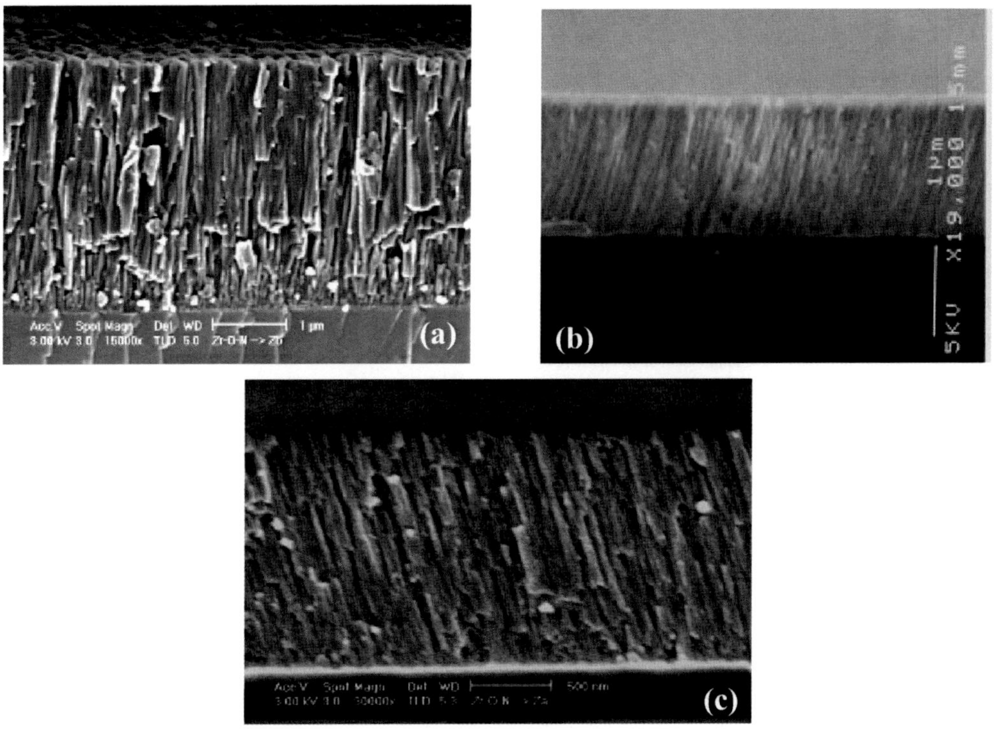

Figure 9: Cross-sections observed by SEM of the films lying in zones III and IV; film prepared with a gas mixture flow of: (a) 17 sccm (zone III) and (b) and (c) 40 sccm (zone IV).

PROPERTIES OF ZIRCONIUM OXYNITRIDE THIN FILMS

Electronic Properties

Following composition and bonding characteristics evolutions, together with the specific types of structural and morphological features, the electrical response of the ZrO_xN_y films show, as well, a wide range of different behaviours, as it can be observed by the evolution of the electrical resistivity at room temperature, as a function of the reactive gas flow (Fig. **10**).

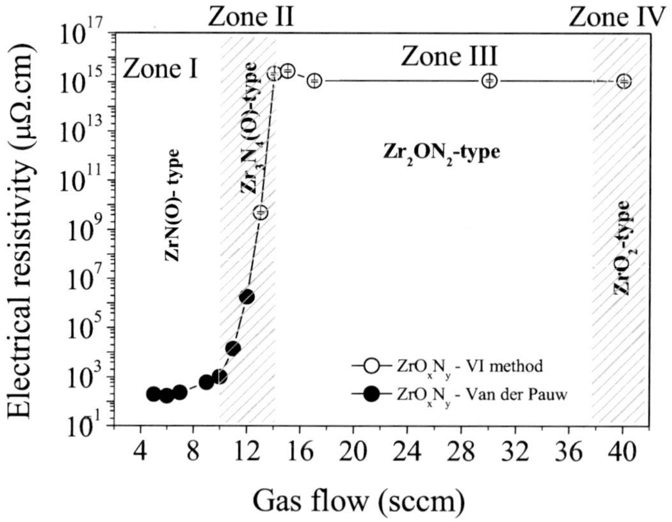

Figure 10: Electrical resistivity variation at room temperature, as a function of the reactive gases, for the zirconium oxynitride system. The error bar was determined by maximum deviation to the average value.

From this plot, it is possible to observe that the evolution of the electrical resistivity provides once again clear evidence of the existence of different zones, although it seems also clear that there are some important differences if the resistivity is directly correlated with the composition and structural features. First of all, it seems that, in terms of electrical resistivity, only three major zones may exist, in comparison to the already claimed four zones for the other properties previous analyzed. In addition, the electrical resistivity of the films prepared within zone II, present a variation of this property of about thirteen orders of magnitude, although the composition within this zone has shown to be varying only slightly (Fig. **2**). Anyway, the XRD patterns showed some important variations, which may induce some explanations for this large variation of resistivity within films with similar composition.

In terms of specific analysis, the films from zone I, (which presented a sub-stoichiometric condition and XRD patterns typical for *fcc*-ZrN), revealed relatively low values of resistivity (around 200 μΩ.cm), which allows to index these films as "roughly" metallic-type. These values are also consistent with those found in the literature for ZrN films [47, 50]. In this zone, the XPS valence band spectrum (Fig. **6**) showed a non-vanishing *d* band near the Fermi energy, with an overlapping between the *d* and *p* bands. This non-vanishing DOS at the Fermi energy should explain the metallic conductivity of these coatings [60]. Schwarz *et al.* [10] claimed that this overlapping of *p* and *d* bands is a sign of metallic-type materials. The films allocated in zone II, where an oxygen doped Zr_3N_4-type phase was indexed (Fig. **4**), associated with the development of over-stoichiometric nitride compounds (see Fig. **2**), show the most interesting behaviour. The resistivity is varying from low metallic-type values (few hundreds of μΩ.cm) at the lower gas flows, to highly insulating-type ones (between 10^{10} and 10^{15} μΩ.cm) at the highest gas flows. The composition analysis for the samples within this second zone (see Fig. **2**) shows only slight changes in the N and Zr content.

The main important change corresponds to the O content, which varies from ~9 at.% to about 15 at.%. The $(C_N+C_O)/C_{Zr}$ varies from 1.22 to about 1.45. The significant increase of resistivity in this zone might be resultant from the progressive change of an oxygen-doped ZrN-type structure towards a nitrogen+oxygen rich phase similar to that one of Zr_3N_4. Several authors reported a significant increase of electrical resistivity of ZrN thin films, when the C_N/C_{Zr} ratio is approximately 1.2–1.3, which results in the change from a metallic-type ZrN structure of the films to an insulating-type Zr_3N_4 phase [50,58]. Anyway, this increase of electrical resistivity cannot be attributed exclusively to the increase of oxygen (Fig. **2**) and the consequent increase of the ionic character of the bonds, since there is a similar oxygen increase in zone I (Fig. **2**), but with almost no effect in the electrical resistivity. Chappé *et al.* [61] claimed, for the TiO_xN_y system, that the increase of oxygen content and the increase of ionic bonding character are responsible for the smooth increase of electrical resistivity in the transition between the nitride and oxide sputtering regimes.

The electronic structure, correlated to the compositional and structural features, may bring some extra insights for the evolution of the resistivity. The XPS valence band plots (Fig. **6**) allowed to identify a decrease of the *d* band near the Fermi level, associated with an increase of the *2p* band. This result, with the incomplete vanishing of the *d* band, indicates an increase of the ionic bond character and thus a tendency to increase the electrical resistivity of the films. Prieto *et al.* [62] claimed that the change from metallic to insulating electrical behaviour results from the complete depopulation of the *d* band just below the Fermi level, due to the large transference of Zr *d* electrons to nitrogen. In the same line of thought, J.M. Sanz and co-workers [58] claimed that the opening of a band gap in over-stoichiometric zirconium nitride compounds, developing the Zr_3N_4-type phase of zone II, could correspond to a percolation transition. The incorporation of nitrogen atoms prevents the formation of Zr-Zr paths, so that the conduction band minimum becomes determined by Zr-N anti-bonding states, and the valence band maximum becomes determined by N *2p* bonding states. Furthermore, these authors also claimed that the intensity of the *d*-band at the Fermi level of zirconium nitride (ZrN_y) depends linearly on the average nitrogen content (y). The density of electrons per atom of zirconium (*i.e.,* nd) could vary linearly according to the formula nd = 4-3y, with a complete suppression of the *d*-band for y = 1.33 (*i.e.,* in the presence of a Zr_3N_4-type compound).

In the case of zirconium oxynitride system, the complete depopulation of the *d* band is not observed. Nevertheless the observed incomplete depopulation of the *d* band could result from the sputtering cleaning process. This argument is reinforced by the observed negative slope in the Arrhenius plot (Fig. **11**). A "real" incomplete depopulation of the *d* band should lead to a dependence of the electrical resistivity with the temperature with an opposite behaviour to that one observed in Fig. **11**. It should be observed a decrease of electrical conductivity (increase of electrical resistivity) with increasing temperature.

The plot of the electrical resistivity as a function of temperature (Fig. **11**) confirms the particular behaviour of the samples belonging to zone II. Although, it should be emphasized that the sample deposited with a gas mixture flow rate of 14 sccm, with resistivity values nine orders or magnitude higher than the first sample within this zone, shows no significant decrease of resistivity with temperature.

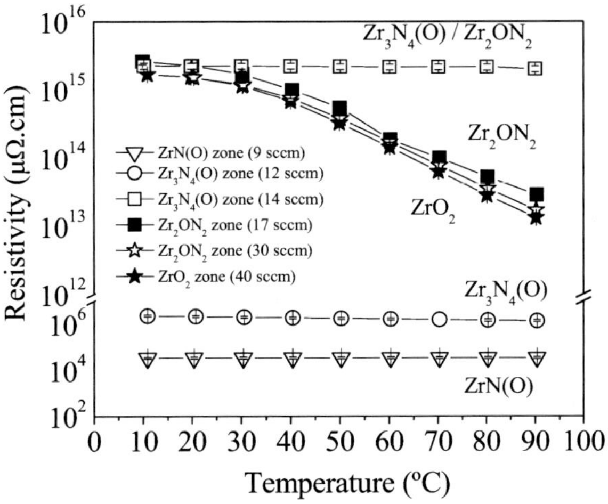

Figure 11: Electrical resistivity evolution as a function of temperature. The error bar was determined by maximum deviation to the average value.

Within zones III and IV, the amount of oxygen is already significant and the tendency to have highly insulating phases, such as Zr_2ON_2 and ZrO_2, is enhanced. Therefore, the films tend to be of insulating type. In these two zones, the samples show relatively high values of electrical resistivity (Fig. **10**), and the conductivity is thermally activated (Fig. **11**), similar to what happens with the zirconium oxide films. With the formation of the insulating phases of Zr_2ON_2 and ZrO_2 and the progressive formation of a gap between the p and d bands, the transport mechanism is thermally activated (Fig. **11**). As already discussed previously, the samples lying in the third and forth zones show a complete depopulation of the d band at the Fermi level, with a continuous shift to higher energies of the p band. This fact explains their particular electrical resistivity behaviour.

OPTICAL PROPERTIES

Following all the previous analyses, and in particular the changes in the electronic behaviour, optical properties of the ZrO_xN_y films show, as well, a wide range of different responses. The variation of the colour coordinates of the coatings, as a function of the reactive gas flow, according the CIELAB 1976 colour space [63,64], are shown in Figs. **12** and **13**. It is important to refer that only the samples that were prepared with gas flows up to 15 sccm are presented in these two plots, since this flow was the higher limit to obtain intrinsic coloured samples.

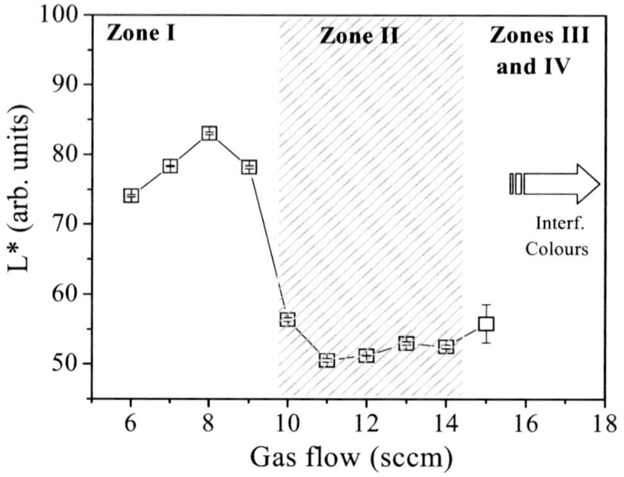

Figure 12: L* coordinate of the CIELAB 1976 colour space under the standard CIE illuminant D_{65}. The error bar was determined by maximum deviation to the average value.

Beyond this range of gas flows the samples revealed a characteristic rainbow-like interference effect. Regarding the brightness values, two major regions can be observed, see Fig. **12**. The first of these regions (corresponding to previous indexed zone I) presents high values of L*, characteristic of the sub-stoichiometric group of samples, with non-metal to zirconium atomic ratios varying from ≈ 0.5 to 0.9. These films develop structural arrangements very similar to those of ZrN-type. The second region includes all the samples that were previously indexed to zone II.

In order to understand these differences between the samples of the first two zones, one must keep in mind that the brightness of metallic materials is determined by interactions between incident photons and free electrons [65]. Thus, the decrease in brightness for the samples of zone II should be associated to a decrease in the density of states at the Fermi level [66]. The samples within zone II showed non-metal to zirconium ratios varying from 1.22 to 1.45. Moreover, XRD measurements were consistent with the development of a structural arrangement where some oxygen could be present in an orthorhombic Zr_3N_4-type structure.

Regarding the chromaticity coordinates (Fig. **13**), the distinction between the two zones is again quite clear. When the flow rate increases from 6 to 9 sccm (samples from zone I) the values of a* show no significant changes, but the positive b* values (yellowness) increase significantly. The colour tones of the films changed from a very bright pale-yellow (sample deposited with a gas flow of 6 sccm) to golden (9 sccm sample) with an increase of the yellowness (increase of the positive value of b* (Fig. **13**)). With further increase of the reactive gas flow, and the consequent changes in composition and the formation of a new structure (Z_3N_4-type one), the samples showed a significant decrease of brightness (L* values) and the colour changed from red-brownish to purple to dark blue.

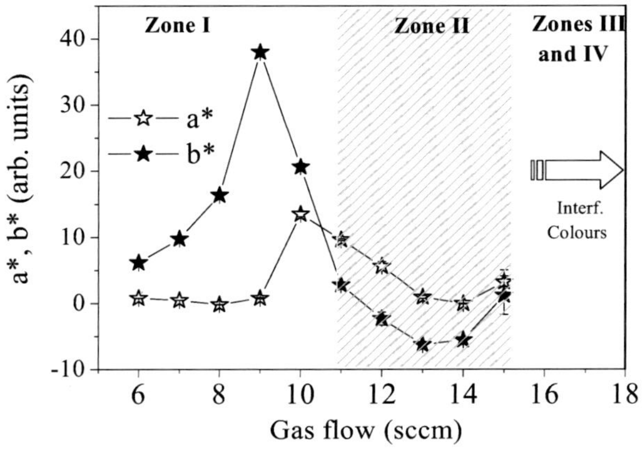

Figure 13: Chromaticity coordinates (a* and b*) of the CIELAB 1976 colour space under the standard CIE illuminant D_{65}, for films prepared with different reactive gas flows. The error bar was determined by maximum deviation to the average value.

The reflectivity spectra of the films deposited with reactive mixture flow values up to 9 sccm is shown in Fig. **14**(a). In this plot, each spectrum is associated to the reactive gas flow of the sample and to the corresponding non-metal to zirconium ratio, $(C_N+C_O)/C_{Zr}$. To establish some possible correlations between the optical and electrical behaviour of the films, Fig. **14**(b) shows also the variation of the electrical resistivity of these films.

Fig. **14**(a) shows that all the films present high values of reflectivity in the near-infrared region (NIR - $\hbar\omega < 1.7$ eV), a continuous decrease of the reflectivity in the visible region ($1.7 < \hbar\omega < 3.1$ eV), and a reflectivity minimum in the UV region ($\hbar\omega > 3.1$ eV). The high values of the reflectivity in the NIR region are

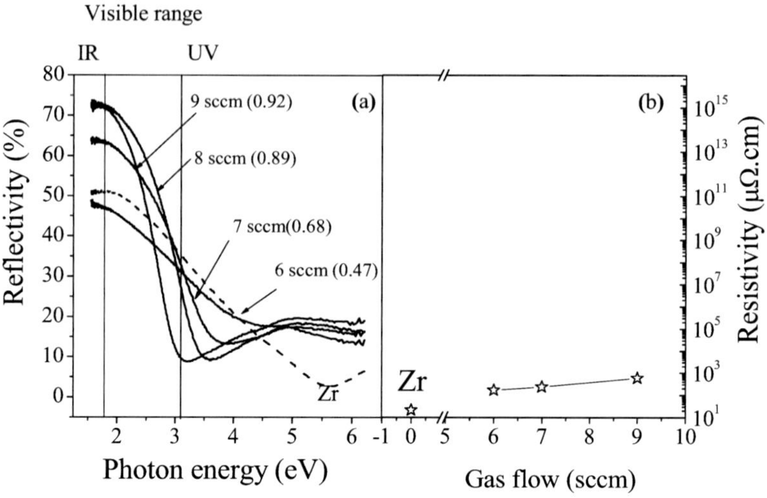

Figure 14: (a) Reflectivity spectra for samples from zone I as a function of the photon energy; (b) corresponding electrical resistivity values as a function of the reactive gas flow.

typical of good conductors and are the result of the free-carrier absorption mechanisms. The high concentration of conduction electrons in the films is explained by an overlap between the *p* and *d*-type bands, so that the Fermi level lies inside the *d*-band [59], as already was discussed previously. This gives rise to electrical and optical "metallic-like" behaviour of the films belonging to this first zone. The results of the electrical resistivity measurements are consistent with this conclusion, showing relatively low values, in the order of few hundreds of μΩ·cm, increasing slightly with the increase of gas mixture flow, in agreement with the gradual shift of the plasma edge seen in Fig. **14**(a). The increase of the non-metal content within this first set of films appears to favour higher values of the reflectivity in the NIR region and the formation of a sharper edge in the reflectivity spectrum. This behaviour of the reflectivity, revealing high values in the NIR, and transparency for shorter wavelengths, puts, in a first approximation, these films as being good candidates to be used, for example, as heat mirrors [67]. Among all the studied films, the highest reflectivity at low photon energies is reached for the nearly stoichiometric sample (8 sccm), which presented a golden yellow colour tone, typical of stoichiometric ZrN. Note that this sample exhibits a minimum reflectivity value of about 10 % at 3.5 eV (within the UV region). The reflectivity starts decreasing in the blue–green region (≈ 2.5 eV), approaching the level of 25 % in the violet wavelength range (≈ 3.1 eV). The result of the increase of the reflectivity in the higher wavelength (low photon energy) region of the visible spectrum gives rise to yellow colour and, since there is low reflectivity in the short wavelength (high photon energy) the colour appears as golden-like.

The samples deposited at lower reactive gas flows (< 8 sccm) show a less pronounced decrease of reflectivity with the increase of the photon energy in the visible region ($1.7 < \hbar\omega < 3.1$ eV). Compared to the sample produced with 8 sccm, these samples present a lower effective reflectivity in the low photon energy region of the visible spectrum, and not so low reflectivity in the high photon energy region. This means that the reflectivity in the visible region doesn't present significant differences and consequently the samples evidence a silver-like colour. Another important fact to be pointed out, for the samples prepared within zone I, is the already mentioned shift of the reflectivity minimum to lower energies with the increase of the gas flow (increase of the non-metallic content, O + N). As first suggested by *B.* Karlsson *et al.* [68], the observed red shift of the reflectivity minimum can be attributed to the decreasing role of the conduction electrons in determining the optical properties and, hence, to the lower metallic character of the coatings [69]. Associated with this shift, it is noticed the increase of the reflectivity maximum at lowest energies and the increase of the slope of the reflectivity in the visible region with the increase of the non-metal content (Fig. **14**(a)).

The plasmon frequency, ω_p, which is characteristic of the plasma edge in the reflectivity spectra is given by [70]:

$$\omega_p^2 = \frac{4\pi N_e e^2}{m_e} \qquad \textbf{(1)}$$

with m_e being the effective mass and N_e the number of conduction electrons per unit volume (*i.e.,* the concentration). The concentration of the conduction electrons depends on the metal content in the coating, so that an increase in the non-metal content will decrease N_e and, accordingly, the shift of the plasmon energy, $\hbar\omega_p$, towards lower photon energy values as may be observed in the reflectivity plot (Fig. **14**(a). *Bendavid et al.* [71] pointed out that the strong red shift of the Plasmon energy may also suggest a change of the conduction band nature. However, in the present case, the variation of $\hbar\omega_p$ with increasing reactive gas flow (particularly in the case of the samples deposited with reactive gas flows of 7, 8 and 9 sccm (Fig. **14**(a)) is relatively smooth and, probably, should be taken into account for the transfer of the *d* electrons into the *p*-band (decrease of N_e). The sample deposited with a reactive gas flow of 6 sccm shows no well-defined plasma edge. One possible explanation is the high amount of structural defects in the film, which results from its considerably sub-stoichiometric condition. The high concentration of defects decreases the electron mobility and, consequently, increases the plasmon damping, leading to an overdamping of the plasma oscillations of the conduction electrons. Later in the text, a detailed analysis of surface defects will be carried out.

The observed variation of the reflectivity spectra agrees qualitatively with previously reported results for *fcc*-ZrN [50,51], measured with increasing nitrogen contents. However, the zirconium oxynitride system, with similar metalloid contents, presents always a shift of approximately 0.5 eV of the reflectivity minimum towards lower energies. In order to understand this, one has to keep in mind that the electronic structure of ZrN is characterized by the existence of a partially filled *d*-band (where the Fermi level is located) and a second band, located about 6 eV below, formed mainly by N 2*p* states. For the ZrN(O) system, the *p*-band is composed by N 2*p* and O 2*p* sub-bands. The transfer of the *d*-electrons occurs also to the O 2*p* sub-band, and probably this process influences the reflectivity spectrum. In fact, the replacement of one atom of nitrogen by an oxygen one, in the unit cell, leads to the formation of a hole in the valence band [52]. Consequently, the position of the Fermi level in the *d*-band could change, affecting the Plasmon frequency. Moreover, and since the oxidation state of oxygen is -2 and the one of nitrogen is -3, the difference in electronegativity between O-Zr (2.1) and N-Zr (1.7), results in a higher ionic character of the former. According to the Pauling theory [72], the ionicity of a single bond increases with the difference of the electronegativity values between the two elements forming the bond. The higher ionic character could explain the 0.5 eV shift of the reflectivity minimum for the ternary system ZrN(O), through the different correlation contribution to the Fermi energy.

Beyond this, the metallic behaviour of the material can explain the high values of the reflectivity in the NIR region. The Drude model must be completed by taking into account the interband absorption in order to explain the reflection minimum in the UV region [68,69]. This behaviour is quite similar to that one observed in the noble metals, where interband transitions are responsible for their characteristic colour [73]. For photon energies higher than the reflectivity minimum, the interband transitions start to dominate [74]. This can be taken into account phenomenologically by introducing a Lorentz oscillator, centred at ≈ 5 eV ($\lambda \approx 250$ nm), into the dielectric function, which can be taken into account for the optical properties of the material in the UV range [47]. Note, however, that this is a rather crude approximation, possible only for narrow bands. Fig. **15**(a) shows a summary of the reflectivity behaviour of the zirconium oxynitride samples lying in zone II, as a function of the photon energy. Again, and to establish a parallel between both optical and electrical behaviours, this figure shows also the electrical resistivity values of this group of samples, as a function of the reactive gas flow.

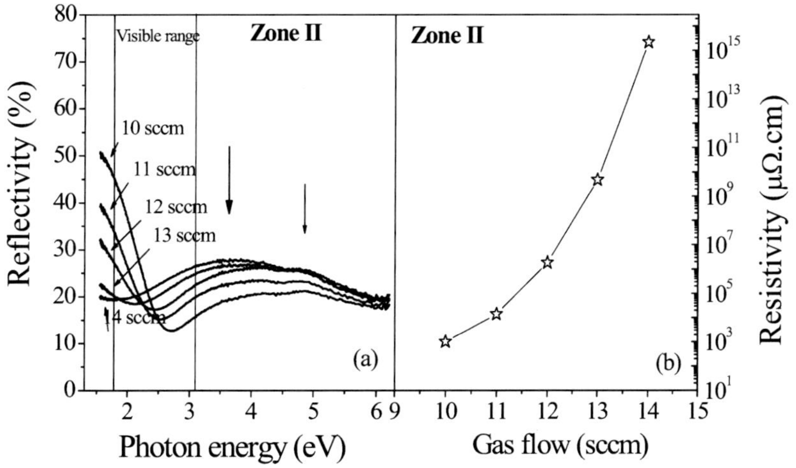

Figure 15: (a) Reflectivity spectra for samples from zone II as a function of the photon energy; b) corresponding electrical resistivity values as a function of the reactive gas flow.

The first observation to be noticed is the clear reduction of the reflectivity in the NIR region and in the visible region close to the IR region, and the continuous shift of the reflectivity minimum to lower photon energy values. Let us keep in mind that this shift has already been observed for the set of samples indexed to zone I, which was attributed to the decrease of the role of free electrons. Regarding the values of the reflectivity maximum in the IR region, there is a clear reduction within the samples indexed to this zone when compared to those from zone I (from ~75 to 50 % for zone I and from 50 to 20 % for this second zone). This significant decrease of the reflectivity maximum near de IR region indicates a clear reduction of the role of free electron absorption mechanism for the optical properties of the coatings. Moreover, a consistent reduction is noticed within all samples belonging to this second zone, reaching values as low as 20 % for the samples prepared with the highest gas flows. This evolution is well correlated with the evolution of the results obtained by XPS valence spectra, discussed previously in this chapter, where a continuous decrease of the Plasmon loss (1-3 eV) and a continuous depopulation of the *d*-band were observed, when going from films within zone I to those in zone II (Fig. **6**).

Finally, and to complete the parallel between both optical and electrical behaviours, Fig. **16** shows the reflectivity and electrical resistivity of the samples that were indexed to the last two zones, III and IV. In these plots it is possible to observe that the samples of these last two zones present semi-transparent optical behaviour (interference fringes), and very high electrical resistivity values, in straight agreement with the already reported compositional, crystalline and electronic structure results. Further, and as already mentioned during the valence band XPS discussion, these coatings present a complete depopulation of the *d* band at the Fermi level, and a parallel growth of a broad band at energies between 3 and 8 eV, and thus the films became of semi-transparent-/insulating-type. Another point that is important to reinforce is the evolution of the optical energy band-gap of the samples, where this parameter may be calculated: zones II, III and IV. Direct or indirect optical transitions are possible in a polycrystalline material depending on the band structure [75].

The nature of the transition involved (direct or indirect), during the optical absorption process, can be determined by studying the dependence of the absorption coefficient α on the photon energy $(\hbar\omega)$ as [76]:

$$\alpha = K\left(\hbar\omega - E_g\right)^{a/2} / \left(\hbar\omega\right) \quad \textbf{(2)}$$

where K is a constant, E_g is the gap between the valence band and conduction band, a is a constant equal to 1 for direct gap materials, and 4 for indirect gap material. The usual method to determine band gap is to plot $(\alpha\hbar\omega)^{1/a}$ against $(\hbar\omega)$ and get value of a, which gives the best linear tendency in the band edge region.

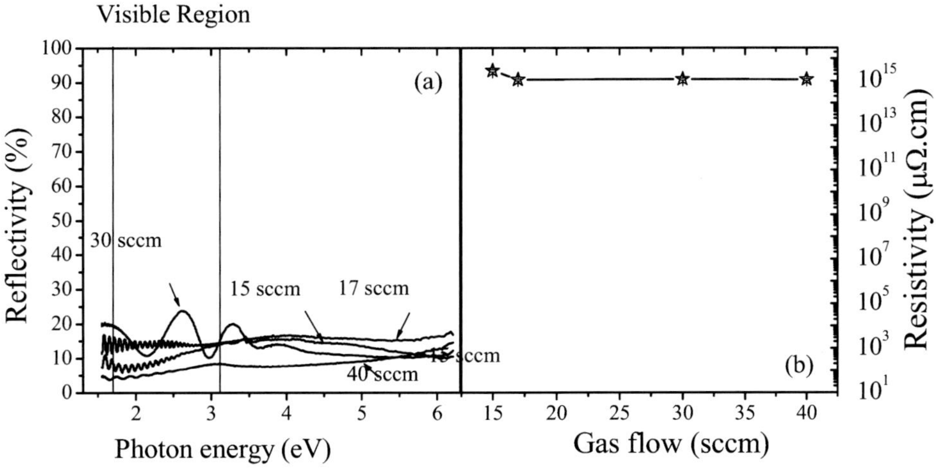

Figure 16: (a) Reflectivity spectra for samples from zone III and IV as a function of the photon energy; (b) corresponding electrical resistivity values as a function of the reactive gas flow.

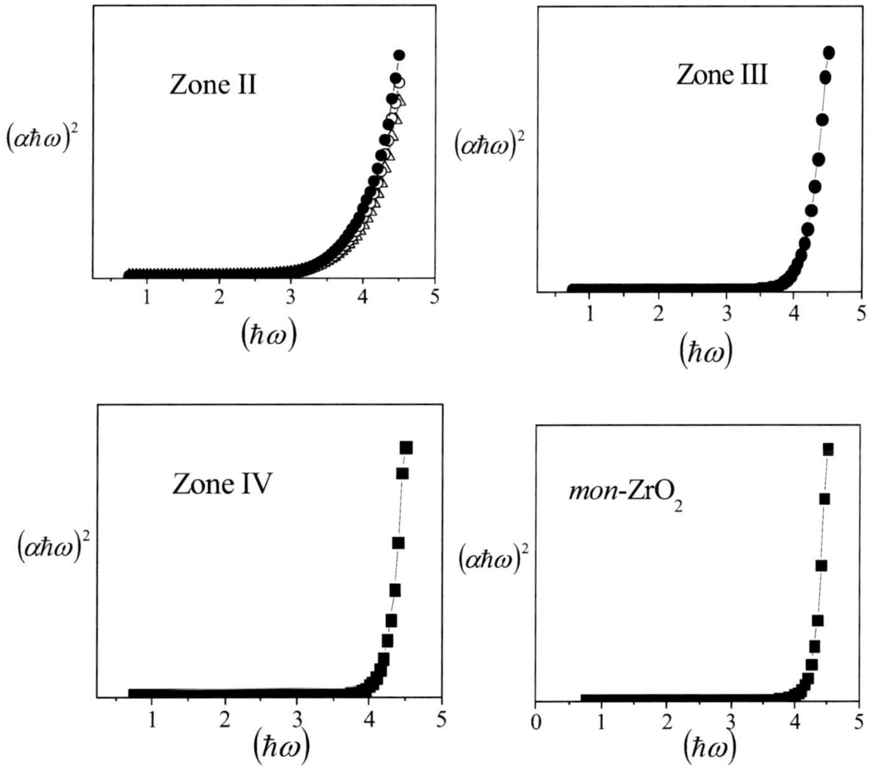

Figure 17: $(\alpha\hbar\omega)^2$ *vs.* $(\hbar\omega)$ for representative samples from zones II, III and IV. For comparison purposes, the result from a "pure" ZrO_2 sample is also shown.

Plots of $(\alpha\hbar\omega)^2$ *vs.* $(\hbar\omega)$ and $(\alpha\hbar\omega)^{1/2}$ *vs.* $(\hbar\omega)$ were analyzed for all the samples of the last three zones. Fig. **17** shows plots of $(\alpha\hbar\omega)^2$ against $(\hbar\omega)$ for representative samples lying in zones II, III and IV.

The observed steep increase of the absorption on the plots implies that there is a direct transition nature for zirconium oxynitride films. From the previous plots, the band-gap energy was determined by extrapolating the linear part of the curve to zero absorption.

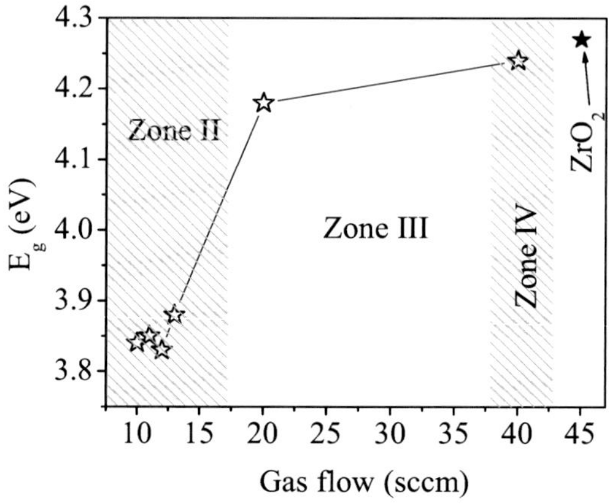

Figure 18: Optical gap for representative samples lying in zones II, III and IV. For comparison purposes, the result from a "pure" ZrO_2 sample is also shown.

From this plot it is possible to observe an abrupt increase of the direct transition energies, when going from samples within zone II to those in zone III. Regarding the increase of the E_g for the samples of the last two zones, the main reason should result from the decrease of narrowing of the *p band*, associated to the decrease of N 2*p* bands contribution for the *p band*, already described in the XPS section. Since for the nitrides and oxides, the energy difference between the Fermi level and the *p* group of bands has to be overcome for the interband transitions to start, the decrease of N 2*p* bands contribution to the *p band* should explain the increase of E_g. Asahi *et al.* [77] found that the substitutional doping of N in TiO_2 leads to the band gap narrowing by mixing N 2*p* states with O 2*p* states. In the present case, with the increase of oxygen content, there is a decrease of N 2*p* bands contribution for the *p band*. Thus, the overlap between N 2*p* states and O 2*p* states decreases, leading to an increase of the band gap.

FUNCTIONAL PROPERTIES

Hardness and Young's Modulus

In the previous section it was shown that the variation of the reactive gas mixture flow has a significant effect on the electronic states, composition and structural/microstructural features of the films, which were then correlated with the changes on the electrical and optical properties of the films. In the present section, the changes in the films characteristics will be also correlated with the functional properties of the films, namely with the evolution of the mechanical properties, as illustrated in Fg. 19, which illustrates particularly the changes in the films' hardness (H_v).

Typical loading/unloading curves measured for representative coatings belonging to each of the four identified zones are presented. By quantifying the "area" located between the loading and unloading curves one obtains the energy dissipated in the coating due to plastic deformation. The "area" under the unloading curve allows obtaining the value of the energy associated to the elastic deformation. All the coatings present loading and unloading curves that are approximately parabolic. This is a good indication that the elastically accommodated deformation is significant. Anyway, it seems clear that the coatings prepared within the last two compositional zones (zone III and IV) present a more abrupt decrease on the unloading curve (Fig. **19**). This could indicate a decrease of the elastic strains compared with the plastic strains, resulting in a decrease of the elastic recovery of the indentation process. The zirconium oxynitride films with intrinsic colorations (Zone I and II), present similar hardness values and, in some cases, even higher than those of zirconium nitride. Furthermore, the increase in hardness of the films belonging to the first compositional/structural zone (samples that presented a *fcc*-ZrN-type structure) is also evident - varying from ~23 up to ~42 GPa.

This initial increase is then followed by a systematic decrease of this film property, reaching a value of about 12 GPa for the sample prepared with the highest gas flow (40 sccm), which revealed an oxide-type composition and structure. Anyway, and in spite of this decrease in hardness being also observed throughout zones II to IV, there are few differences that are worth to point out. In fact, it is possible to observe that, within zone II, there is a much more pronounced decrease in the hardness values, when compared to what happens in zones III and IV.

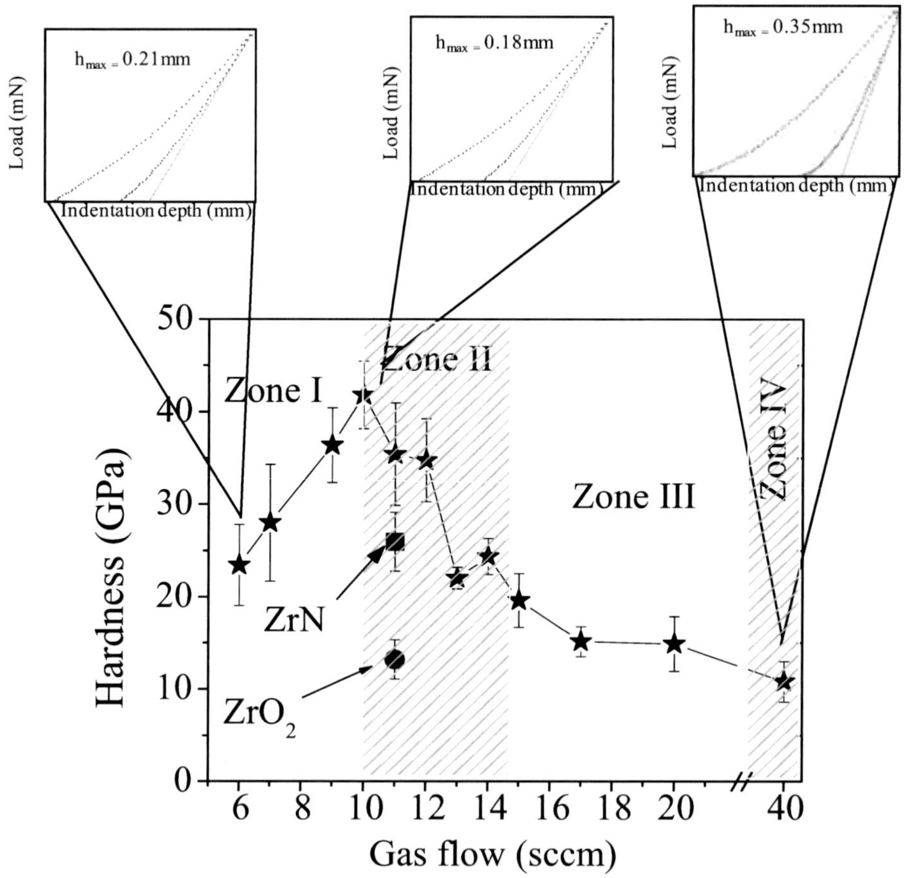

Figure 19: Hardness values as a function of reactive gas flow. The loading and unloading curves are also shown as a function of indentation depth. The error bar was determined by maximum deviation to the average value.

In terms of the main features that may explain this particular behaviour of hardness, it is important to state that, first of all, this evolution is well correlated with the structural (and thus, its precursor – the composition) evolution. In fact, a closer look to Fig. **19** shows that the previous indexed zones have particular trends that corroborate different behaviours according to those different zones. It is clear that the change from zone I to zone II corresponds exactly to an inversion in the hardness evolution tendency. The continuous decrease of the hardness within the last three zones should result from the decrease of the covalent bonding character of the coatings [78]. Covalent nitrides like ZrN are better candidates for high hardness values than ionic compounds like ZrO_2 because electrostatic interaction are omnidirectional and yield low bond-bending force constants, which result in low shear moduli. Fig. **20** shows the Young's modulus (elastic modulus) values (E_f) as a function of the reactive gas flow. Similarly to what has been shown for the hardness evolution, the Young's modulus presents also an initial increase in the first compositional/structural zone, followed by a two-stage decrease. In zone II, where the structural analysis revealed a progressive formation of a $Zr_3N_4(O)$-type structure, the Young's modulus values present a significant decrease, while within the last two zones (those corresponding to oxynitride - γ-Zr_2ON_2 and oxide *mon*-ZrO_2 type structures) this parameter decreases in a much smoother way.

The notorious coherence in the variation tendencies of both properties shows that there is an approximately linear dependency between both quantities. Fig. **21** shows a plot of hardness *vs.* the Young's modulus. The main conclusion that can be drawn is that the samples lying in the first two compositional/structural zones appear to present a good linear evolution of hardness with Young's modulus, which is an indication of a brittle-like behaviour [79].

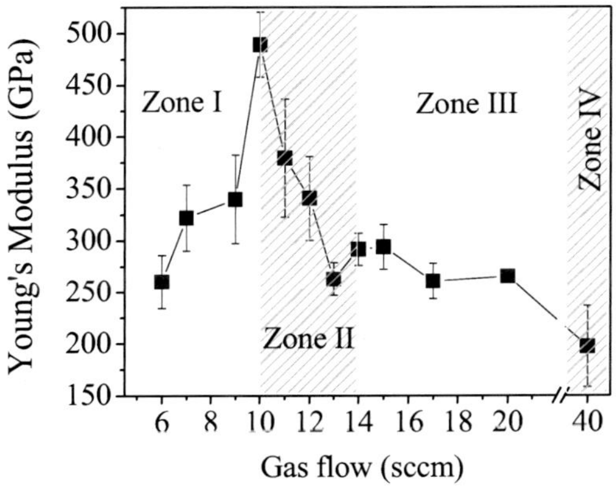

Figure 20: Young's Modulus values as a function of reactive gas flow. The error bar was determined by maximum deviation to the average value.

Regarding zones III and IV, the reduced numbers of points could not give any information about the hardness/Young's modulus ratio. At best, this small group of samples could eventually also follow a linear trend.

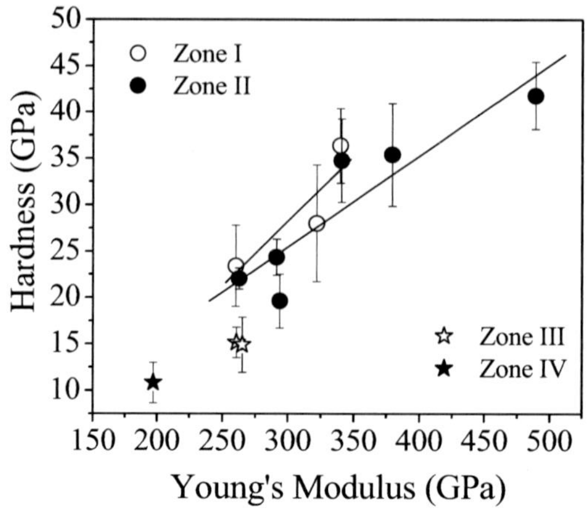

Figure 21: Hardness values as a function of Young's Modulus. The error bar was determined by maximum deviation to the average value.

The Effect of Oxygen Addition

In transition metal compounds, the chemical bond and cohesion energy influences significantly their mechanical properties. In this domain the Young's modulus and the hardness are the most well-known properties directly correlated with bonding/cohesion characteristics [80]. High cohesion energy and high hardness values normally correlate with a higher covalent character of the bonding [81]. In this case, and although the composition and structural changes, already described, seem to dictate the evolution of the hardness and Young's modulus, the particular behaviour in each zone suggests that other parameters may be playing important roles as well. In fact, several works may be found in the literature that try to explain how hardness evolves, taking into account not only the particular crystalline structure of the films, but also other parameters such as the lattice parameter variations, the peak intensity ratio and grain size, as well as other related parameters such as the macrostress states and bonding conditions [82].

Regarding the initial increase of hardness values for the samples deposited within zone I, the solid solution hardening mechanisms should be an important factor to take into account. When preparing ternary nitrides, based on the introduction of a third element in a well-known binary nitride, such as the case here by adding oxygen to ZrN, the concentration of the added element is seen as a major parameter with a significant role in the observed behaviour [41]. In fact, and reporting to the case expressed here, the addition of low oxygen amounts to the films leads to an increase of hardness until a maximum of about 40 GPa, which can result from an increasing oxygen doping of the ZrN lattice, giving rise to a solid solution hardening mechanism. In fact, the sub- and close stoichiometry of the films within this region is certainly a factor that may facilitate this doping of the nitride lattice. With oxygen insertion, an increased strength of the material may arise as a consequence of the resultant lattice distortions [83]. The lattice distortions inhibit the mobility of the dislocations, thus, an enhancement in hardness is expected.

In fact, *solid solution hardening* should play an important role in the hardness evolution for the samples lying in the first compositional zone. Several authors claim that the hardness and the elastic modulus of group IVb nitrides such TiN_x, ZrN_x and HfN_x should decrease as the concentration of the non-metal vacancy increases. Seung-Hoon Jhi *et al.* [84], in the study of vacancy hardening and softening in transition metal carbides and nitrides, claimed that occupied metallic *dd* bonding states in TiN and ZrN, which give a negative contribution to the shear modulus, are enhanced as the vacancy concentration increases, resulting in a decrease of the shear modulus (and correspondingly the hardness). Since the coatings lying in the first compositional zone present an increase of metalloid content until the stoichiometric condition is achieved, the reduction of the vacancy concentration could also play an important role in the increase of the hardness values.

The Effect of Structural Parameters

Previous studies have reported that preferred orientation is a very important factor to the film hardness enhancement. Major achievements in these studies have shown that the $\langle 111 \rangle$ orientation is the hardest orientation in the TiN films due to geometrical strengthening [85,86]. The hardness is expected to increase as the $\langle 111 \rangle$ texture coefficient increases. Since ZrN has the same NaCl crystal structure as TiN, $\langle 111 \rangle$ is thus supposed to be the hardest orientation of the ZrN film.

So far, it has been shown in the previous sub-sections that none of the parameters *per se* may explain the overall behaviour of the coatings. It is quite clear that the conjugation of different parameters can account for the hardness (and Young's modulus) variation. In order to check for other possible parameters that may have a role in hardness variation, plots of hardness *vs.* lattice parameter and grain size were produced (Fig. **22**). Due to the particular structural features (developed phases and high degree of peak broadening), only samples from zone I will be checked in this sub-section.

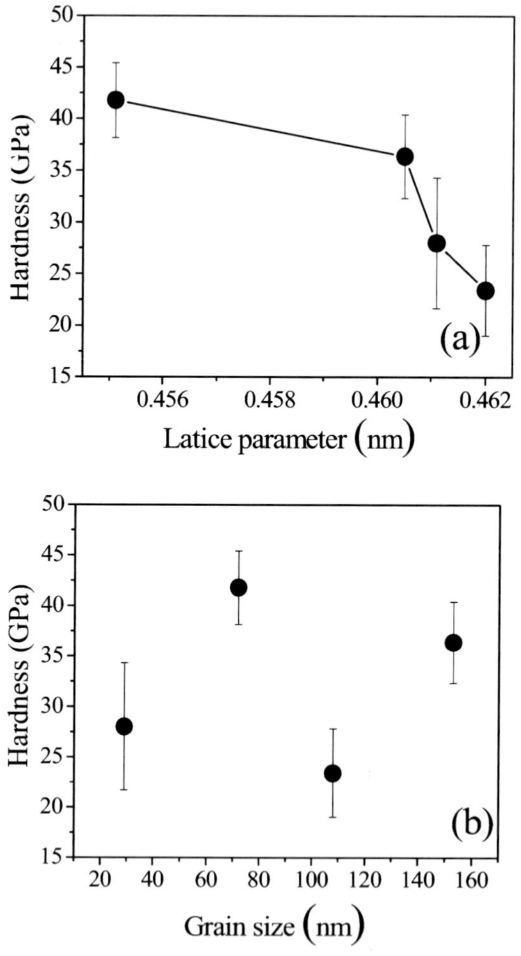

Figure 22: Hardness evolution as a function of: (a) lattice parameter and (b) grain size; for samples revealing a *fcc* ZrN-type structure. The error bar was determined by maximum deviation to the average value.

Regarding the evolution of hardness as a function of lattice parameter (Fig. **22**(a)), it is possible to observe that these samples present a decrease of hardness values with increasing lattice parameter. This fact should be correlated with the slight decrease of the lattice parameter with the increase of the non-metal to zirconium atomic ratio. A better arrangement of the adatoms decreases the lattice disorder, provoking, as consequence, a reduction of the lattice parameter. The increase of the metalloid content is directly correlated with the increase of the covalent bonding character, with the consequential increase of hardness (Fig. **22**(b)) shows that there is no correlation between the hardness evolution and grain size values, indicating that no Hall-Petch effect (direct or inverse) is observed. This non-dependence of the Hall-Petch effect should result from the composition variation, which change of $(C_N+C_O)/C_{Zr}$ from close to 0.5 through ~0.9.

The Effect of Macrostresses

Fig. **23**(a) shows the macrostress evolution as a function of reactive gas flow. For easier better comprehensive analysis, the plot of hardness evolution is also shown in Fig. **23**(b).

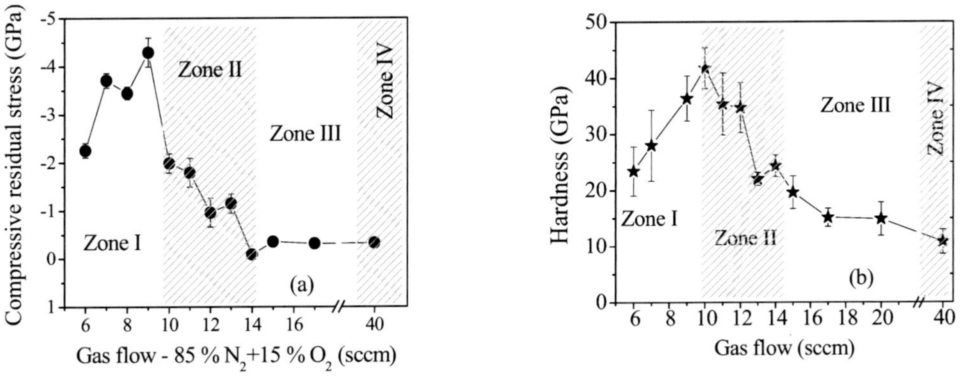

Figure 23: (a) Residual stress and (b) hardness evolution; as a function of reactive gas flow. The error bar was determined by maximum deviation to the average value.

In sputter-deposited hard polycrystalline films, stresses are commonly observed. The compressive stresses are typically generated by the atomic peening mechanism (the creation of interstitials or implanted argon during the film growth) [87]. The first important feature observed in the two plots of Fig. **23** is that both properties reveal similar variation tendencies, meaning that beyond the already mentioned structural parameters, the compressive macrostress states of the samples are a key factor which has to be taken into account. Nevertheless, there are also some particularities that may be worth mentioning. In fact, if these two properties are correlated in the same plot, these particularities may be easier to be observed, see Fig. **24**.

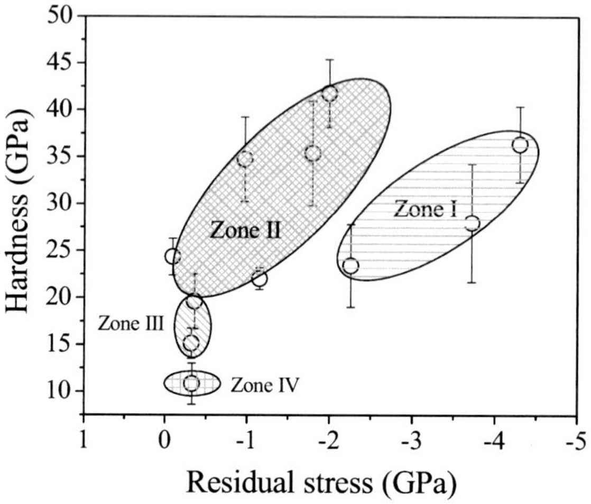

Figure 24: Hardness evolution as a function of macrostress values. The error bar was determined by maximum deviation to the average value.

This figure reveals the tendency that, within zones I and II, the hardest films are also the ones with the highest macrostresses. The relatively high macrostresses of the samples from the first compositional zone is most likely to result from the already mentioned incorporation of oxygen within the ZrN lattice, which would induce lattice defects, thus increasing the stress levels. However, the hardest samples (those prepared with reactive gas flows varying from 9 to 12 sccm, and having the hardness values ranging from ~35 to ~40 GPa), show macrostress states that actually decrease from ~ -4.5 to about -1 GPa. This clearly shows that stresses are not the only important parameter for the interpretation of the hardness data. Moreover, a closer look to Fig. **24** shows that there are films with hardness values ranging from 20 to ~10 GPa (zones III and IV), but with similar macrostress values - around -0.5 GPa.

Adhesion

The knowledge about how coatings fail on coated parts and objects is essential to assess not only the mechanical behaviour of the system (coating/substrate), but above all, the coating suitability for a particular purpose. In addition, the magnitude of the failure event is often more decisive to discriminate between different coatings than the difference in critical load values. Fig. **25** shows the "typical" failure events evolution on the zirconium oxynitride system with increasing load, for the samples lying in the first three compositional zones.

The evolution of the failure events, with increasing load, is similar in all the analysed coatings (zone I, II and III). In the first part of the scratch track (lower loads), the samples behaviour is characterised by a plastic deformation of the coating, without showing any failure event (Fig. **25**(a)). This first plastic deformation is then followed by a *chevron*-type cracking, with tensile-type Hertzian cracks (Fig. **25**(b)). With increasing load, *conformal*-type *buckling cracking* starts to appear (Fig. **25**(c)). Finally, significant interfacial *shell-shaped spallation* (Fig. **25**(d)) occurs, leading to what one could call the "end of the coating". Fig. **26** shows the critical loads of representative samples as a function of reactive gas flow. Lc_1 and Lc_2 represent the critical adhesion loads for the first cohesive and first adhesive failures, respectively. Lc_3 represents the load at which more than 50 % of the coating is removed from the substrate.

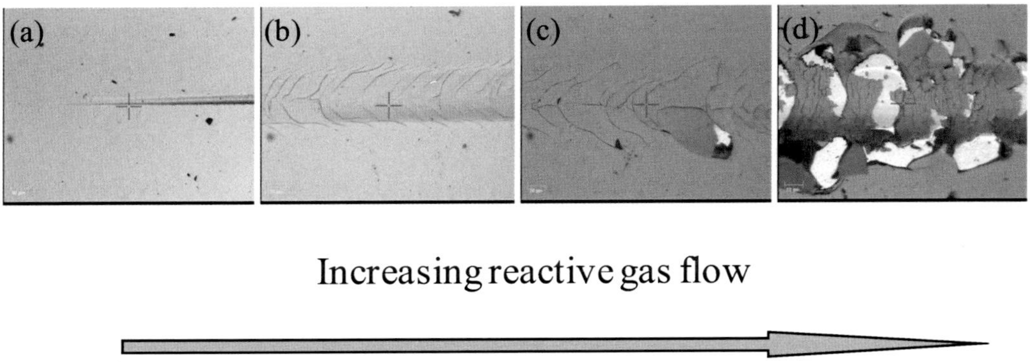

Figure 25: Typical failure events evolution on the zirconium oxynitride system, with increasing load.

Regarding the evolution of critical load values (Fig. **26**), the first general observation that one might deduce is that the adhesion of the samples to the substrate tends to decrease with the increase of the reactive gas flow, and the consequent changes in composition and structure.

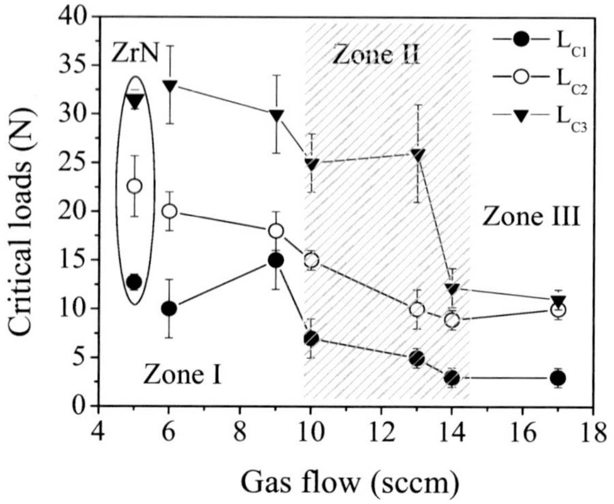

Figure 26: Critical loads values as a function of reactive gas flow regarding zone I, II and III. The error bar was determined by maximum deviation to the average value.

Banakh et al. [88] studied the adhesion of magnetron sputtered tantalum oxynitride on high speed steel substrates, and also reported the decrease of coatings adhesion with increase of metalloid content.

Regarding the samples lying in the second compositional zone, Fig. **26** shows a decreasing tendency of all critical load values. In spite of this tendency, the sample deposited with a reactive gas flow of 14 sccm seems to present a small increase of the L_{C3} values. This increase may, however, not be real, but the result of the measurement resolution itself is real, since in this particular situation the error bar is much larger than in the case of the other samples. The third zone representative sample presents critical load values similar to those obtained for the last sample of zone II. Although composition and structure appear to present a dominant role in the critical loads evolution, other factors were also taken into account. Regarding the correlation between adhesion and macrostresses, Fig. **27** presents L_{C1}, L_{C2} and L_{C3} values as a function of the macrostresses. This figure shows that critical load values trend to increase with macrostress. It is expected that coatings with higher residual stresses present higher tendency to flake or debond from the substrate with applied exterior loads. A close example of this behaviour is the sample deposited with a reactive gas flow of 9 sccm, which does not present the second critical load value (L_{C2}), and the first adhesion failure was characterized to be a *gross interfacial* *spallation*-type. This sample presents the highest values of the residual stress state and the highest hardness values (~4.3 GPa and ~42 GPa respectively). In brittle films, if cohesive strength is higher than adhesion, the coating tends to debond or flake away at the edges of the groove [89].

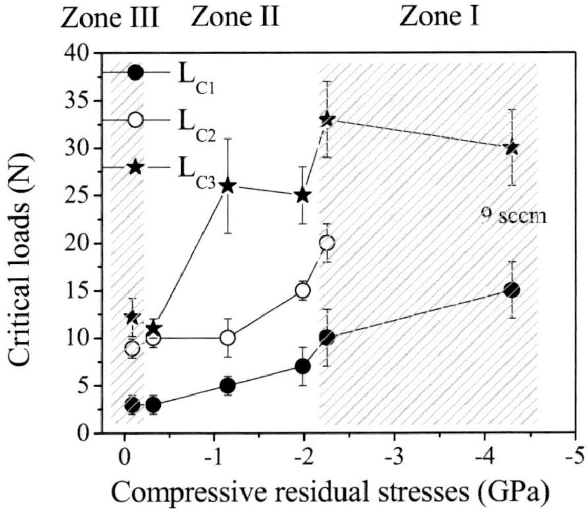

Figure 27: Critical load values a function of the macrostresses of the coatings. The error bar was determined by maximum deviation to the average value.

The apparent correlation between adhesion and residual stresses should be related with the specific type of material involved, and the consequent changes in the interface strength. As a conclusion, one may claim that the adhesion of the coatings present a clear relationship with the compositional/structural zones. In spite of the basic deposition conditions of these coatings (no bias, no adhesion layer, no external heating), it appears that they are adequate to give rise to coatings with enough adhesion to be used in decorative applications. However, to improve adhesion, several methods are possible, namely the use of negative substrate bias, heating of the substrates, the use of ion gun bombardment or to use an interlayer [90].

Corrosion Behaviour

Fig. **28** shows the potentiodynamic polarization curves of representative samples lying in the first two composition zones for different reactive gas flow values. These samples were the only ones selected because of their intrinsic colours and thus, the most important ones for decorative applications. The ZrO_xN_y films were used as plates positioned horizontally and mounted in an acrylic electrochemical cell (20 ml electrolyte volume) with the working surface of the film facing upwards. An artificial sweat solution (pH = 4.5), containing 7.5 g l−1 NaCl; 1.2 g l−1 KCl; 1 g l−1 CH_4N_2O (urea) and 1 g l−1 $C_3H_6O_3$ (lactic acid), was used as electrolyte. All potentials were measured and expressed with reference to a standard calomel electrode (SCE), and a platinum wire with an area of 1 cm^2 was used as a counter electrode. For comparison proposes, the polarization behaviour of the substrate (M2 steel) is also plotted. The plot clearly shows that the corrosion resistance and the pitting of the coated M2 steel is clearly enhanced regarding the uncoated M2 steel [91].

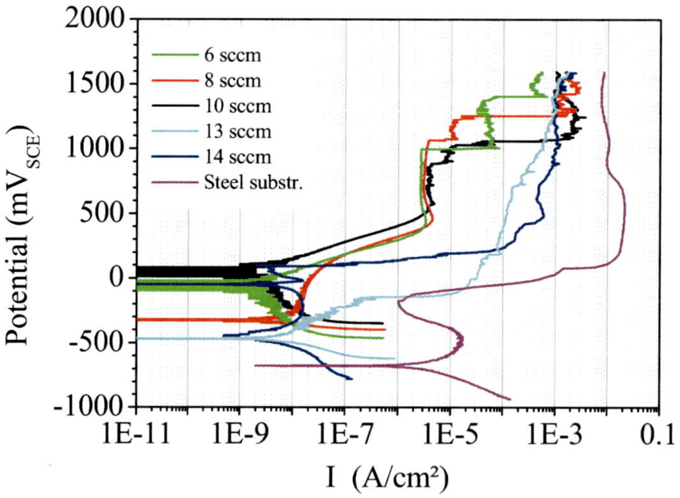

Figure 28: Potentiodynamic polarization curves of representative of samples.

The first observation that can be drawn is the existence of two different potentiodynamic polarization behaviours, for the samples deposited with reactive gas flow higher and lower than 10 sccm. The samples that still present the *fcc*-ZrN structure (prepared with a gas flow ranging from 6 to 10 sccm) revealed pitting potentials of about 1000 mV (SCE), showing a well-defined large potential independent region, indicative of a passive plateau. These samples present a corrosion current density varying from ~0.1 to about 0.3 μA/cm^2, three orders of magnitude lower than the uncoated M2 steel [91].

Regarding the samples deposited with reactive gas flows of 13 and 14 sccm, results show poorer corrosion behaviour, with pitting potential values of ~200 and ~400 mV, respectively, and corrosion current densities close to 0.7 $\mu A/cm^2$. Although these samples present worse corrosion resistance, they still present a clear higher corrosion resistance when compared with the M2 steel.

Although composition and structural features appear, once again, to have an important influence in this property, in order to further understand the origin of the different corrosion behaviours, some possible correlations between coatings morphology and surface analysis, by optical microscopy, were checked.

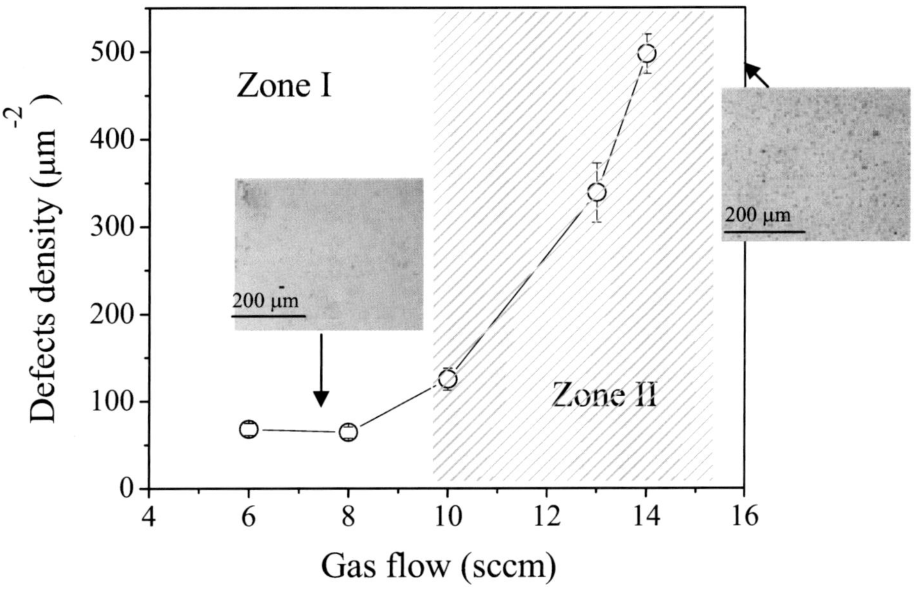

Figure 29: Density of surface defects of representative samples lying in the first composition zones. The error bar was determined by maximum deviation to the average value.

One important factor concerning the corrosion resistance is the quality of the surface of the films. In fact, if one intends to use a particular part with a coating material, the number/density of surface defects will have a major role on the part oxidation/corrosion resistance, and thus a unique importance for the in-service behaviour and durability of the coated part. E. Ariza *et al.* [91] claimed that several features, such as the thickness of the film, the density of superficial defects and/or the chemical composition and morphology of the coatings are expected to influence the corrosion behaviour of zirconium oxynitride thin films. Fig. **29** shows the surface defect density of representative samples from the first two compositional zones.

From this figure it is possible to observe an increase of the density of defects for the films deposited with reactive gas flow between 10 and 14 sccm. From the analysis of the density of defects and the potentiodynamic polarization curves (Figs. **29** and **28**, respectively), it is possible to observe a close correlation between the corrosion behaviour and the surface quality. The increase of density of defects is followed by a decrease of the corrosion resistance, indicating that the surface quality is a very important parameter for the corrosion resistance of the coatings. This could be explained by the fact that superficial defects can act as preferential routes for the penetration of the solution through the coating and reach the substrate [92, 93].

Comparing the morphological features (see section "morphology") with the corrosion behaviour, it is also possible to observe some correlations. The samples presenting better corrosion behaviour (zone I) present also a compact fibrous microstructure, with no clear development of columnar-type morphology (Fig. **7**). The samples deposited with reactive gas flows of 13 and 14 sccm, which present a clear decrease in corrosion resistance in comparison to the samples lying in the first compositional zone, show a change from quasi-featureless morphology (Fig. **8**).

CONCLUSIONS

Structural characterization of the ternary system ZrO_xN_y revealed a strong dependence of the film texture on the oxygen content, exhibiting a change from a *fcc* B1 NaCl-type films for low gas flows (low oxygen contents), to a structure that might be similar to that of *orth*-Zr_3N_4, with oxygen inclusions, at intermediate gas flows (intermediate oxygen contents). For relatively high gas flows (high oxygen amounts), the films exhibited only crystalline structures that were indexed as oxynitride (γ-Zr_2ON_2) and oxide phases (*mon*-ZrO_2).

XPS results showed that the electronic structure of zirconium oxynitride is ruled by the Zr 4*p* levels, but the O 2*s*, N 2*s* and molecular orbitals associated with O 2*p*/N 2*p* and

Zr 4*d* valence levels can also be observed. With the increase of the oxygen content, the films present a progressive transfer of zirconium 4*d* valence electrons to the O 2*p* and N 2*p* bands, with a shift to lower energies of the "*p band*", until the complete depopulation of the *d-band* where the films become an insulator-type material.

The zirconium oxynitride films electrical behaviour presented a straight correlation between the crystalline and electronic structure. The low values of resistivity were found to be related with the roughly metallic nature of the films, with compositions between that of Zr and sub-stoichiometric ZrN(O). For higher reactive gas flows (zone II) the resistivity of the films increased, which can be explained by the decrease the metallic bond contribution and the formation of oxygen-doped ZrN crystalline phase towards a nitrogen+oxygen rich phase, similar to that of Zr_3N_4. Particularly noticeable were the results obtained for the electrical resistivity of films for intermediate gas flows (oxygen content between about 8 to 15 at.% - zone II films), where in spite of almost no changes in the Zr and N contents, the resistivity values increased about twelve orders of magnitude ($\sim10^3$ to $\sim10^{15}$ $\mu\Omega$.cm). The structural and bonding changes were found to be in accordance with these changes. Regarding the last two zones, the coatings present high electrical resistivity and the conductivity is thermally activated.

The ternary system zirconium oxynitride is characterized by three major different optical behaviours, directly correlated with the compositional/structural features of the coatings. The small optical response variations in each zone result from small variations in composition. The coatings with an *fcc*-ZrN structure present an optical behaviour similar to zirconium nitride, showing high reflectivity near the infrared region, an edge in the visible region and a reflectivity minimum in the UV region. The colour of these coatings varies from bright yellow to golden yellow. The coatings which presented a Zr_3N_4(O) structure showed low and almost constant values of reflectivity, with colour tones varying from red-brownish to purple and dark blue. The coatings where the crystalline structure changed from Zr_2ON_2 to ZrO_2, presented a semi-transparent optical behaviour, with optical band gaps close to 4 eV.

The coatings which show intrinsic coloration present good mechanical properties. The hardness values were always above 20 GPa and Young's Modulus higher than 250 GPa. Regarding adhesion, the coatings with composition close to ZrN presented, as expected, critical load values similar to those of "pure" zirconium nitride. The samples presenting a $Zr_3N_4(O)$ structure showed a decrease of the critical load values, indicating a decrease of the adhesion of the coatings to the substrate. Regarding the corrosion behaviour, a similar evolution was observed.

ACKNOWLEDGEMENTS

Authors acknowledge the Portuguese Science Foundation - FCT "Fundação para a Ciência e Tecnologia" for the PhD Grant SFRH/BD/31907/2006 and project nº POCTI/CTM/38086/2001, co-financed by European community fund FEDER. Also to acknowledge the European Union through the NMP3-CT-2003 505948 project "HARDECOAT". These two institutions have also made possible this work.

CONFLICT OF INTEREST

The author(s) confirm that this chapter content has no conflict of interest.

REFERENCES

[1] H. Heidsieck, Surf. Coat. Technol., vol. 112, pp. 324, 1999.
[2] K.H. Jack, Non-Oxide Technical and Engineering Ceramics, Elsevier, Amsterdam, 1, 1986.
[3] S. Komarrneni, J. Mater. Chem., vol. 2, pp. 1219, 1992.
[4] A. Navrotsky, J. Alloys Compd., vol. 321, pp. 300, 2001.
[5] S. Dreer, R. Krismer, P. Wilhartiz, G. Friedbacher, Thin Solid Films, vol. 354,pp. 43, 1999.
[6] S. Horita, T. Tujikawa T.H. Akahori, M. Kobayashi, T. Hata, J. Vac. Sci. Technol., vol. A 11, pp. 2452, 1993.
[7] S.T. Oyama, Chemistry of Transition Metal Carbides and Nitrides, Blackie Academic & Professional, (New York, 1996) 215-309.
[8] J.-L. Calais, Adv. Phys., vol. 26, pp. 847, 1977.
[9] A. Neckel, P. Rastl, R. Eibler, P. Weiberger, K. Schwarz, J. Phys. C, vol. 9, pp. 579, 1976.
[10] K. Schwarz, H. Ripplinger, A. Neckel, Z. L. Phys. B, vol. 48, pp. 79, 1982.
[11] D.F. Arias, Y.C. Arango, A. Devia, Appl. Surf. Sci., vol. 253, pp. 1683, 2006.
[12] T. Yotsuya, M. Yoshitake, T. Kodama, Cryogenics, vol. 37, pp. 817, 1997.
[13] K. Schwarz, A.R. Williams, J.J. Cuomo, J.H.E. Harper, H.T.G. Hentzell, Phys. Rev. B, vol. 32, pp. 8312, 1985.
[14] M. Wautelet, J.P. Dauchot, F. Debal, S. Edart, M. Hecq, J. Mater. Res., vol. 11, pp. 825, 1996.
[15] Y. Igarashi, T. Yamaji, S. Nishikawa, Jpn. J. Appl. Phys., vol. 29, pp. L2337, 1990.
[16] B. M. Takeyama, T. Itoi, E. Aoyagi, A. Noya, Appl. Surf. Sci., vol. 190, pp. 450, 2002.
[17] B. M. Takeyama, A. Noya, K. Sakanishi, J. Vac. Sci. Technol., vol. B 18, pp. 1333, 2000.

[18] M. Burghartz, G. Ledergerber, H. Hein, R.R. Van der Laan, R.J.M. Konings, J. Nucl. Mater., vol. 288, pp. 233, 2001.

[19] R. A. Andrievski, J. Mater. Sci., vol. 32, pp. 4463, 1997.

[20] B. O. Johansson, H. T. G. Hentzell, J. M. E. Harper, J. J. Cuomo, J. Mater. Res., vol. 1, pp. 442, 1986.

[21] R. P. Netterfield, P. J. Martin, D. R. McKenzie, J. Mater. Sci. Lett., vol. 9, pp. 972, 1990.

[22] H. K. Pulker, Coatings on Glass, Thin Films Science and Technology, vol. 6, pp. 343-417, 1984.

[23] H. Bach and N. Neuroth, The properties of Optical Glass, Springer, Heidelberg, pp. 263-388, 1995.

[24] P.T. Gao, L.J. Meng, M.P. dos Santos, V. Teixeira, M. Andritschky, Vacuum, vol. 56, pp. 143, 2000.

[25] S. Venkataraj, O. Kappertz, H. Weis, R. Drese, R. Jayavel, M. Wuttig, J. Appl. Phys., vol. 92, pp. 3599, 2002.

[26] W. Kriven, M.W.L. Fraser, S.W. Kennedy, in: A.H. Heuer, L.W. Hobbs (Eds.), Science and Technology of Zirconia, The American Ceramic Society, Inc., Columbus, OH 82, 1981.

[27] R.A. Miller Surf. Coat. Technol., vol. 30, pp. 1, 1987.

[28] W.J. Brindley, R.A. Miller. Adv Mat and Proc., vol. 8, pp. 29, 1989.

[29] S.M. Meier, D.K. Gupta. J Eng Gas Turbines Power., vol. 116, pp. 250, 1994.

[30] J.T. DeMasi-Marcin, K.D. Sheffler, S. Bose, J Eng Gas Turbines Power, vol. 112, pp. 521, 1990.

[31] N.P. Barradas, C. Jeynes, R.P. Webb, Appl. Phys. Lett., vol. 71, pp. 291, 1997.

[32] A.F. Gurbich, Nucl. Instr. and Meth. B, vol. 268, pp. 1703, 2011.

[33] QUASES-Tougaard Inc. Available from: www.quases.com.

[34] M. Del Re, R. Gouttebaron, J.-P. Dauchot, P. Leclère, G. Terwagne, M. Hecq, Surf. Coat. Technol., vol. 174 –175, pp. 240, 2003.

[35] I. Miloŝev, H. H. Strehblow, M. Gaberscek, B. Navinsĕk, Surf. Interface Anal., vol. 24, pp. 448, 1996.

[36] J.M. Sanz, L. Soriano, P. Prieto, G. Tyuliev, C. Morant, E. Elizalde, Thin Solid Films., vol. 332, pp. 209, 1998.

[37] S. Tsunekawa, K. Asami, S. Ito, M. Yashima, T. Sugimoto, Appl. Surf. Sci., vol. 252, pp. 1651, 2005.

[38] P. Carvalho "Development of new decorative coatings based on zirconium oxynitrides": Ph.D. Thesis, Universidade do Minho, 2008.

[39] D.R. Lide, Handbook of Chemistry and Physics, 85th Edition ed. CRC Press, Florida, 2004.

[40] H. J. Goldschmidt, Interstitial Alloys, London, Butterworths, 1967.

[41] P. Carvalho, F. Vaz, L. Rebouta, L. Cunha, C.J. Tavares, C. Moura, E. Alves, A. Cavaleiro, Ph. Goudeau, E. Le Bourhis, J.P. Rivière, J.F. Pierson, O. Banakh, J. Appl. Phys., vol. 98, pp. 023715, 2005.

[42] P. Carvalho, J.M. Chappé, L. Cunha, S. Lanceros-Méndez, P. Alpuim, F. Vaz, E. Alves, C. Rousselot, J.P. Espinós, A.R. González-Elipe, J. Appl. Phys., vol. 103, pp. 104907, 2008.

[43] N. Martin, R. Sanjines, J. Takadoum, F. Lévy, Surf. Coat. Technol., vol. 142–144, pp. 615, 2001.

[44] S.H. Mohamed, O. Kappertz, J.M. Ngaruiya, T. Niemeier, R. Drese, Wakkad, M. Wuttig, Phys. Stat. Sol., vol. 201, pp. 90, 2004.

[45] Francisco Macedo, Pedro Carvalho, Luís Cunha, Filipe Vaz, Juergen Gibkes, Bruno K. Bein Josef Pelzl, Plasma Process. Polym., vol. 6, pp. S760–S766, 2009.
[46] Chuan-Pu Liu, Heng-Ghieh Yang, Thin Solid Films, vol. 444, pp. 111, 2003.
[47] D. Pilloud, A.S. Dehlinger, J.F. Pierson, A. Roman, L. Pichon, Surf. Coat. Technol., vol. 174-175, pp. 338, 2003.
[48] J.V. Ramana, S. Kumar, C. David, A.K. Ray, V.S. Raju, Mater. Lett., vol. 43, pp. 73, 2000.
[49] L. Pichon, T. Girardeau, A. Straboni, F. Lignou, P. Guérin, J. Perrière, Appl. Surf. Sci., vol. 150, pp. 115, 1999.
[50] H. M. Benia, M. Guemmaz, G. Scmerber, A. Mosser, J.C. Parlebas, Appl. Surf. Sci., vol. 200, pp. 231, 2002.
[51] R. Lamni, E. Martinez, S.G. Springer, R. Sanjinés, P.E. Schmid, F. Lévy, Thin Solid Films, vol. 447-448, pp. 316, 2004.
[52] D.I. Bazhanov, A.A. Knizhnik, A.A. Safonov, A.A. Bagatur'yants, M.W. Stoker, A.A. Korkin, J. Appl. Phys., vol. 97, pp. 044108, 2005.
[53] M. Lerch, J. Mater. Sci. Lett., vol. 17, pp. 441, 1998.
[54] S.J. Clarke, C.W. Michie, M.J. Rosseinsky, J. Sol. Stat. Chem., vol. 146, pp. 399, 1999.
[55] E. Fuglein, R. Hock, M. Lerch, Z. Anorg, Allg. Chem., vol. 623, pp. 304, 1997.
[56] C. Morant, J. M. Sanz, L. Galán, L. Soriano, F. Rueda, Surf. Sci., vol. 218, pp. 331, 1989.
[57] C. Palacio, J.M. Sanz, J.M. Martnez Duart, Surf. Sci., vol. 191, pp. 385, 1987.
[58] J.M. Sanz, L. Soriano, P. Prieto, G. Tyuliev, C. Morant, E. Elizalde, Thin Solid Films, vol. 332, pp. 209, 1998.
[59] A. Delin, O. Eriksson, R. Ahuja, B. Johansson, M. S. S. Brooks, T. Gasche, S. Auluck, J.M. Wills, Phys. Rev. B, vol. 54, pp. 1673, 1996.
[60] I. Milošev, H.-H. Strehbtow, Navinsek, Thin Solid Films, vol. 303, pp. 246-254, 1997.
[61] J.-M. Chappé, N. Martin, J.F. Pierson, G. Terwagne, J. Lintymer, J. Gavoille, J. Takadoum, Appl. Surf. Sci., vol. 225, pp. 29, 2004.
[62] P. Prieto, L. Galán, J. M. Sanz, Phys. Rev. B, vol. 47, pp. 1613, 1993.
[63] Recommendations on uniform colour spaces, colour-difference equations, psychometric colour terms, CIE publ. 15, Suppl. 2– 70 CIE, Paris, 1978.
[64] Colorimetry, CIE Publ. (Commission Internationale de L'Éclairage) 15, 1971.
[65] M. Ohring, The Materials Science of Thin Films, Academic Press, San Diego CA, 1992.
[66] J.F. Pierson, F. Bertran, J.P. Bauer, J. Jolly, Surf. Coat. Technol., vol. 142-144, pp. 906, 2001.
[67] M. Veszelei, K. E. Andersson, A. Roos, Opt. Mater., vol. 2, pp. 257, 1993.
[68] B. Karlsson, R.P. Shimshock, B.O. Seraphin, J.C. Haygarth, Solar Energy Mater., vol. 7, pp. 401, 1983.
[69] M. Veszeilei, K. Andersson, C.G. Ribbing, K. Järrendahl, h. Arwin, Appl. Opt., vol. 33, pp. 1993, 1994.
[70] S. Logothetidis, I. Alexandrou, J. Stoernenos, Appl. Surf. Sci., vol. 86, pp. 185, 1995.
[71] A. Bendavid, P.J. Martin, R.P. Netterfield, T.J. Kinder, Surf. Inter. Anal., vol. 24, pp. 627, 1996.
[72] L. Pauling, The Nature of the Chemical Bond, Chap. 3, Cornell University, Ithaca, New York, 1960.
[73] J. Taut, The Optical Properties of Solids, Proceedings of the International School of Physics (ENRICO FERMI), course XXXIV, Academic, New York, 106, 1966.
[74] A. Schlegel, P. Wachter, J.J. Nickl, Lingg, J. Phys. C: Solid State Phys., vol. 10, pp. 4889-4896, 1977.

CHAPTER 5

Gradual Evolution of the Properties in Titanium Oxynitride Thin Films

Jean-Marie Chappé and Nicolas Martin*

Institut FEMTO-ST, 32, Avenue de l'observatoire, 25044 Besançon, France

Abstract: Titanium oxynitride thin films were deposited by dc reactive magnetron sputtering. The reactive gas pulsing process, namely RGPP, was used to modify the oxygen and nitrogen concentrations in titanium oxynitride thin films TiO_xN_y. Nitrogen gas was constantly introduced into the sputtering process, whereas oxygen mass flow rate was periodically changed *vs.* time. A rectangular pulsing pattern was implemented and the duty cycle α, defined from the introduction time of oxygen, was systematically modified from 0 to 100 % of the pulsing period T maintained at 45 s. In this way, the chemical composition of TiO_xN_y sputter deposited coatings was tuned from pure metallic titanium nitride TiN to titanium dioxide TiO_2 through various metalloid concentrations, *i.e.,* x varied from 0 to 2.0 and y inversely reduced from 1.0 to 0. Such tuneable chemical compositions led to a gradual and smooth transition of optical and electrical properties of the films. Optical transmittance in the visible range was null for oxygen contents lower than 30 at. % and was higher than 75 % for the highest oxygen contents (> 60 at. %). A progressive evolution of the transmittance was observed for intermediate oxygen contents. Similarly, electrical conductivity was close to 10^5 S. m^{-1} for the lowest oxygen concentrations and dropped down to few S m^{-1} as oxygen concentration increased. This gradual evolution from metallic to semiconducting and finally dielectric behaviours correlates well with the occurrence of f.c.c. NaCl-type phase in an amorphous TiO_2 matrix. This f.c.c. phase vanishes as a function of the oxygen concentration in the films leading to an amorphous structure and typical dielectric characteristics.

Keywords: Titanium oxynitride, reactive magnetron sputtering, reactive gas pulsing process (RGPP), duty cycle, rectangular pulsing pattern, tuneable chemical composition, ternary diagram, gradual transition, metallic titanium nitride, titanium dioxide, crystalline structure, dielectric, semiconducting behaviour, spectroscopic ellipsometry, optical transmittance, refractive index, extinction coefficient, electrical conductivity, temperature coefficient of resistance (TCR), activation energy.

INTRODUCTION

In the last ten years, a growing attention has been paid on metallic oxynitride thin films [1-6]. A significant concentration of oxygen in nitride compounds or

***Address correspondence to Nicolas Martin:** Institut FEMTO-ST, 32, avenue de l'observatoire 25044 Besançon, Cedex, France; Tel: +33 (0)3 81 85 39 69; Fax: +33 (0)3 81 85 39 98; E-mail: nicolas.martin@femto-st.fr

Filipe Vaz, Nicolas Martin and Martin Fenker (Eds)

inversely, the addition of nitrogen in oxide materials leads to unexpected and promising range of functional materials. Thus, oxynitrides have encountered an emerging interest since they may reveal behaviours included between metallic nitrides, which are well known for their mechanical properties, and their corresponding oxides, especially employed for optical and insulating performances [7-11]. Rapidly, a gap of knowledge appeared between the standard nitride and oxide thin films. Even if silicon oxynitride coatings have lately deserved a great attention due to their technological importance in micro-electronics [12-15], several works have been devoted to transition metal oxynitrides [16-23]. Among these types of films, titanium oxynitrides have been one of the most intensively studied systems [24-30]. Recent investigations on TiO_xN_y thin films with various oxygen and nitrogen concentrations, showed that these compounds exhibit a complex structure and revealed some intermediate characteristics between metallic TiN and insulating TiO_2 [31-33]. Indeed and as many metallic oxynitride compounds, titanium oxynitrides beneficiate from properties of metallic oxides (colours, chemical resistance, optical properties ...) and nitrides (hardness, wear resistance ...).

In order to synthesize such titanium oxynitride coatings with a wide range of oxygen and nitrogen concentrations, various deposition processes have been developed [34-39]. Among these processes, some chemical and physical vacuum deposition methods have been shown to be attractive ways to achieve adjustable chemical compositions and consequently, a variety of physico-chemical properties [40-44]. In the present work, dc reactive magnetron sputtering has been implemented to growth titanium oxynitride thin films. However, one of the main drawbacks of the reactive sputtering process is the typical hysteresis loop of some experimental parameters against the reactive gas supply in the system. This non-linear effect has extensively been studied for many years [45-48]. Some improvements of the conventional reactive sputtering have been proposed so as to avoid instability phenomena and to better control the process. Because of such instabilities, some compositions become difficult to achieve without using particular devices [49-53]. Aronson *et al.* [54] and other authors [55, 56] proposed an original method for the deposition of pure nitride and oxide coatings in which the reactive gas is pulsed according to periodic on and off intervals at regular times. This method was first applied to binary compounds. Few years ago, Martin *et al.* [57-60] developed the reactive gas pulsing process (RGPP) for the preparation of titanium oxynitrides. The present study involves RGPP to get a well controllable and tuneable N/O ratio in the films, and thus a wide range of properties. The evolution of the chemical composition, the crystallographic structure and the optical as well as electrical properties of the deposited titanium oxynitrides are discussed as a function of the oxygen concentration in the films.

MATERIALS AND METHODS

The titanium oxynitride coatings were sputter deposited by dc reactive magnetron sputtering. A stainless-steel vacuum reactor (60 L) was evacuated with a turbomolecular pump backed by a mechanical pump giving an ultimate vacuum of 10^{-5} Pa. A metallic titanium target (purity 99.6 at. %, 50 mm diameter) was sputtered in a reactive atmosphere composed of argon, nitrogen and oxygen gases. A constant current density $J_{Ti} = 200$ A m^{-2} was applied to the target. The substrate holder was maintained at room temperature and grounded during the deposition. The target-to-substrate distance was 50 mm. The pumping speed was set constant at S = 10 L s^{-1}. Ar and N_2 mass flow rates were kept at 2.30 and 0.80 sccm, respectively. Such operating conditions correspond to Ar and N_2 partial pressures of 0.24 and 0.09 Pa, respectively and consequently, to the full nitridation of the titanium target surface. Oxygen gas was introduced using a home-made computer controlled system called the reactive gas pulsing process (RGPP). Such a system allows a well-controlled injection of oxygen. Several types of pulses can be generated *vs.* time and some temporal parameters can be adjusted. A rectangular modulation pattern of the oxygen mass flow rate was chosen in this work (Fig. **1**). These pulses allow an alternation of the reactive process between the oxidized and nitrided sputtering modes. The maximal oxygen flow rate was fixed at q_{O2Max} = 2.00 sccm during the t_{ON} time and the minimal oxygen mass flow rate was fixed at q_{O2min} = 0 sccm during the t_{OFF} time (*i.e.*, oxygen injection was completely stopped). For all depositions, the titanium target was pre-sputtered for few minutes in pure argon atmosphere. Afterwards, nitrogen gas was supplied to set the process at its steady state equilibrium (nitrided sputtering mode). Oxygen was finally introduced following rectangular patterns described in Fig. **1**.

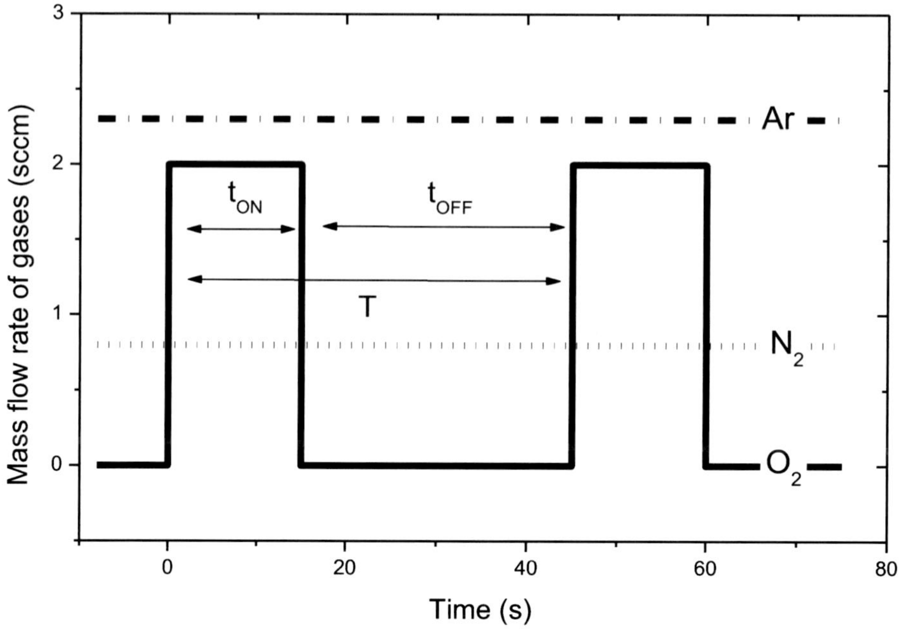

Figure 1: Argon, nitrogen and oxygen mass flow rates as a function of time. Argon and nitrogen gases are kept constant for all depositions, whereas oxygen is pulsed according to rectangular patterns. Duty cycle α defined as the t_{ON}/T ratio is systematically changed.

The duty cycle α, defined as $\alpha = t_{ON}/T$, where T is the pulsing period kept constant at 45 s, was systematically changed from 0 to 100 %. Thin films were deposited on (100) silicon wafer and glass substrate with a thickness close to 400 nm.

Optical transmittance of the films deposited on glass was measured at $\lambda = 633$ nm. X-ray diffraction patterns (XRD) were recorded in the Bragg-Brentano θ-2θ configuration using monochromatized Cu K_{α} radiation. The chemical composition was measured by Rutherford backscattering spectroscopy (RBS) and nuclear reaction analysis (NRA). Combining both techniques, it is possible to perform the elemental analysis of all elements present in the coatings. For the heavy element (Ti), a 2 MeV α particle beam used for the backscattering experiment was produced by a 2 MV Tandetron accelerator. The scattered particles were detected in a PIPS detector placed at 165° relative to the incident beam. The analysis of light elements in the films was performed with a 5.2 MeV α particles beam produced with the same accelerator. The particles due to $^{14}N(_{\alpha,p0})^{17}O$ and $^{14}N(_{\alpha,p1})^{17}O$ nuclear reactions were detected in a PIPS detector placed at 90° relative to the incident beam. The scattered α particles were stopped in a 24 µm Mylar foil filter placed in front of the detector. Simultaneously, the scattered α particles were detected in a PIPS detector at backward angle (165°). The refractive index and the extinction coefficient were calculated from data recorded by spectroscopic ellipsometry. A phase-modulated ellipsometer in the energy range 1.5 – 5.0 eV at an incidence angle of 70° was used. A classical Lorentz-Drude dispersion function was applied to model the optical constants of the metallic-like films. The Cauchy polynomial function was used in the case of semi-transparent films. For transparent coatings, the Sellmeier dispersion relation was applied in the transparent region (1.5 to 3.0 eV) whereas a Cauchy equation served as a model in the region from 3.0 to 5.0 eV, where the films are absorbent. The electrical conductivity of the films deposited on glass substrates was measured in the van der Pauw geometry in the temperature range 300 – 500 K. For metallic films the temperature coefficient of resistance (TCR) was determined, whereas activation energy of the conduction mechanism was calculated for semiconducting films.

TUNEABLE PROPERTIES

Chemical Composition and Crystallographic Structure

Oxygen and nitrogen concentrations (x and y) in TiO_xN_y films are inversely affected by the duty cycle α (or t_{ON} time) as shown in Fig. **2**.

At first, it is worth to note that no abrupt transition of the metalloid concentrations is measured, as commonly observed in reactive sputtering due to non linear effects. However, for duty cycles α lower than 1 % (t_{ON} times close to few seconds), oxygen and nitrogen concentrations in the film both significantly change from x = 0 to 0.82 and inversely from y = 1.0 to 0.63. It proves again the very high reactivity of oxygen towards titanium in spite of an initial nitrided state of the target surface. A further increase of the duty cycle up to α = 10 % leads to a continuous and moderated evolution of the oxygen and nitrogen concentrations since x and y similarly change from x = 0.82 to 1.26 and from y = 0.63 to 0.47.

On the other hand, the most significant variation of composition can be noticed for duty cycles in-between α = 10 to 15 % where x increases from 1.26 to 1.98 whereas y drops from 0.47 to 0. For duty cycles higher than 15 %, stoichiometric or oxygen-rich TiO_2-like thin films are always produced with nitrogen traces. This atypical evolution of TiO_xN_y chemical composition will be later correlated with the evolution of the crystallographic structure *vs.* duty cycle.

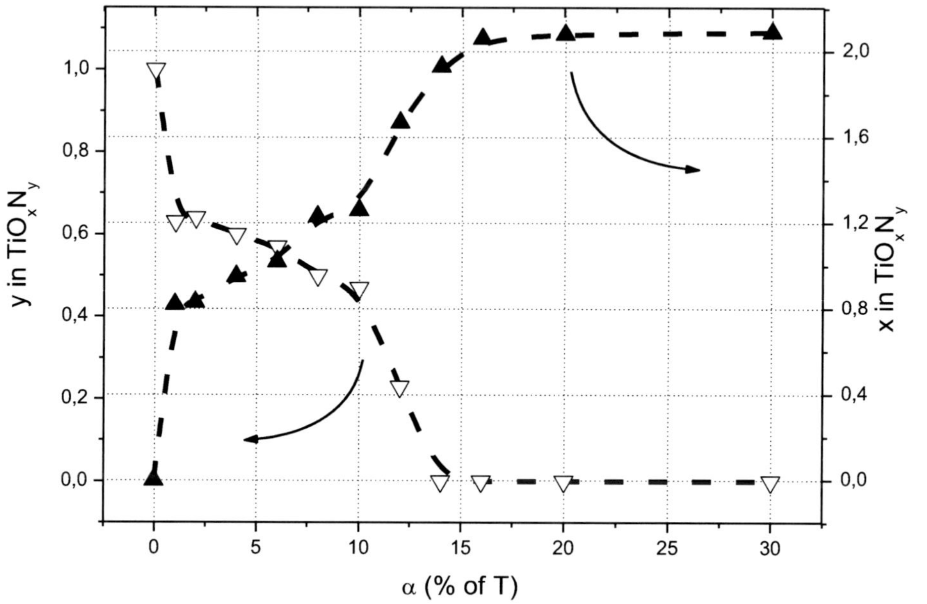

Figure 2: Gradual evolution of oxygen and nitrogen concentrations (x and y) in TiO_xN_y thin films as a function of the duty cycle α (or t_{ON} time). A reverse evolution of metalloid contents can be noticed as the duty cycle increases from 0 to 30 %.

The concentration of each element in TiO_xN_y films can also be represented as a function of an increasing duty cycle in a ternary diagram (Fig. **3**).

By conventional reactive sputtering, some metalloid concentrations cannot be reached without using some innovative feedback control systems [61, 62]. So, classical methods (*e.g.* constant supply of oxygen and nitrogen gases) lead to titanium nitride with oxygen as doping element or titanium dioxide with some traces of nitrogen (regions defined by the dotted lines in Fig. **3**). In addition, the transition from nitride to the oxide-like coatings is so abrupt that the process cannot be controlled to accurately adjust the chemical composition of the deposited films. On the other hand, it is worth to notice that a wide range of metalloid concentrations included between pure titanium nitride and pure titanium dioxide materials is achieved using the reactive gas pulsing process. A systematic rise of the duty cycle α from 0 to 100 % leads to a smooth evolution of the chemical composition in the films. In addition, one can remark that in the ternary diagram (Fig. **3**), the transition from the pure nitride to the pure oxide is not a linear function of the duty cycle. Besides, for the lowest duty cycles (*e.g.* $\alpha < 1$ %), oxygen concentration is below 30 at. %. As-deposited oxynitrides exhibit a poorly crystalline f.c.c. NaCl-type phase for this low range of duty cycles (Fig. **4**).

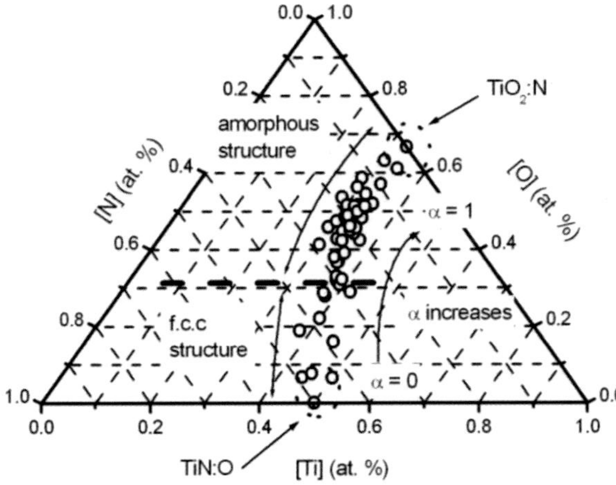

Figure 3: Evolution of the chemical composition in titanium oxynitride thin films as duty cycle rises from 0 to 1. Structural changes from f.c.c. to amorphous structure are also indicated as well as a virtual evolution of the chemical composition for a complete substitution of N by O atoms in the f.c.c. structure.

It is in agreement with previous studies focused on TiO_xN_y films synthesized by activated reactive evaporation [63] or by dc reactive sputtering [64, 65]. Since TiO and TiN compounds both develop the NaCl-type of structure with very close lattice constant values (a_{TiO} = 0.4190 nm and a_{TiN} = 0.4240 nm), it remains very ambiguous and inaccurate to state whether the titanium oxynitride films are single phase (*i.e.,* TiN:O) or contain mixed phases of sub-stoichiometric TiO and TiN materials. The f.c.c. TiN stays chemically stable over a wide N/Ti ratio in-between 0.5 to 1.2. The main point defects are both titanium vacancies and nitrogen interstitials. For TiO phase, both titanium and oxygen vacancies can exist over a wide range of O/Ti ratios in-between 0.8 to 1.3. Nevertheless, for the lowest duty cycles (*i.e.,* $\alpha < 10$ %), one can notice a shift of the diffracted peaks of the f.c.c. phase to higher 2θ angles, which corresponds to a decrease of the interplanar d-spacing. In addition, due to the strong affinity of oxygen with regards to titanium, it suggests that the formation of TiO phase is rather favoured with increasing oxygen content in TiO_xN_y films.

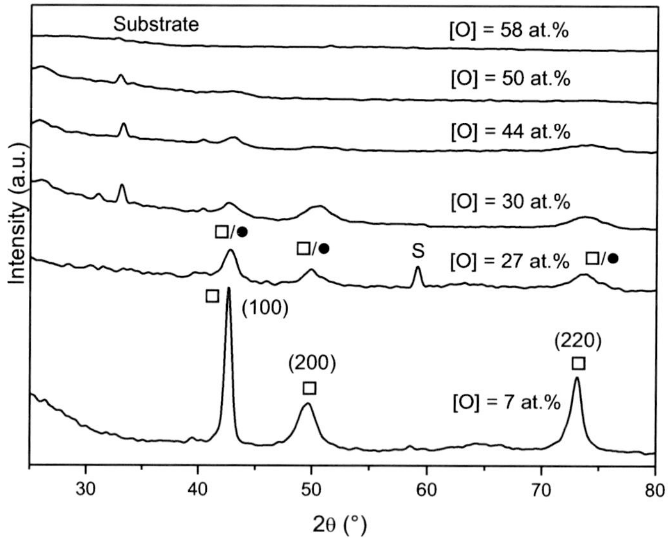

Figure 4: X-ray diffraction patterns of titanium oxynitride thin films as a function of a growing oxygen concentration. Diffracted signals corresponding to the f.c.c. NaCl-type phase are measured for the lowest oxygen concentrations, whereas an amorphous structure is observed for oxygen contents higher than 30 at. % (• = f.c.c. TiO phase; □ = f.c.c. TiN phase).

Thus, the lowest duty cycles ($\alpha < 0.10$) can lead to the periodic alternation of the nitride layer by an oxide layer at the surface of the sputtered target since the reactive gas pulsing process is used to produce TiO_xN_y films. The formation of mixed solid phases occurs by growth competition between TiO and TiN phases at the surface of depositing films. Besides, the stronger reactivity of oxygen than nitrogen towards titanium, and due to the ability of TiN to contain high concentrations of point defects, a progressive substitution of the nitrogen atoms by the oxygen ones in the metalloid f.c.c. sub-lattice during the film growth cannot be ruled out completely. However, it is worth noticing that the substitution phenomenon is not the only one as the oxygen concentration increases in TiO_xN_y films since the evolution of the elemental concentrations deviates to the hypothetic full substitution straight line defined by f.c.c. TiN and TiO phases in the ternary diagram (virtual straight line linking TiN and TiO stoichiometric compounds in Fig. **3**). A complete substitution should lead to the formation of TiO_xN_{1-x} compounds with a continuous modification of the f.c.c. phase towards the TiO one. Thus, the atomic ratio (O+N)/Ti should be close to unity. From analyses of the chemical composition presented in Fig. **2**, (O+N)/Ti ratio changes from 1.30 to 1.70 when the oxygen content in the films increases from 0 to 30 at. % (duty cycle α is lower than 10 %). So, the right structure of titanium oxynitride thin films could be a mixture of f.c.c. nanocrystallized Ti (N, O) + amorphous TiO_2 phases as suggested by Guillot *et al.* [66], but it still remains an open question.

For oxygen concentrations higher than 30 at. % (duty cycles higher than 10 %), the composition of TiO_xN_y films tends to stoichiometric TiO_2 compound with nitrogen traces and an amorphous structure is produced for the deposition performed at room temperature. When the substrates are heated to 473 K during the deposition, the growth of tetragonal TiO_2 phase prevails. The single polycrystalline anatase TiO_2 compound can be produced for the lowest oxygen concentrations, whereas mixed anatase + rutile TiO_2 phases are formed for the highest oxygen concentrations [57].

Optical Behaviours

Optical transmittance spectra of the TiO_xN_y films prepared on glass substrate measured in the visible region 4.5 to 0.5 eV (*i.e.,* 276 to 2483 nm), exhibit different features as a function of the oxygen concentration (Fig. **5**).

The spectra of TiN and TiO_xN_y films containing the lowest oxygen content up to 35 at. % (x = 0.95) are completely or strongly absorbent in the measured energy range, which is representative of a metallic-like behaviour. To the opposite, oxynitride thin films with an oxygen amount higher than 60 at. % (x = 1.90) exhibit the best optical transmittance (average transmittance in the visible region is close to 80 %) with interference fringes typical of dielectric materials. These results support that this kind of films behave as titanium dioxide. It is also interesting to note that the absorption edge is shifted to higher energies as the oxygen concentration increases. This blue shift is commonly observed in transparent conducting oxides. It is mainly attributed to the Burstein-Moss effect and correlates with an enhancement of the carriers concentration [67] and then, with the oxygen composition.

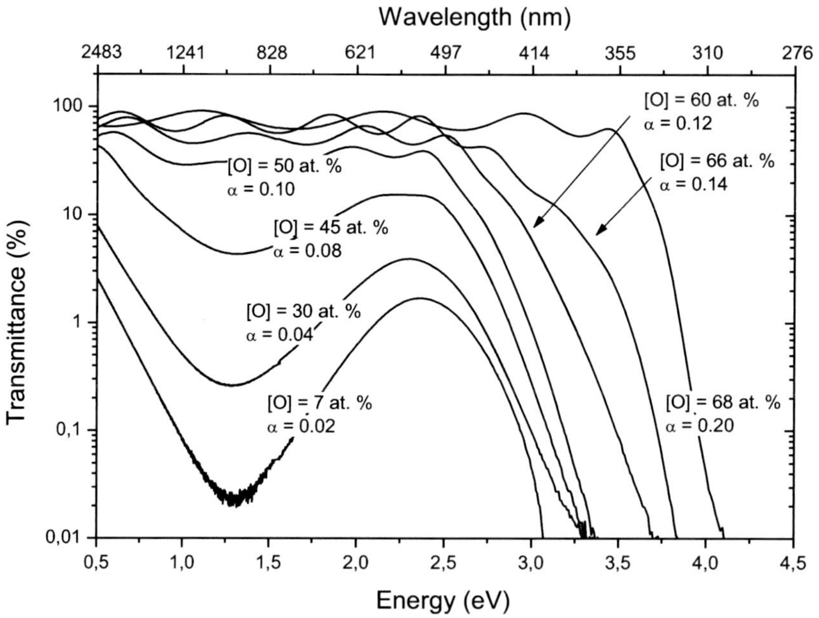

Figure 5: Optical transmittance spectra in the visible region of titanium oxynitride thin films 400 nm thick sputter deposited on glass substrates and for various oxygen concentrations in the films. Absorbent films are prepared for oxygen contents lower than 30 at. %, whereas typical and regular interferential fringes are observed for the films containing more than 60 at. % of oxygen.

For titanium oxynitride thin films containing intermediate oxygen concentrations, *i.e.,* from 35 to 60 at. % (x = 0.95 to 1.90), the transmittance spectra gradually increases from few % to more than 60 % in the visible range. In addition, some irregular fringes can be observed for the oxygen-rich TiO_xN_y films, whereas a strong absorption band is measured in the energy range between 1.0 and 1.5 eV

(1000 to 800 nm). This absorption can be attributed to a resonance phenomenon of the dielectric function corresponding to a weakly allowed electronic transition of Ti 3d to O/N 2p levels [64].

The smooth transition from metallic f.c.c. structure to typical dielectric compound TiO_2 is even more illustrated in Fig. **6** showing the average optical transmittance measured at 630 nm as a function of the oxygen concentration.

Up to 35 at. % of oxygen concentration, the optical transmittance in the visible region of TiO_xN_y films 400 nm thick is null. It corresponds to the occurrence of the NaCl-type of structure. The latter vanishes as the oxygen content rises. The films become poorly crystallized and an amorphous structure is adopted for the highest oxygen concentrations. Transparent TiO_2-like coatings are produced for oxygen concentrations higher than 60 at. %. The gradual change of the optical behaviours from metallic and absorbent TiN to dielectric transparent TiO_2 compound is clearly observed and supports that the most attracting range of compositions is included between 35 and 60 at. % of oxygen.

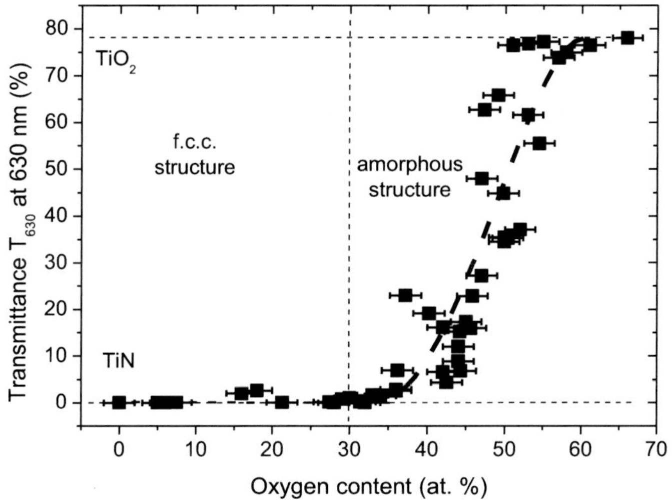

Figure 6: Optical transmittance at 633 nm of TiO_xN_y thin films 400 nm thick deposited on glass substrates *vs.* oxygen content. Up to 30 at. % of oxygen, films are absorbent and correlate with the occurrence of the f.c.c. structure. A gradual transition from absorbent to transparent oxynitrides can be noticed as the oxygen content changes from 30 to 60 at. %.

Refractive index n and extinction coefficient k at 633 nm of TiO_xN_y films are deduced from spectroscopic ellipsometry measurements *vs.* the metalloid concentrations (Fig. 7).

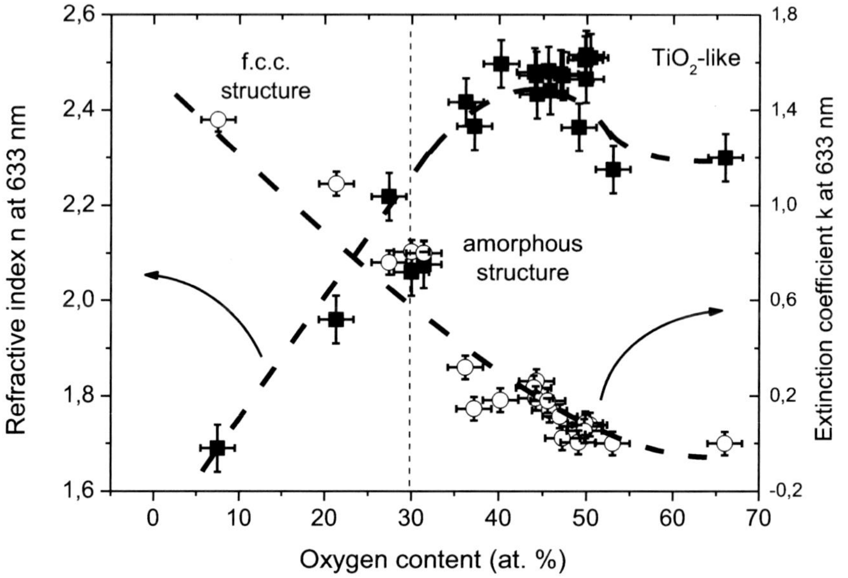

Figure 7: Refractive index n and extinction coefficient k of titanium oxynitride thin films calculated at 633 nm from spectroscopic ellipsometry as a function of the oxygen concentration.

For the lowest oxygen contents (< 10 at. %), refractive index values tend to that of typical TiN compound at 633 nm, *i.e.,* n_{633} = 1.47. A sharp increase of the index from n = 1.69 to 2.40 is observed when the oxygen concentration increases from 7 to 35 at. %. This range of compositions corresponds to the occurrence of the f.c.c. type of structure. In spite of the remaining open question about the right structure of titanium oxynitride thin films, *i.e.,* a single Ti(N, O) phase or a mixed TiN and TiO phases, the rise of refractive index as a function of oxygen concentration may be assigned to the formation of a more complex structure. Since TiN and TiO exhibit quite different refractive indices (n_{TiN} = 1.47 and n_{TiO} = 2.25 at 633 nm), a simple combination of each f.c.c. phase cannot be stated. As previously suggested by Guillot *et al.* [66], the coexistence of two phases: a nanocrystallized titanium oxynitride f.c.c. phase and an amorphous titanium dioxide TiO_2 one, may be formed *vs.* the oxygen incorporation in the material. In addition, it can be suggested that the f.c.c. patterns observed by XRD are composed of TiO and/or TiN phases. As a result, an increasing oxygen content in TiO_xN_y films favours the formation of the amorphous TiO_2 compound and reduces the amount of the f.c.c. structure. Due to the high value of TiO_2 refractive index (bulk values of anatase and rutile are n_{Anatse} = 2.49 and n_{Rutile} = 2.90 at 589 nm, respectively), the global refractive index of titanium oxynitride thin films is enhanced. Oppositely, the extinction coefficient k at 633 nm chiefly drops *vs.* oxygen content and tends to be negligible in the amorphous zone. For oxygen concentrations close to 45 at. %, the refractive index reaches a maximum value close to 2.50 and reduces down to 2.30 for TiO_2-like thin films (oxygen concentration reaches 66 at. %). This reduction of the refractive index may be assigned to a decrease of defects in the amorphous structure and due to the formation of a nearly stoichiometric TiO_2 compound. The refractive index is smaller than the anatase or rutile bulk value since the density of thin sputter deposited films is commonly below that of the bulk material.

Electrical Properties

Dc electrical conductivity measured at room temperature as a function of the oxygen concentration corroborates the structural and optical properties (Fig. **8**).

Films prepared with the lowest oxygen concentrations (< 30 at. %), exhibit an electrical conductivity σ at 300 K close to 10^5 $S.m^{-1}$. It slightly reduces as the oxygen content reaches 30 at. %. This range of compositions corresponds to the occurrence of the NaCl-type phase recorded previously by XRD and also to the most significant changes of the optical properties, *i.e.,* an enhancement of the refractive index, a drop of the extinction coefficient and a transition of the optical transmittance from absorbent to transparent films. Thus, up to 30 at. % of oxygen, the conductivity is weakly influenced by the metalloid concentrations and remains in the same order of magnitude consistent with the metallic-like behaviour.

The dc electrical conductivity was also measured as a function of the temperature between 300 and 500 K by the van der Pauw method (not shown here). These measurements support the metallic-like characteristics of the films up to an oxygen concentration close to 20 at. % since the conductivity reduces with an increasing temperature. Then, the TCR (temperature coefficient of resistance) can be determined at 300 K (Fig. **9**). Positive values lower than 5×10^{-4} K^{-1} are always obtained with a trend to decrease with the oxygen content. Such TCRs are lower than typical values of metallic compounds (for metallic single crystals, TCR is higher than 10^{-3} K^{-1}) because of the poorly crystallized structure of TiO_xN_y films (broad and small diffracted peaks due to the f.c.c. structure) and certainly the presence of defects associated to metalloid (both oxygen and nitrogen) and titanium deficiencies in the films. As the oxygen content reaches 30 at. %, one can assume that the films tend to the limit of the metallic-like properties. The electrical conductivity is still in the order of magnitude close to 10^5 S m^{-1}, but the TCR becomes negative. It drops down to -4×10^{-3} K^{-1} for an oxygen content of 29 at. %. Assuming that the f.c.c. structure is kept but is reduced up to 30 at. % and since the long range order (crystallite size decreases below few nanometres) is lost with a growing concentration of defects as the oxygen content increases in TiO_xN_y films, the low or negative TCR values can be attributed to the decrease of the mean free path of electron per grain size [68].

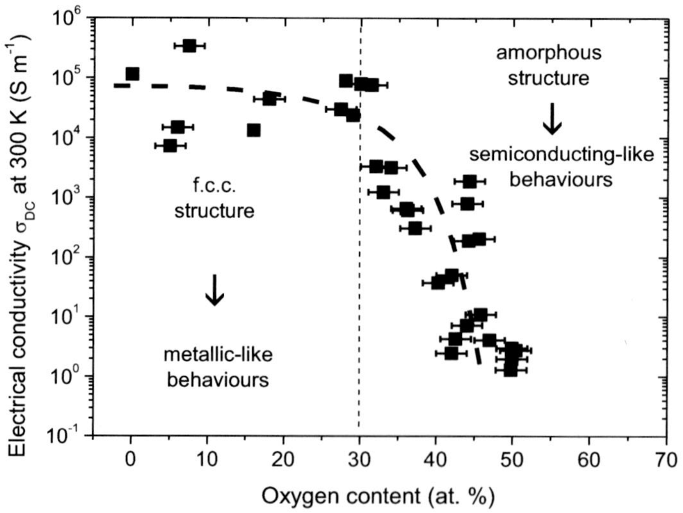

Figure 8: Dc electrical conductivity *vs.* oxygen content of titanium oxynitride thin films. A metallic-like behaviour is noticed up to 30 at. % of oxygen. Afterwards, films behave a semi-conductor and become too much resistive to be measured by the van der Pauw method. The drop of conductivity also correlates with the change of the long range order, *i.e.,* from f.c.c. to an amorphous structure.

A further increase of the oxygen concentration from 30 to 55 at. % leads to an abrupt drop down to few S m^{-1} of the dc electrical conductivity measured at room temperature (Fig. **9**).

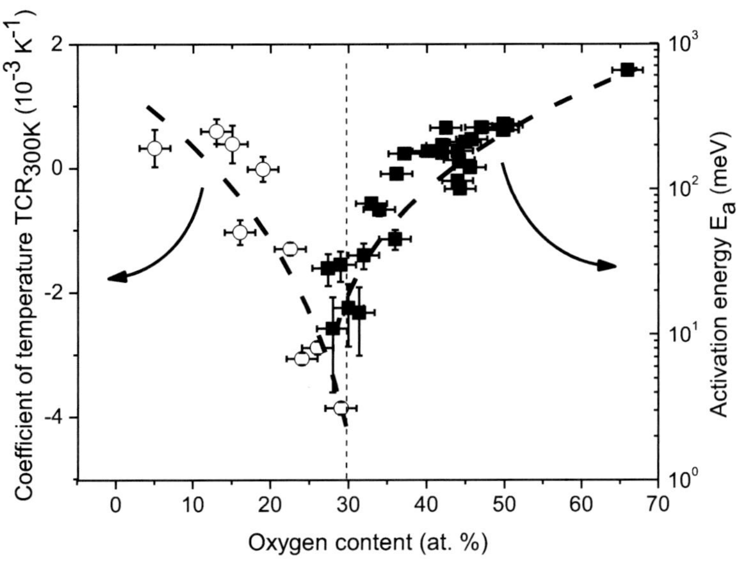

Figure 9: Temperature coefficient of resistance TCR at 300 K (metallic behaviour) and activation energy E_a (semi-conducting behaviour) as a function of oxygen content for titanium oxynitride thin films.

For higher oxygen contents, electrical resistivity becomes too high to be measured by the four-point probe technique, and the coatings tend to behave as pure titanium dioxide, *i.e.,* an insulator. Here again, resistivity correlates well with optical and structural properties discussed previously. The smooth transition from metallic to semi-conducting and finally insulating-like behaviours occurs as the oxygen concentration rises. In addition, conductivity *vs.* temperature measurements (not shown here) support that TiO_xN_y films produced with oxygen contents in-between 30 to 60 at. %, exhibit a semi-conducting characteristic. A negative slope of $\log(\sigma) = f(1000/T)$ is measured (linear evolution in the Arrhenius plot), which supports that the electrical conductivity σ as a function of temperature T follows an exponential behaviour according to $\sigma \propto \exp(-E_a/kT)$, where E_a is the activation energy and k the Boltzmann's constant. This activation energy is systematically calculated as a function of the oxygen concentration (Fig.

9). It continuously increases from tens of meV up to few hundreds meV as the oxygen concentration is maximum at 66 at. %.

Therefore, the transition of the electrical conductivity from metal to semiconductor of titanium oxynitride thin films is clearly illustrated as similarly demonstrated for optical properties in the previous section. The right structure of TiO_xN_y compounds is not completely known. It still remains an open discussion and further investigations are required to rightly understand this transition. However, the difference of ionicity between the metal – O and metal – N bonds and taking into account that the structure of TiO_xN_y films is composed of a nanocrystallized f.c.c. metallic phase in an amorphous matrix, one can suggest that such a gradual transition from metal to semi-conductor, originates from chemical and structural changes. Supposing that titanium oxynitride thin films are formed by a mixture of Ti – O and Ti – N bonds and assuming that the amount of Ti – O bonds increases (whereas that of Ti – N bonds decreases) with the oxygen content, the progression from metal to semiconductor can be ascribed to a larger ionicity of TiO_xN_y films. In addition, the decreasing amount of the metallic phase in the mixture f.c.c. + amorphous films also contributes to the formation of an insulating compound.

CONCLUSION

The reactive sputter deposition of titanium oxynitride thin films was successfully developed by the reactive gas pulsing process – RGPP. In order to prepare coatings with various metalloid concentrations, nitrogen gas was continuously supplied into the sputtering chamber, whereas the oxygen mass flow rate was pulsed. A rectangular shape was implemented using a variable duty cycle α and a constant pulsing period T = 45 s. A systematic change of the duty cycle α from 0 to 100 % led to a continuous and reverse evolution of oxygen and nitrogen concentrations in TiO_xN_y films as a function of α: from x = 0 to 2.0 and reversely, from y = 1.0 to 0.

For oxygen concentrations lower than 30 at. %, the films exhibited a poorly crystalline structure. The f.c.c. NaCl-type phase was produced for this range of compositions and vanished to give an amorphous structure for the highest oxygen contents. Since both TiO and TiN compounds adopt the same f.c.c. structure with very close lattice parameters, and due to the high metalloid contents in the films (oxygen + nitrogen concentrations higher than that of the titanium ones), the substitution of nitrogen atoms by the oxygen atoms cannot be simply considered to describe the right structure of titanium oxynitride thin films. It is strongly suggested that a mixture of two or three phases (TiO, TiN, amorphous TiO_2) can compose such oxynitrides. However, the accurate structure of TiO_xN_y compounds *vs.* the metalloid concentrations still remains an open field of discussions and further investigations are required.

Similarly, optical properties like optical transmittance in the visible region or refractive index strongly depend on the metalloid contents. Absorbent films are obtained for oxygen concentrations lower than 30 at. %. The refractive index changes from 1.69 to 2.40, whereas the extinction coefficient drops from 1.40 to negligible values. Afterwards, the films become transparent and dielectric compounds are rather deposited, which correlates with the occurrence of amorphous compounds. In addition, the electrical conductivity corroborates the optical and structural behaviours. A metallic-like characteristic was measured for TiO_xN_y films containing the lowest amount of oxygen (*i.e.,* oxygen concentration < 30 at. %) with TCRs lower than 10^{-3} K^{-1} or with negative values. An increasing oxygen concentration gave rise to a smooth transition towards a semi-conducting behaviour with activation energies tending to that of TiO_2-like compound.

ACKNOWLEDGEMENTS

The authors gratefully acknowledge the financial support of the region of Franche-Comté, France and the European Union through the NMP3-CT-505948 project "HARDECOAT".

CONFLICT OF INTEREST

The author(s) confirm that this chapter content has no conflict of interest.

REFERENCES

[1] E. Martinez-Ferrero, Y. Sakatani, C. Boissire, D. Grosso, A. Fuertes, J. Fraxedas, C. Sanchez, Advanced Functional Materials, vol. 17, pp. 3348-3354, 2007.

[2] E. Aperathitis, M. Bender, V. Cimalla, G. Ecke, M. Modreanu, J. Appl. Phys., vol. 94, pp. 1258-1266, 2003.
[3] R. Franchy, Surf. Sci. Rep., vol. 38, pp. 199-204, 2000.
[4] S. Venkataraj, O. Kappertz, R. Jayavel, M. Wuttig, J. Appl. Phys., vol. 92, pp. 2461-2466, 2002.
[5] C.A. Jong, T.S. Chin, Mater. Chem. Phys., vol. 74, pp. 201-209, 2002.
[6] M. Radecka, E. Pamula, A. Trenczek-Zajac, A. Zakrzewska, A. Brudnik, E. Kusior, N.T.H Kim-Ngan, A.G. Balogh, Solid State Ionics, vol. 192, pp. 693-698, 2011.
[7] R. Arvinte, J. Borges, R.E. Sousa, D. Munteanu, N.P. Barradas, E. Alves, F. Vaz, L. Marques, Appl. Surf. Sci., vol. 257, pp. 9120-9124, 2011.
[8] M. Grafouté, C. Petitjean, C. Rousselot, J.F. Pierson, J.M. Greneche, J. Phys. Cond. Mat., vol. 19, pp. 226207-12, 2007.
[9] O. Banakh, P.A. Steinmann, L. Dumitriscu-Buforn, Thin Solid Films, vol. 513, pp. 136-141, 2006.
[10] M. Fenker, H. Kappl, O. Banakh, N. Martin, J.F. Pierson, Surf. Coat. Technol., vol. 201, pp. 4152-4157, 2006.
[11] N.M.G. Parreira, Y.T. Pei, D. Galvan, T. Polcar, J.T.M. De Hosson, A. Cavaleiro, Microsc. Microanal., vol. 14, pp. 27-30, 2008.
[12] W.L. Scopel, M.C.A. Fantini, M.I. Alayo, I. Pereyra, Thin Solid Films, vol. 413, pp. 59-64, 2002.
[13] H. Wong, V.A. Gritsenko, Microelectron. Reliab., vol. 42, pp. 597-605, 2002.
[14] M.A. Quevedo-Lopez, S.A. Krishnan, P.D. Kirsch, G. Pant, B.E. Gnade, R.M. Wallace, Appl. Phys. Lett., vol. 87, pp. 262902-3, 2005.
[15] M. Muraoka, K. Kurihara, H. Satake, J. Appl. Phys., vol. 94, pp. 2038-2045, 2003.
[16] P. Carvalho, J.M. Chappé, L. Cunha, S. Lanceros-Mendez, P. Alpium, F. Vaz, E. Alves, C. Rousselot, J.P. Espinos, A.R. Gonzalez-Elipe, J. Appl. Phys., vol. 103, pp. 104907-15, 2008.
[17] S. Agouram, F. Bodart, G. Terwagne, Surf. Coat. Technol., vol. 180, pp. 164-168, 2004.
[18] C.S. Kang, H.J. Cho, R. Choi, YH. Kim, C.Y. Kang, S.J. Rhee, C.W. Choi, M.S. Akbar, J.C. Lee, IEEE T. Electron Dev., vol. 51, pp. 220-227, 2004.
[19] A. Von Richthofen, R. Domnick, R. Cremer, D. Neuschutz, Thin Solid Films, vol. 317, pp. 282-284, 1998.
[20] M. Futsuhara, K. Yoshioka, O. Takai, Thin Solid Films, vol. 317, pp. 322-325, 1998.
[21] I.P. Parkin, G.S. Elwin, J. Mater. Chem., vol. 11, pp. 3120-3124, 2001.
[22] V.A. Matylitskaya, W. Block, K. Thoma, B.O. Kolbesen, Appl. Surf. Sci., vol. 252, pp. 205-210, 2005.
[23] L. Castaldi, D. Kurapov, A. Reiter, V. Shklover, J. Patscheider, J. Appl. Phys., vol. 109, pp. 053720-7, 2011.
[24] S.K. Rawal, A.K. Chawla, R. Jayganthan, R. Chandra, Appl. Surf. Sci., vol. 257, pp. 8755-8761, 2011.
[25] F. Vaz, P. Cerqueira, L. Rebouta, S.M.C Nascimento, E. Alves, P. Goudeau, J.P. Rivière, K. Pischow, J. De Rijk, Thin Solid Films, vol. 447, pp. 449-454, 2004.
[26] S.H. Mohamed, O. Kappertz, J.M. Ngaruiya, T. Niemeier, R. Drese, R. Detemple, M.M. Wakkad, M. Wuttig, Phys. Status Solidi A, vol. 201, pp. 90-102, 2004.
[27] S. Venkataraj, D. Severin, S. Mohamed, J. Ngaruiya, O. Kappertz, M. Wuttig, Thin Solid Films, vol. 502, pp. 228-234, 2006.

[28] J.M. Chappé, N. Martin, G. Terwagne, J. Lintymer, J. Gavoille, J. Takadoum, Thin Solid Films, vol. 440, pp. 66-73, 2003.
[29] G. Hyett, M.A. Green, I.P. Parkin, J. Am. Chem. Soc., vol. 129, pp. 15541-15548, 2007.
[30] E. Alves, A.R. Ramos, N.P. Barradas, F. Vaz, P. Cerqueira, L. Rebouta, U. Kreissig, Surf. Coat. Technol., vol. 180, pp. 372-376, 2004.
[31] D. Herman, J. Sicha, J. Musil, Vacuum, vol. 81, pp. 285-290, 2006.
[32] J.B. Metson, K.E. Prince, Surf. Interface Anal., vol. 28, pp. 159-162, 1999.
[33] C. Nunes, V. Teixeira, M.L. Prates, N.P. Barradas, A.D. Sequeira, Thin Solid Films, vol. 442, pp. 173-178, 2003.
[34] A. Bittar, D. Cochrane, S. Caughley, I. Vickeridge, J. Vac. Sci. Technol., vol. A15, pp. 223-229, 1997.
[35] X.G. Yang, C. Li, B.J. Yang, W. Wang, Y.T. Wian, Chem. Phys. Lett., vol. 383, pp. 502-506, 2004.
[36] F. Fabreguette, L. Imhoff, M. Maglione, B. Domenichini, M.C.M. De Lucas, P. Sibillot, S. Bourgeois, M. Sacilotti, Chem. Vapor Depos., vol. 6, pp. 109-114, 2000.
[37] M. Drygas, C. Czosnek, R.T. Paine, J.F. Janik, Chem. Mater., vol. 18, pp. 3122-3129, 2006.
[38] X.M. Song, D. Gopireddy, C.G. Takoudis, Thin Solid Films, vol. 516, pp. 6330-6335, 2008.
[39] S.K. Pradhan,.J. Reucroft, J. Cryst. Growth, vol. 250, pp. 588-594, 2003.
[40] O.I. Yaskiv, I.M. Pohrelyuk, V.M. Fedirko, D.B. Lee, O.V. Tkachuk, Thin Solid Films, vol. 519, pp. 6508-6514, 2011.
[41] I.N. Martev, Vacuum, vol. 67, pp. 261-265, 2002.
[42] F. Maury, F.D. Duminica, Surf. Coat. Technol., vol. 205, pp. 1287-1293, 2010.
[43] V. Atuchin, V. Kruchinin, A. Kalinkin, V. Aliev, S.V. Rykhlitskii, V. Shvets, E. Spesivtsev, Opt. Spectrosc., vol. 196, pp. 72-77, 2009.
[44] M.H. Chang, F.H. Lu, Thin Solid Films, vol. 517, pp. 5006-5009, 2008.
[45] S. Berg, H.O. Blom, T. Larsson, C. Nender, J. Vac. Sci. Technol., vol. A5, pp. 202-207, 1987.
[46] D. Depla, R. De Gryse, Surf. Coat. Technol., vol. 183, pp. 184-189, 2004.
[47] L. Pekker, Thin Solid Films, vol. 289, pp. 140-146, 1996.
[48] P.J. Kelly, R.D. Arnell, J. Vac. Sci. Technol., vol. A17, pp. 945-953, 1999.
[49] V. Bellido-Gonzalez, B. Daniel, J. Counsell, D. Monaghan, Thin Solid Films, vol. 502, pp. 34-39, 2006.
[50] H. Bartzsch, P. Frach, Surf. Coat. Technol., vol. 142, pp. 192-200, 2001.
[51] T. Nyberg, C. Nender, S. Berg, J. Vac. Sci. Technol., vol. A16, pp. 1868-1872, 1998.
[52] A. Billard, C. Frantz, Surf. Coat. Technol., vol. 86-87, pp. 722-727, 1996.
[53] N. Danson, G.W. Hall, R.P. Howson, Thin Solid Films, vol. 289, pp. 99-106, 1996.
[54] A.J. Aronson, D. Chen, W.H. Class, Thin Solid Films, vol. 72, pp. 535-540, 1980.
[55] H. Sekiguchi, A. Kanzawa, T. Imai, T. Honda, J. Vac. Sci. Technol., vol. 12, pp. 3176-3179, 1994.
[56] D. Gall, R. Gampp, H.P. Lang, P. Oelhafen, J. Vac. Sci. Technol., vol. A14, pp. 374-379, 1996.
[57] N. Martin, O. Banakh, A.M.E. Santo, S. Springer, R. Sanjinès, J. Takadoum, F. Lévy, Appl. Surf. Sci., vol. 185, pp. 123-133, 2001.

[58] N. Martin, J. Lintymer, J. Gavoille, J.M. Chappé, F. Sthal, J. Takadoum, F. Vaz, L. Rebouta, Surf. Coat. Technol., vol. 201, pp. 7720-7726, 2007.

[59] N. Martin, J. Lintymer, J. Gavoille, J.M. Chappé, F. Sthal, J. Takadoum, F. Vaz, L. Rebouta, Surf. Coat. Technol., vol. 201, pp. 7727-7732, 2007.

[60] N. Martin, J. Lintymer, J. Gavoille, J.M. Chappé, F. Sthal, J. Takadoum, F. Vaz, L. Rebouta, Surf. Coat. Technol., vol. 201, pp. 7733-7738, 2007.

[61] W.D Sproul, D.J. Christie, D.C. Carter, S. Berg, T. Nyberg, 46th Annual Technical Conference Proceedings, vol. 98, 2003.

[62] W.D. Sproul, D.J. Christie, D.C. Carter, Thin Solid Films, vol. 491, pp. 1-17, 2005.

[63] M. Lazarov, P. Raths, H. Metzger, W. Spirkl, J. Appl. Phys., vol. 77, pp. 2133-2137, 1995.

[64] E. Vogelzang, J. Sjollema, H.J. Boer, J.T.M. De Hosson, J. Appl. Phys., vol. 61, pp. 4606-4611, 1987.

[65] M.H. Kazemeini, A.A. Berezin, N. Fukuhara, Thin Solid Films, vol. 372, pp. 70-77, 2000.

[66] J. Guillot, J.M. Chappé, O. Heintz, N. Martin, L. Imhoff, J. Takadoum, Acta Mater., vol. 54, pp. 3067-3074, 2006.

[67] G. Sanon, R. Rup, A. Mansingh, Phys. Rev. B, vol. 44, pp. 5672-5680, 1991.

[68] G. Reiss, J. Vancea, H. Hoffmann, Phys. Rev. Lett., vol. 56, pp. 2100-2103, 1986.

Send Orders of Reprints at reprints@benthamscience.net
Metallic Oxynitride Thin Films by Reactive Sputtering and Related Deposition Methods, 2013, 133-162

CHAPTER 6

Growth and Characterization of Chromium Oxynitride Thin Films Prepared Using Reactive Unbalanced Magnetron Sputtering in Presence of Air as Reactive Gas

Saïd Agouram[1,*], Guy Terwagne[2] and Franz Bodart[2]

[1]Department of Applied Physics and Electromagnetism, Universitat de Valencia, C./ Dr. Moliner N50, 46100 – Burjassot, Spain and [2]Department of Physics, LARN, FUNDP, 61 rue de Bruxelles, 5000 – Namur, Belgium

Abstract: Chromium nitride and chromium oxynitride thin films were successfully deposited onto carbon, silicon and low carbon steel substrates by DC reactive magnetron sputtering using a chromium metallic target. Argon-nitrogen and argon-air mixtures were used as reactive gases to deposit CrN and $Cr_xN_yO_z$ films, respectively. The total flow (argon + reactive gas) was kept constant during all deposition processes. Relative humidity of air (water vapour pressure) was systematically changed from 0 to 90 % RH. Behaviour of the cathode was studied as a function of reactive gas flow and water vapour pressure. Evolution of the deposition rate and composition of the deposited films with increasing of reactive gas flow and amount of water vapour injected into the process was studied by Rutherford Backscattering Spectroscopy (RBS) and Resonant Nuclear Reaction Analysis (RNRA). The chemical bond was determined by Energy Electron Induced X-ray Spectroscopy (LEEIXS) and X-ray Photoemission spectroscopy (XPS). Hydrogen and nitrogen profiles were determined by RNRA. RBS and LEEIXS results have shown that in metallic mode, the composition can be fitted by CrN and Cr_2O_3 stoichiometries, but in compound mode the CrO_2 stoichiometry predominates. XPS results have shown the coexistence of another phase with $(CrO_2)_3N$ stoichiometry. Its proportion depends on the N content in thin films.

Keywords: Thin films, chromium nitride, chromium oxide, chromium oxynitride, PVD, magnetron sputtering, RBS, RNRA, XPS, LEEIXS, hysteresis effect, relative humidity.

INTRODUCTION

Transition metal nitrides and oxides such as titanium nitride, chromium nitride and chromium oxides have widely been used as protective and hard coatings to

***Address correspondence to Saïd Agouram:** Department of Applied Physics and Electromagnetism, Univesitat de Valencia, C./ Dr. Moliner N 50, 46100 Burjassot, Spain; Tel: +34 963543337; Fax: +34 96 354 3146; Email: said.agouram@uv.es

Filipe Vaz, Nicolas Martin and Martin Fenker (Eds)

increase lifetime and performance of cutting and forming tools. Many hard coatings that can be used at high temperatures have been studied. Similar to TiN, CrN has good mechanical properties [1, 2] and improves the corrosion protection [1, 3] because it has a high oxidation resistance and a good chemical stability. A great advantage of CrN is that the internal stresses are very low, as evident by coatings with thickness greater than 40 μm could be obtained on a variety of substrates. Like CrN, Cr_2O_3 thin films also present interesting properties such as a high hardness [1, 4] and colours intended for decoration [1, 5].

Recently there has been increased interest in transition metal oxynitrides. This has emanated from the possibility of attaining attractive properties such as high thermal, good electrical, mechanical and chemical stability properties [6]. These compounds are, in fact, promising and it is possible to obtain a wide range of materials ranging from oxide to nitride compound. On the other hand, transition metal oxynitrides have been found to possess properties intermediate between oxides and nitrides and present promising industrial applications which justify the increased interest in these compounds. Among the metallic oxynitrides compounds that are being developed there are chromium oxynitride and titanium oxynitride. They have rarely been described in the literature but their presence is observed, particularly in the development of chromium nitride and titanium nitride thin films. Indeed, despite the use of oxygen-free precursors, despite the drastic conditions of vacuum deposition, many authors observe the presence of oxygen in nitride compounds [7, 8]. Oxygen presence in the films has not been clarified and it is very difficult to control the residual oxygen content in a vacuum chamber. However, few people attribute this oxygen to the existence of oxynitrides compound. Chromium oxynitride is a potential candidate in many applications which include high hardness and a low friction coefficient [4]. It presents different colours, which vary with thickness and composition. In comparison with Cr_2O_3, the chromium oxynitride has a higher corrosion resistance and the layer has a better adhesion and a uniform structure [9]. In comparison with CrN, chromium oxynitride films have superior properties and they could be produced by only a Cr target and N_2 and O_2 gases as reactive materials. In Table **1**, we summarise some remarkable properties and technological potentialities of some metal oxynitride thin films.

Table 1: Crystalline structures and applications of some transition metal oxynitride films [10].

Oxynitride	Crystalline Structure	Applications
AlN_xO_y	Cubic spinel	Optics (transparent, UV, IR) Electrical and mechanic properties. Food packaging
CrN_xO_y	cubic (type NaCl)	Mechanical properties (hardness, abrasion and corrosion resistance)
SiN_xO_y	quartz	Microelectronics Optics sensors, detectors Electromechanical systems
TaN_xO_y	monoclinic	Microelectronics Optics sensors (pH, temperature)
TiN_xO_y	cubic (type NaCl)	Microelectronics, mechanics, optics Biocompatible materials
VaN_xO_y	cubic (type NaCl)	Semi-conductors technology Catalyst
ZnN_xO_y	type ZnO	Catalyst
ZrN_xO_y	cubic	Thermal, mechanical properties

PRODUCTION OF CHROMIUM OXYNITRIDE THIN FILMS

Reactive Sputtering

Among the various deposition methods, reactive sputtering has been used to produce metallic oxynitride thin films and enables us to control the composition of the grown films. Sputtering is a process where atoms are ejected from a solid target. Physical sputtering is driven by momentum exchange due to collisions between energetic particles and atoms in the materials. A negative dc voltage is applied to the target -cathode-, which is a metallic material to be deposited. The effect of the negative voltage is to accelerate the ions, mostly argon, towards the cathode. The incident ions produce collision cascades in the target and leads to the ejection of target atoms by Ar^+ ion bombardment. The emission of electrons from the cathode leads to amplification of the ionization process of argon gas and hence to increase sputtered material. The average number of atoms ejected from the target per incident ion is called sputtering yield. It is mentioned for the first time by Sigmund [11], and depends on ion incident angle, energy of ions, masses of ions and target atoms, and surface binding energy of atoms in the target. According to Sigmund [11] the sputtering yield, Y, is simply reciprocal to the surface binding energy. An algorithm that simulates sputtering based on a quantum mechanical treatment including electrons stripping at high energy is implemented in the program TRIM [12].

Deposition of compound films by means of sputtering is performed by adding a reactive gas such as oxygen and nitrogen for deposition of oxides and nitrides, respectively, with the inert gas into the deposition chamber. In basic terms, reactive sputtering results from the sputtering of elemental targets in a reactive atmosphere and it promotes the reaction between sputtered atoms and reactive gas (Fig. **1**). It is a widely used technique to deposit compounds thin films and it is particularly suitable to modulate the chemical composition and so, the final properties of the films. Oxides, nitrides, oxynitrides of single or multi-metallic elements can be synthesized by this method [13-20]. For the deposition of the metal-oxynitride films, both gases oxygen as well as nitrogen could be used simultaneously. The variation of the reactive gas flow enables the deposition of films with different atomic compositions. Nevertheless, and in spite of numerous cases of success, there are some typical instability phenomena of experimental parameters, associated to the reactive process involving one metallic target with reactive gases [21-25]. Some peculiarities are observed in the reactive sputtering mode compared to the sputtering of pure metal films.

Unfortunately, in reactive sputtering, reactive gas does not react only with the ejected atoms but it also reacts with the target (Fig. **1**) and compound formation on the target surface will reduce the sputter rate, originating from the higher binding energy of the oxide (in case of O_2 as a reactive gas) compared to the corresponding metal. The deposition rate will remain constant, if the target surface is totally oxidized, which is called the compound mode. Increasing the reactive gas flow, the compound formation on the target surface raises until a total coverage of the target occurs (the target poisoning effect). This value of reactive gas flow is defined as the critical flow where transition between metallic and compound mode occur at the target. For reactive gas flows lower than the critical value, the target works in a metallic mode, while for reactive gas flows higher than that critical value, it works in the compound mode. The transition from the metallic mode to compound mode also called "poisoned mode", most of the times, exhibits a hysteresis [26, 27] that can be reduced by increasing the pumping speed [28].

Usually when the reactive gas flow is higher that the critical limit, the deposition rate decreases due to several phenomena among which the more relevant are:

i) the lower sputtering yield of the compound formed on the surface of the target;

ii) in some cases, the increase of secondary electron emission coefficient when the target surface is poisoned [29].

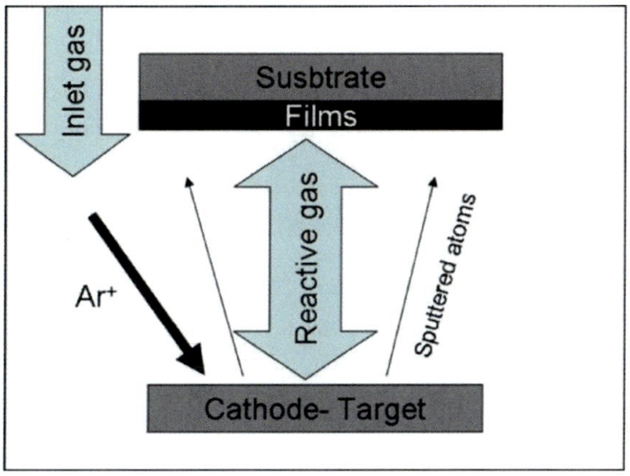

Figure 1: Simplified schematic view of reactive sputtering process.

State of the Art of Chromium Oxynitride Films

Since 1982, Granier *et al.* [30], are interested in metal- oxygen-nitrogen interactions. They developed and studied titanium oxynitrides, and described in the ternary Ti-O-N diagram a wide region of solid solution between cubic TiO and TiN. Other investigations followed and the results could be found in references [31-36].

Chromium oxynitrides were mentioned twenty years ago for the deposition of CrN on steel for corrosion applications. Compared with Cr_2O_3, chromium oxynitrides exhibit high corrosion resistance and the layer of compound product shows good adhesion and uniform structure [37]. Cr-O-N films have been synthesized by various methods (r.f.-sputtering [38], pulsed laser deposition [39], and cathodic arc evaporation [40]). The Cr-O-N films consisted of a single-phase solid solution based on deformed CrN or Cr_2O_3, or a two-phase mixture of CrN and Cr_2O_3 [1, 39, 41-44], or a single Cr-oxynitride phase [45]. Baborowski *et al.* [46, 47] have grown chromium oxynitride layers by reactive sputtering in the presence of nitrogen and water vapour. The characterization of these layers by XPS showed that the composition is close to the $Cr_2O_4N_{0.3}$ stoichiometric formula and O and N contents depend on the amount of water vapour in the plasma. Collard *et al.* [1] also produced thin films of Cr-N-O by sputtering in a reactive atmosphere containing two reactive gases: nitrogen and oxygen. From their studies two major structures appear: the CrN for low contents of oxygen and Cr_2O_3 for higher concentrations of oxygen. They concluded that the structure of the chromium-oxynitride films was based on a deformed Cr_2O_3. In contrast, Suzuki *et al.* [39] observed a single phase, which crystallizes according to the fcc structure. They suggested that the Cr-O-N solid-solution phase had fcc CrN structure due to the non-equilibrium condition in the plasma environment and oxygen atoms replace those of nitrogen in CrN lattice. Mientus *et al.* [48] prepared CrO_xN_y films by reactive magnetron sputtering from a chromium target by using a $Ar/O_2(N_2O)/N_2$ gas mixtures. They observed that by changing the gas composition, the film stoichiometry was continuously varied from CrN to Cr_2O_3 and the films consist of CrN nanocrystals in an amorphous oxide matrix. Recently, Shirahata *et al.* [49] synthesized chromium oxynitride thin films by radio-frequency reactive unbalanced magnetron sputtering at various O_2 flow rates and they observed that hardness is increased with increasing oxygen content.

Other deposition methods were also applied to grow chromium oxynitride films, among we can list: Castaldi *et al.* [42] produced chromium oxynitride films by reactive cathodic arc deposition at different N_2/O_2 flow ratios. Their results suggest the formation of a substitutional CrO_xN_{1-x} solid solution and hardness and the Young's modulus increase with increasing O content up to a maximum value for the $Cr_2(O_{0.97}N_{0.03})_3$ stoichiometry coatings, followed by a progressive decrease for pure Cr_2O_3. Gautier *et al.* [44] used a cathodic arc method to deposit Cr-N-O by adding a flow of oxygen into the vacuum deposition chamber during deposition of chromium nitride. Cr-N-O films prepared by this method contain two different phases: CrN and Cr_2O_3. Wierzchon *et al.* [37] synthesized Cr (N, O) films by annealing chromium nitride layers at 560° C in an atmosphere of $N_2 + H_2$ with 10% of the total flow of air. They showed that the Cr (N, O) layers produced during this process contain 30 at. % oxygen and that the corrosion resistance was improved compared to CrN.

Regarding the process of formation of oxynitrides, in 1993 Carlsson *et al.* [50] tried to create a model that predicts the reactive sputtering process and deposition conditions of oxynitride films with a defined composition. They showed that, due to different reasons, the control of the deposition process was very complicate when two reactive gases were used. One reason is the difference in sticking coefficients of oxides and nitrides, also the formation of a complex compound as a mixture of phases with different structures changes the target behaviours. For the reactive sputtering system, the difference in reactivity of these gases will give new restrictions in the processing control. In addition, a complication in controlling the partial pressures of both reactive gases will appear. Varying the partial pressure of one of these gases will affect the partial pressure of the other gas. Therefore quite advanced feed-back control systems are needed to obtain fully control of such a reactive sputtering process. Oxygen reactivity is more important than that of nitrogen. We must introduce a small amount of oxygen in the plasma to produce oxynitrides. Carlsson *et al.* [50] suggested that the introduction of water vapour in the deposition chamber could be used to produce oxynitride thin films and the proportion of nitride and oxide depends on relative humidity. Based on this work on Cr-N-O films, it seemed interesting to go further forward than to deposit thin layers of chromium oxynitride using two separate reactive gases (nitrogen + oxygen, nitrogen + water vapour or nitrogen in the presence of residual oxygen in the deposition chamber) but to produce Cr-N-O films from a plasma of argon and air with various concentrations of water vapour. It is important to note that chromium oxynitride layers were obtained for the first time by reactive sputtering with air as a reactive gas in 2000 [51] and the first results showed that by using lower air flow the composition was CrN and Cr_2O_3, while with higher air flow the target is poisoned and the CrO_2 stoichiometry predominates. The subject of this chapter is to study the water vapour effect on the deposition rate and the composition of chromium oxynitride thin films deposited onto carbon, silicon and low carbon steel substrates by DC magnetron reactive sputtering in an argon + air mixture containing various concentrations of water vapour.

Experimental Conditions

Chromium oxynitride films were produced by dc reactive sputtering using an unbalanced magnetron system in an Ar and air gases mixture. The vacuum chamber (0.25 m^3) was equipped with a chromium (99.99 at. % purity) sputtering target disc of 19.6 cm^2 area and a 400 L min^{-1} turbomolecular pump. Before starting the sputtering process, the pressure in the chamber was lower than 5×10^{-5} Pa. The deposition working pressure and plasma power density were maintained at 0.56 Pa and 5.86 W cm^{-2}, respectively. For these operating conditions, the temperature measured by a thermocouple fixed at the rear end of the substrate during the sputtering process was about 100 °C. Silicon, polished carbon and low carbon steel were used as substrates. Before depositing, pure argon plasma was used to clean the chromium target in order to eliminate the oxide layer formed on the surface of the target. After this the substrate was etched for 4 min in a pure argon plasma atmosphere. High-purity argon and dry air were used and both gas flows were controlled. The reactive gas (air) and argon were admitted separately through a MKS flow control system, keeping a constant total gas flow of 35 sccm for all deposition runs. The reactive gas flow varied from 0 % (0 sccm) to 70 % (25 sccm). In this work, different reactive gases such as dry air, air at 20, 40, 65 and 95 % relative humidity (RH) were used.

The thickness and composition of chromium oxynitride films were measured by Rutherford backscattering spectroscopy (RBS) and resonant nuclear reaction analysis (RNRA) techniques at LARN using the 2 MV Tandetron accelerator. The RBS technique was performed with 2 MeV α-particles and the scattered particles were detected in a passivated implanted planar silicon (PIPS) surface barrier detector, located at an angle of 175° to the incident beam. The experimental spectra were fitted with the RUMP code [52, 53]. The nitrogen and hydrogen profiles were measured by using the narrow resonance nuclear reaction and TOF-SIMS. The ^{15}N (p, αγ) ^{12}C reaction at E_p = 429 keV for nitrogen and H (^{15}N, αγ) ^{12}C reaction at 6.385 MeV for hydrogen profiling were used. The 4.43 MeV γ-radiation of $^{12}C^*$ was measured with a NaI detector as a function of the incident particle energy to obtain the concentration profile. The chemical bond and the composition of the films were studied by low energy electron-induced X-ray spectroscopy and X-ray photoemission spectroscopy (LEEIXS). The LEEIXS is based on the ionization of the inner-shell of the analyzed element by electrons of a few keV. In this work we investigate the intensity measurement of the Cr Lα with a TAP crystal. The energy of the incident electron beam is 5 keV, the gas pressure is 0.23 Torr and the current is 0.12 mA. With 5 keV of electron beam, the thickness of samples analyzed was between 120 and 240 nm. XPS measurements were recorded with a SSX-100 spectrometer using the monochromatised X-ray Al Kα (1486.4 eV).

Study of the Cathode Behaviours

In this paragraph in order to illustrate better reactive sputtering process, we study the variation of parameters controlling the plasma discharge, which will allow us to compare the behaviour of the cathode when we introduce argon and the reactive gas as nitrogen or air.

Evolutions of Cathode Voltage Discharge

Fig. **2** presents the evolution of the cathode voltage as a function of air and nitrogen flow. Plasma power density and total flow were maintained at 5.86 W cm^{-2} and 30 sccm, respectively. The total pressure was fixed at 0.28 Pa.

As a global effect, an increasing of the reactive gas flow results in a rise of the discharge voltage. This leads us to verify that the current generated by secondary electrons emitted from chromium compound formed on the target is less than metallic chromium target. The probability of ionization of argon by electron impact is more than 10 times higher than that of nitrogen or air [54].

From results of Fig. **2**, in case of dry air as reactive gas, we observe linear and reversible evolution of the discharge voltage. From 20 sccm dry air flow, the voltage reaches a stable value at 400 V. As said before, in reactive sputtering, two rival phenomena take place on the cathode: reactive gas reacts with the surface of the cathode by forming an oxide/nitride compound when air is introduced. The formed compound on the cathode affects surface properties of the cathode and

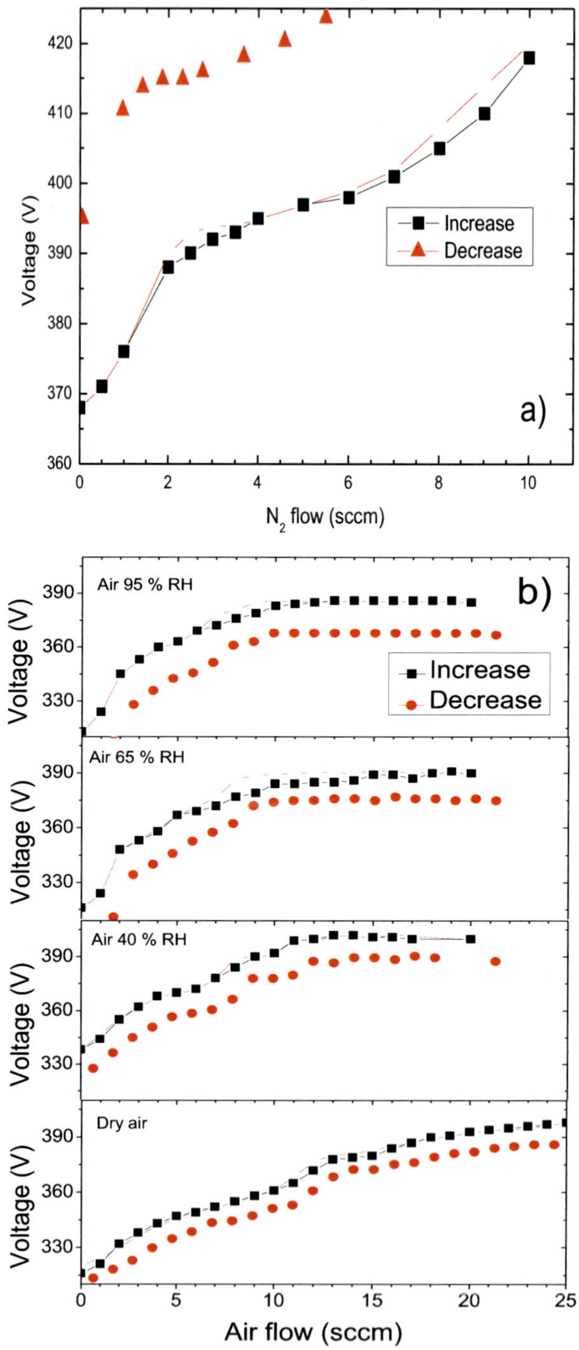

Figure 2: Cathode voltage evolution as a function of a) nitrogen and b) air flow (dry, 40, 65 and 95 % RH). Plasma power density was 5.86 W cm^{-2} and the total flow was 30 sccm.

influences directly the emission of secondary electrons. Emission of secondary electrons by ion impact with metal or oxide/nitride compound is not identical, the voltage and the current of discharge are then going to evolve towards new stable regime. In case of humidified air, we observe that the discharge voltage increases continuously and stabilizes when the air flow reaches a critical value (12 sccm for air at 40% RH and 10 sccm for air at 65 and 95 % RH). It can be related to the content of water vapour, which accelerates the oxidation of the cathode. This can be explained by the fact that molecules (that make up air like nitrogen and hydrogen) impacting on the target surface produce little secondary electrons. We also observe that the evolution curves exhibit a hysteresis voltage of the cathode. Reducing the air flow does not immediately lead to return to the metallic mode. This is because the cathode is saturated and it is necessary to return to very low values of flow for metallic return mode. A hysteresis phenomenon is observed in voltage discharge evolution and it could be attributed to lower sputtering yield of chromium nitride on the surface in comparison with metallic chromium. The same behaviour was observed for a power density of 12 W cm^{-2}: a linear increase of the discharge voltage followed by stabilization.

Evolution of the Partial Pressures in the Plasma

Fig. **3** illustrates the evolution of partial pressures of H_2, H_2O, N_2 and O_2 present in argon-nitrogen and argon-air plasma maintaining power density and total pressure at 5.86 W cm^{-2} and 0.28 Pa, respectively. We observe a gradual increase of nitrogen partial pressure as a function of nitrogen flow. This rapid increase is due to the small amount of nitrogen consumed to form the chromium nitride, and each addition of nitrogen flow results in an increase of the nitrogen partial pressure and the formation of nitride on the cathode surface, which reduces the number of sputtered chromium atoms. The same behaviour is observed in case of argon-air plasma discharge. An increase of oxygen partial pressure as a function of dry air flow is observed. This increase is due to the fact that oxygen molecules are not totally consumed by chromium to form chromium oxide on the substrate, and each addition of the flow of air causes oxidation of the chromium target until its saturation. The low reactivity of nitrogen causes a rapid and linear increase in its partial pressure. Between 0 and 6 sccm of air flow H_2O remains constant. A rapid increase is observed between 6 and 10 sccm and then stabilizes when the flow reaches 10 sccm.

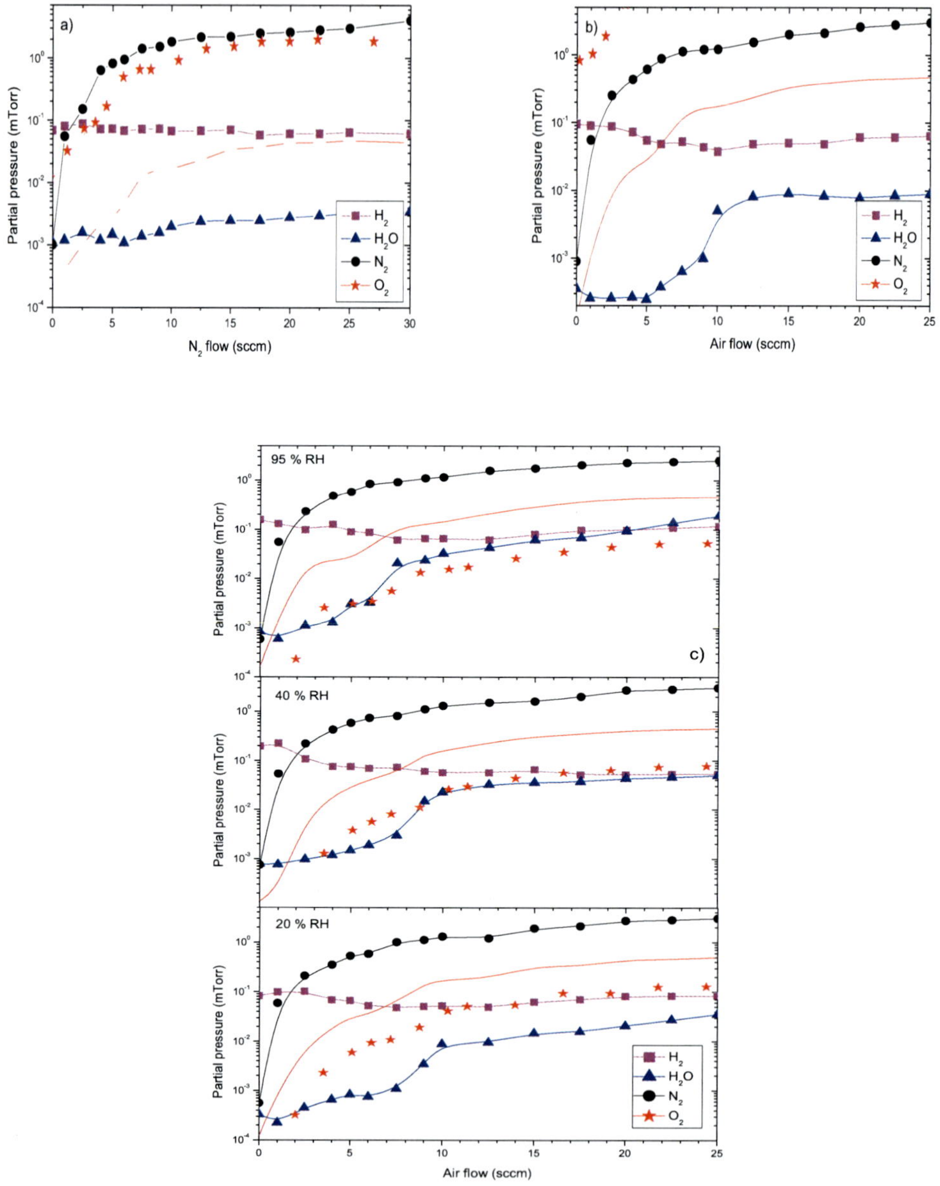

Figure 3: Evolution of partial pressures of N_2, O_2, H_2O and H_2 as a function of a) nitrogen; b) dry air and c) air flow. Plasma discharge is maintained at constant power density 5.86 W cm^{-2} and total pressure 0.28 Pa.

The same trends were observed in case of air with different relative humidities. It is observed that partial pressures increase rapidly when RH is increased. It can be explained by the fact that oxidation of the chromium cathode increases with relative humidity and the transition between metallic mode to compound mode occurs in a short range in case of air at 95 % RH than with dry air.

In order to get a close look on the evolutions of water vapour partial pressure, it was represented in Fig. **4** as a function of air flow and relative humidity. Two zones of partial pressure could be observed. The first one is between 0 and 6 sccm of air in which the water vapour partial pressure is very low. Another zone is observed when air flow reaches 7 sccm. It is expected to see a gradual increase of the partial pressure of H_2O as a function of the air flow relative humidity. A probable phenomenon could be considered: after ionization and dissociation of H_2O molecules in the plasma, a recombination effect of ions (H^+, OH^-, O^{2-}) could occur when air flow and relative humidity are increased. This phenomenon could explain the small decrease in hydrogen partial pressure. The probability of recombination of these ions increases with air flow and relative humidity and as a result, increases H_2O partial pressure.

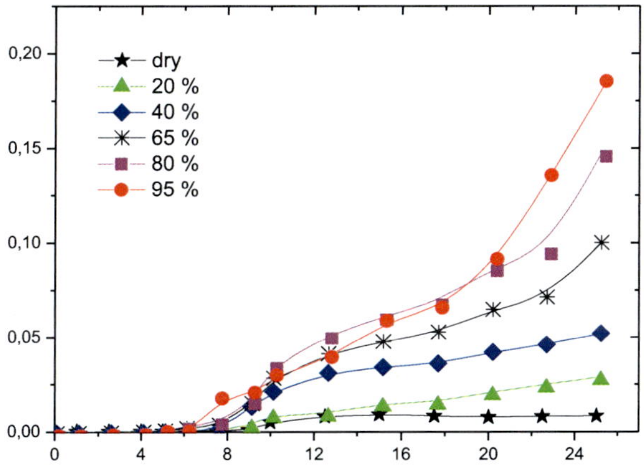

power. The voltage variations were illustrated in Fig. **2** for argon and air plasma at different relative humidities. These changes in plasma exhibit a hysteresis phenomenon, which corresponds to a transition between metallic and compound mode of the cathode. This transition between the two regimes could be illustrated by the variations of partial pressures, which reflect the consumption of reactive gas. The evolution of the partial pressures of N_2 and O_2 when air flow is increased or decreased in the deposition chamber are presented in Fig. **5**.

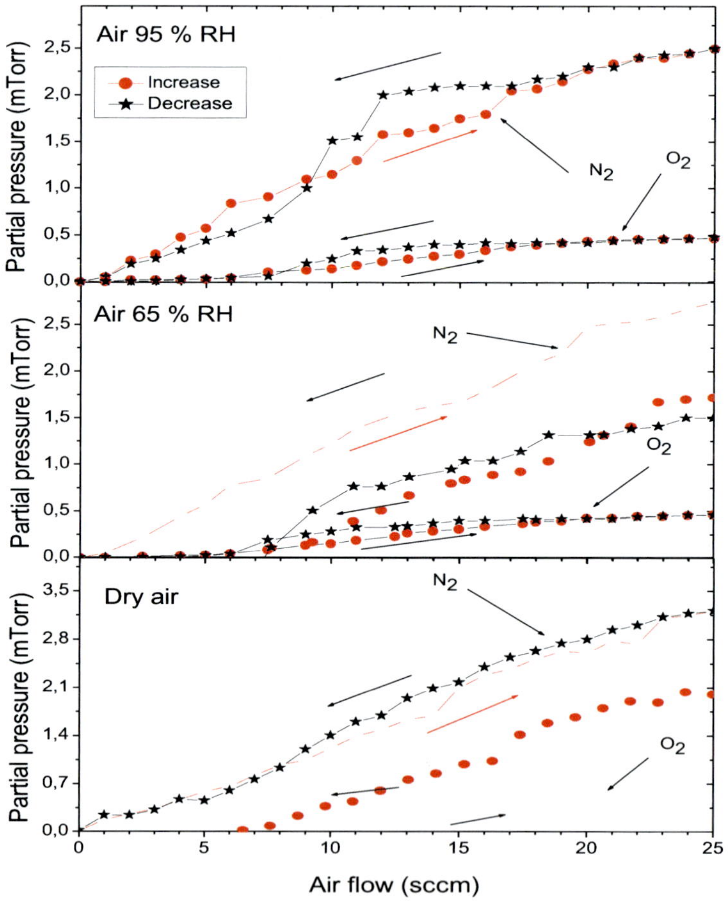

Figure 5: Evolution of nitrogen and oxygen partial pressure as a function of air flow.

A typical feature of the reactive sputtering process is the nonlinear evolution of some experimental parameters *vs.* the reactive gas injected in the deposition chamber. This behaviour is clearly identifiable in the hysteresis loop. When the reactive gas is introduced into the sputtering chamber, it reacts with the sputtered atoms to produce the growing film, and with the target surface as well. A total compound coverage of the target occurs when reactive gas flow exceeds a critical flow value where the transition between metallic and compound mode occurs. For reactive gas flows lower than the critical value, the target works in a "metallic mode", while for reactive gas flows higher than that critical value, it works in the "compound mode" and the consumption of reactive gas drops (Fig. **5**).

Fig. **5** also shows the behaviour of the oxygen and nitrogen partial pressures as function of air flow. We observe that the partial pressures are varying in an irreversible manner. An unstable behaviour is detected and the partial pressure of nitrogen and oxygen exhibit a hysteresis phenomenon. When we reduce the air flow, the partial pressure does not reach the same values obtained when the flow is increased in the deposition chamber. When a compound is formed on the cathode, we have a decrease of sputtering rate and the amount of N_2 and O_2 adsorbed by the deposited film decreases. Reducing the air flow does not immediately lead to return to the metallic mode, which explains the difference of partial pressures increasing and decreasing. This is because the cathode is still in the form of a compound (saturated) and it is necessary to reach very low air flow value to return to the metallic mode. The shift from one mode to another depends on the initial and final states of the cathode and on the nature of reactive gas. In case of dry air, no much difference was observed between the partial pressures corresponding to increasing and decreasing of flow. In the case of humidified air, we observe a difference between the two regimes. This can be explained by the effect of water vapour that accelerates the formation of compound on the cathode.

COMPOSITION AND CHEMICAL STATE OF $Cr_XN_YO_Z$ THIN FILMS

The elemental composition of chromium oxynitride samples were determined by RBS and RNRA.

In Fig. **6**, we present the evolution of the Cr, O, N and H content as a function of air flow and relative humidity. We observe the Cr content decreases, while the O content increases when air flow increases. N content increases and reaches a maximum at 2.5 sccm then decreases and becomes constant. The decrease of Cr content and increase of O content is explained by O incorporations in Cr in order to form a chromium oxide. The low content of N (< 10 at. %) in chromium oxynitride films is due to the low nitriding rate compared to oxidation rate of chromium.

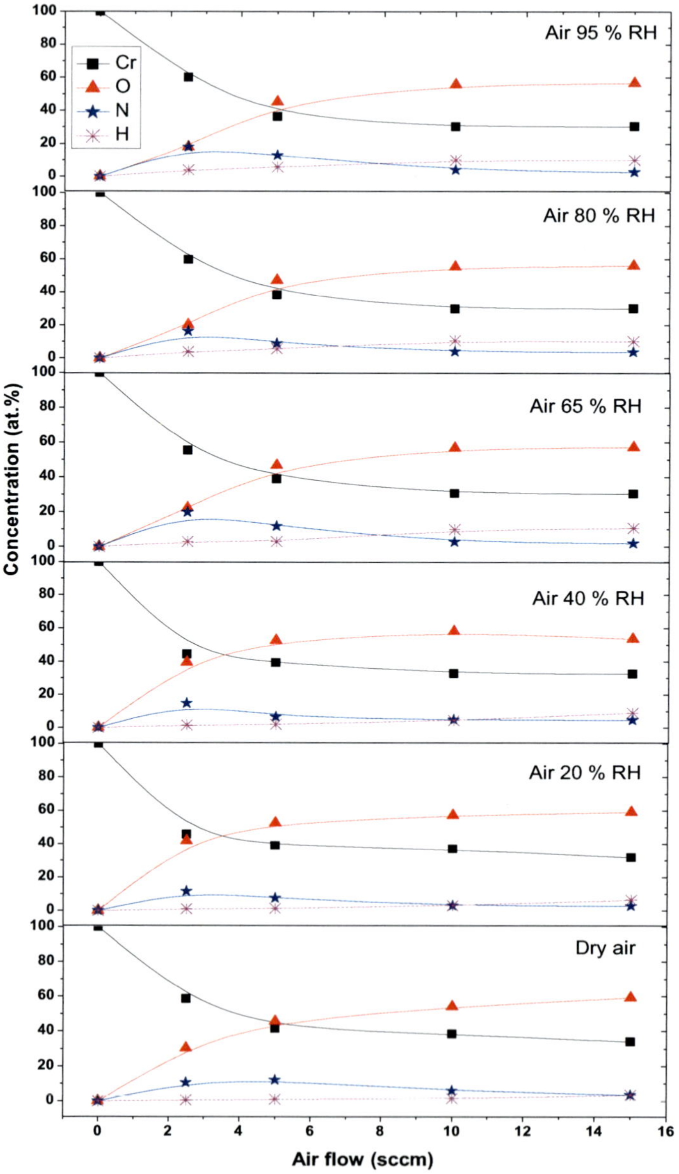

Figure 6: Elementary composition determined by RBS and RNRA of chromium oxynitride thin films deposited on carbon substrate.

Fig. 7a presents H and N depth profiles measured by RNRA for chromium oxynitride sample deposited with 5 sccm of dry air flow. H profile tends to increase in the interface and in the surface of the sample, which can be explained by the deposition of the water vapour present in air on the surface of substrate and the layer. The evolution of H content with air flow and relative humidity is reported in Fig. 7b. It is normal to observe an increase of the H content when flow and relative humidity increase. The augmentation of water vapour in the plasma provokes a rising amount of the hydrogen species (H + OH) and (H^+ + OH^-) coming from the dissociation of H_2O molecules, and consequently there is an increase of H incorporations in the deposited films. This phenomenon increases with air flow and relative humidity.

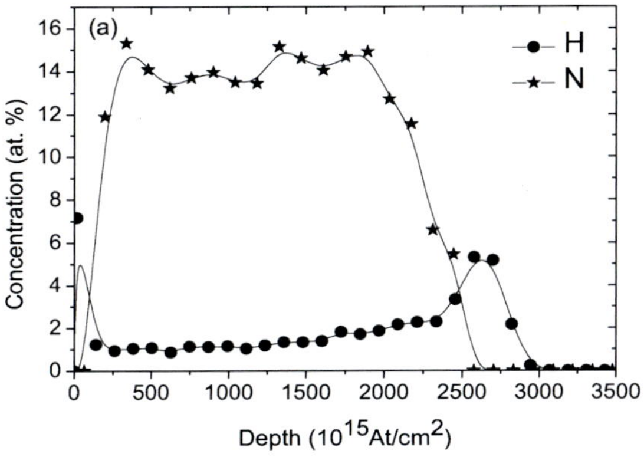

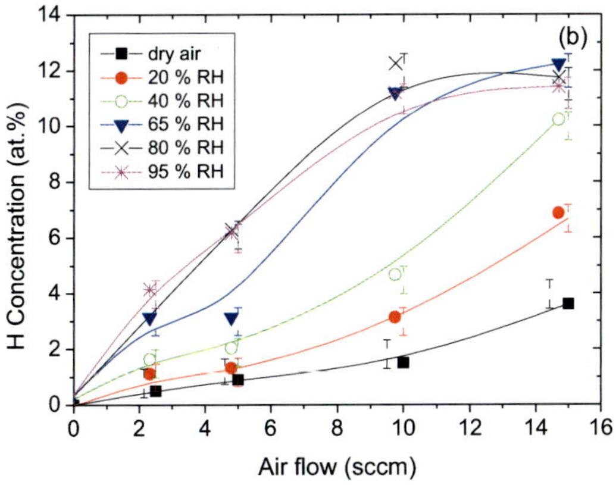

Figure 7: a) H and N profiles determined by RNRA and b) evolution of hydrogen content determined by RNRA as a function of air flow and relative humidity.

In Fig. **8**, we present the evolution of the H, N and O contents in chromium oxynitride samples according to the relative humidity and H_2O partial pressure. The samples were deposited at 2.5 sccm (c & d) and 1 sccm (a & b) of air flow with various relative humidities.

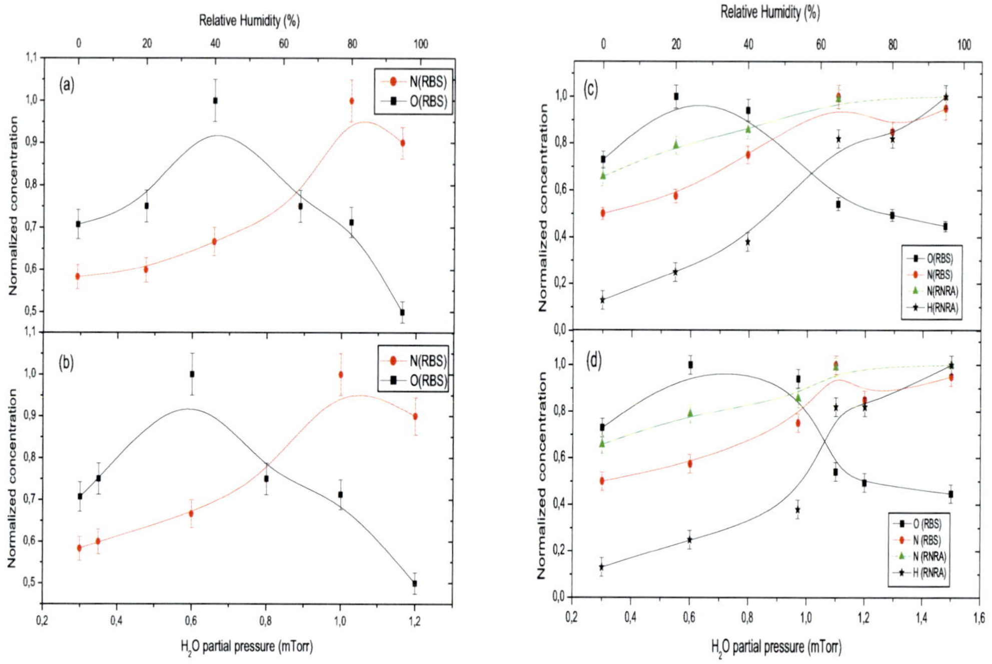

Figure 8: Evolution of O, N and H contents in chromium oxynitride samples deposited at 1 sccm (a & b) and at 2.5 sccm (c & d) of air flow with different relative humidities. (In case of samples deposited at 1 sccm, the H content was not measured).

We observe an increase of H and N with a decrease of O content when the relative humidity increases. This phenomenon is probably due to the reaction of hydrogen, coming from the dissociation of H_2O, with the oxygen to form the H_2O molecules provoking an increase of the H_2O partial pressure. The H_2O partial pressure was measured by a mass spectrometer connected to the deposition chamber. The reconstitution of H_2O molecules reduces the O_2 in the plasma and leads to a decrease of O content accompanied by an increase of N in the films. We also observe that the N and H contents exhibit the same evolution. This phenomenon was observed by various authors. In fact, Terwagne *et al.* [55] showed that the H_2 injection in an Ar+N_2 plasma causes an increase in the nitrogen content in the films of stainless steel deposited by dc magnetron sputtering. Fewell *et al.* [56-58] observed a small increase of the N content for small quantities of H. Some authors [59] explain the increase in the N with H contents by the probability of creation of hydrogenated species (N_xH_y).

As mentioned elsewhere [60], the reactive gas reacts with sputtered atoms to grow a film on the substrate but it also reacts with the target and changes the surface properties of the target, and directly affects sputtering and of course deposition rate and composition of chromium oxynitride. In fact, Fig. **9** presents the evolution of deposition rate and composition of deposited films as a function of deposition time. Samples were prepared with 5 sccm of air flow at 40 % RH. The deposition times were 2, 5, 10, 15 and 20 min. RBS measurements showed the evolution of thickness and composition *vs.* deposition time. However, the deposition rate decreases slightly and seems to be stabilized at 17 nm min^{-1} when the deposition time reaches 15 min. This small decrease can be attributed to the oxidation of the target, which increases under reactive sputtering in presence of air with the water vapour and causes a small decrease of chromium sputtering yield. The hydrogen content determined by $H(^{15}N, \alpha\gamma)^{12}C$ resonant reaction is around 2 at. %.

The Cr-Lα experimental spectra were fitted by a linear combination of data obtained from the standard samples (Cr, CrN and Cr_2O_3). The experimental data obtained from standard sample are given in reference [61]. An example of LEEIXS spectrum is shown in Fig. **10**, where we can observe the peaks characteristic of CrN and Cr_2O_3.

Deposition rate and LEEIXS analyses of chromium oxynitride thin films are given in Fig. **11**. During sputtering in metallic mode, we observe that the Cr_2O_3 stoichiometry is in majority with a small percentage of CrN and CrO_2. When air flow and relative humidity are increased, the content of CrO_2 increases while the Cr_2O_3 content decreases. When the compound mode is installed, the CrO_2 stoichiometry predominates in coexistence with low content of CrN. The low content of CrN compared to that of Cr_2O_3 and CrO_2 can be explained by the fact that enthalpy of formation of chromium nitride at 25 °C is –124.7 kJ mol^{-1}, very low absolute value in comparison with that of Cr_2O_3 *i.e.*, -1134.7 kJ mol^{-1} and that of CrO_2 *i.e.*, -597.6 kJ mol^{-1} [62, 63]. Thermodynamically, the formation of CrN is more favourable than that of Cr_2N.

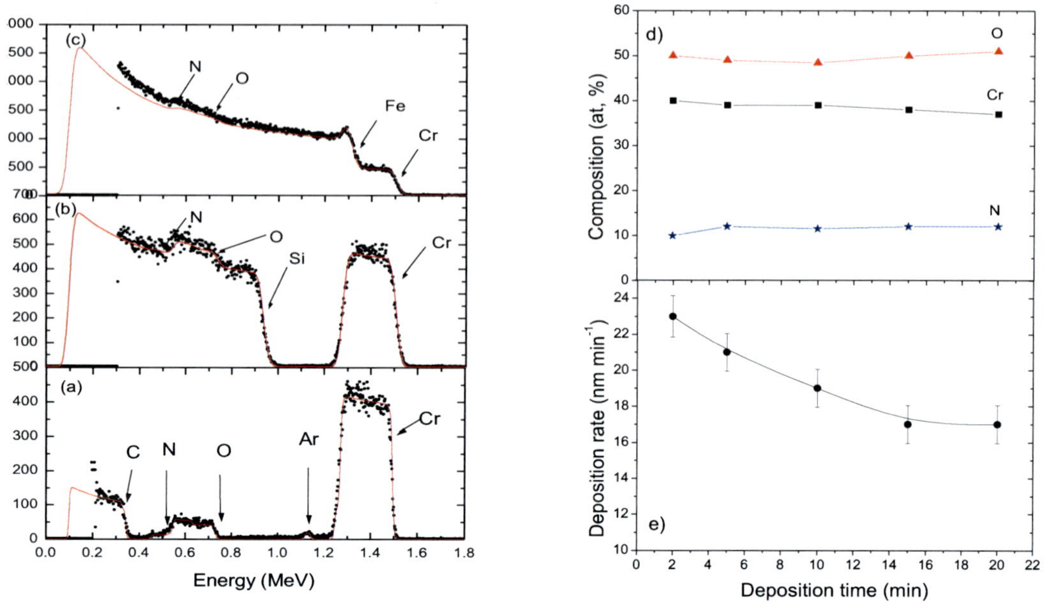

Figure 9: RBS spectra and RUMP simulations obtained from chromium oxynitride thin films deposited on a) carbon, b) silicon and c) low carbon steel. d) Composition and e) deposition rate of chromium oxynitride thin films as a function of deposition time. Air flow is maintained at 5 sccm, power density: 5.28 W cm^{-2} and total pressure at 0.28 Pa.

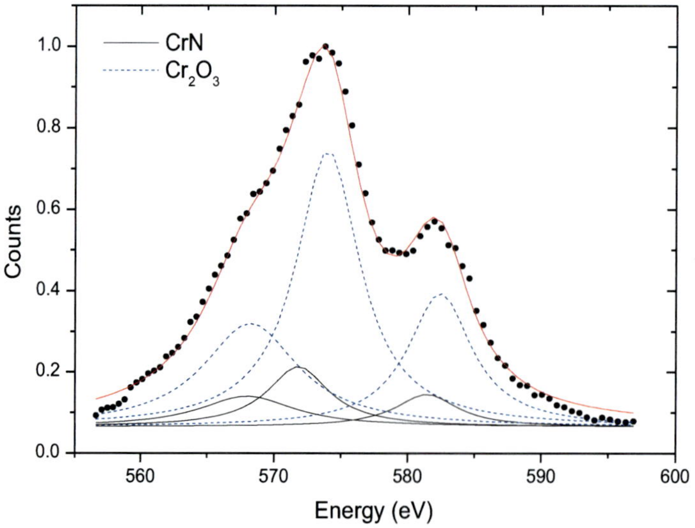

Figure 10: Cr L-spectrum obtained from oxynitride sample deposited on carbon substrate at 5 sccm air flow 40 % RH.

At 2.5 sccm of air flow, the CrN content increases while the Cr_2O_3 concentration decreases with the water vapour content in the plasma but the deposition rate is not affected. RBS results have shown that if the relative humidity of air is less than 40 % RH, the deposition rate is maximal in a large region of air flow (between 2.5 and 10 sccm) with dominance of Cr_2O_3. When the relative humidity reaches 65 % RH, the deposition rate is maximal between 2.5 and 5 sccm of air flow and decreases drastically when the air flow reaches 10 sccm (compound mode) with predominance of CrO_2 stoichiomety. With respect to the thermodynamic data, for a very low content of oxygen or H_2O, the following reactions (1 & 2) [62] are more probable:

$$4Cr + 3O_2 \rightarrow 2Cr_2O_3 \text{ and } 2Cr + 3H_2O \rightarrow Cr_2O_3 + 3H_2 \quad \textbf{(1)}$$

and when the quantities of oxygen or the water vapour increase, the chromium oxidation's results in CrO_2 compound according to the following reactions with augmentation of H_2 partial pressure:

$$Cr + O_2 \rightarrow CrO_2 \text{ and } Cr + 2H_2O \rightarrow CrO_2 + 2H_2 \quad \textbf{(2)}$$

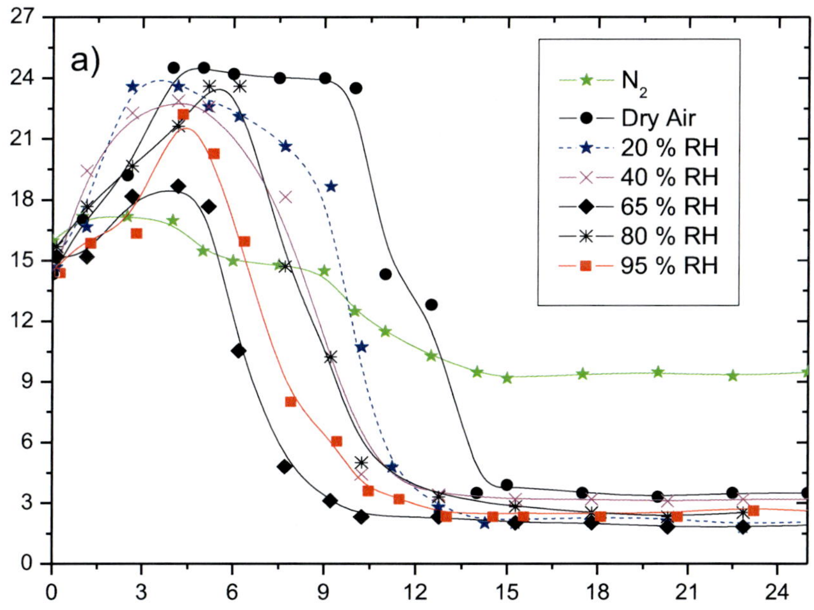

Fig. 11: contd....

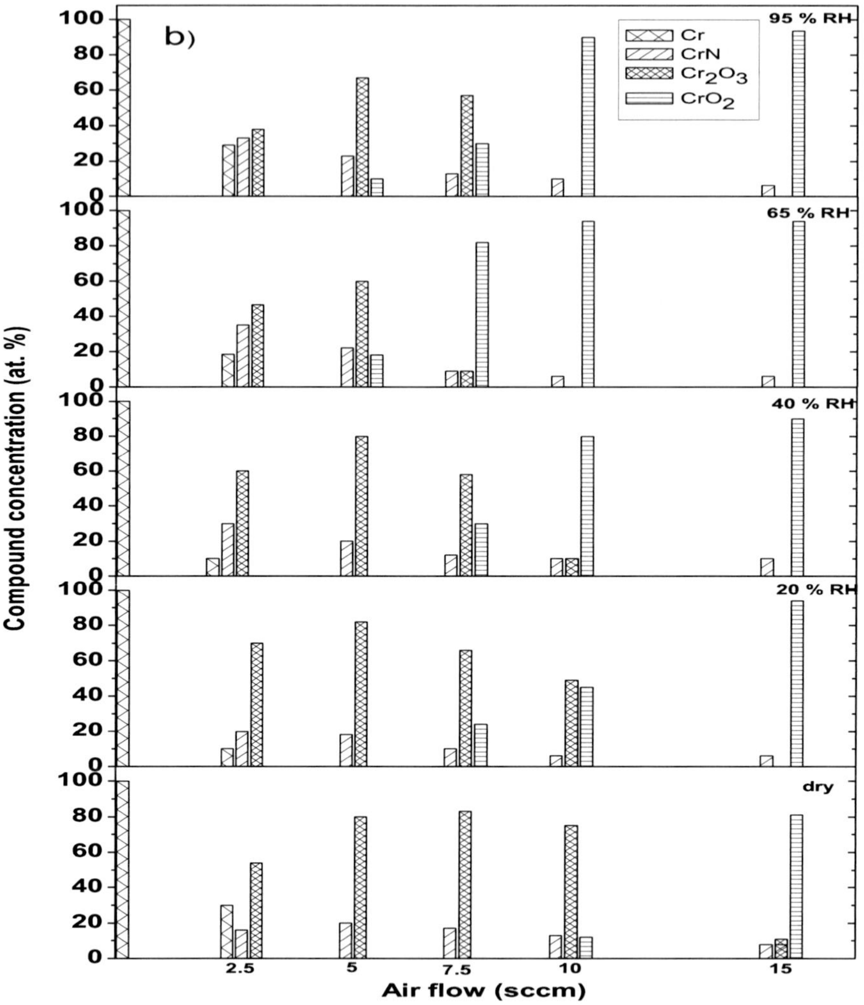

Figure 11: a) deposition rate and b) evolution of Cr, CrN, Cr_2O_3 and CrO_2 contents determined by LEEIX spectroscopy as a function of air flow and relative humidity.

The evolution of the composition of chromium oxynitride thin films can be discussed with deposition rate correlated with sputtering yield (Y). Fig. **11** illustrates the variations of the deposition rate of $Cr_xN_yO_z$ films as a function of dry air flow and relative humidity compared with that of Cr-N. We can observe an increase of the deposition rate when air flow is between 0 to 4 sccm (0 and 11.4 %). The deposition rate is almost constant between 4 and 10 sccm air (11.4 and

28.6 %). Then it decreases when the air flow exceeds 11 sccm, and finally stabilizes at 3 nm min^{-1} when the air flow reaches 14 sccm (40 % of mixture) to result in a sputtering process in the compound mode. The process presents a rapid transition when air flow reaches a critical value (saturation of the cathode), which will decrease the sputtering yield and requires a change in regime discharge by increasing the cathode voltage (Fig. **2**) and partial pressures of nitrogen and oxygen (Fig. **3**). The increase in deposition rate between 0 and 4 sccm of air could be explained by a low oxidation of the cathode, which makes easily emission of electrons and results in an increase of the number of Ar^+ ions, and finally improves the sputtering efficiency. The presence of three zones could be attributed to three different stoichiometries of oxynitride chromium based on the: Cr, Cr_2O_3 and CrO_2 with presence of CrN. The number of particles of chromium sputtered from an oxidized chromium target is lower than that obtained from a pure metallic target. The ratio between maximum and minimum of deposition rate is about 6, while the ratio of the corresponding sputtering yield approaches 5 [12].

We note that the maximum value of the deposition rate decreases when the relative humidity increases. This decrease may be due to the decrease of Ar^+ ions, which are neutralized by OH^-. We also observe a shift of the transition zone from metallic to compound mode of the cathode. This transition zone decreases with relative humidity of the air. The transition between the metallic and the compound mode occurs at 12.5 and 10 sccm in the case of dry air and air at 40% RH, respectively. When relative humidity exceeds 65 %, the transition occurs at 7.5 sccm. During reactive sputtering, in addition to the reaction of O_2 and N_2 with the target, the H^+ ions will be attracted by negative voltage and implanted, modifying the composition of the target and affecting the deposition rate. An augmentation of H in the chromium target causes a reduction of the chromium sputtering yield Y(Cr) [12]. However, a continuous sputtering removes a fraction of the surface oxygen, nitrogen and hydrogen. Indeed, supposing that there is formation of the same compound on the target and on the substrate and of the ions implantation process in the target, calculated sputtering yield by TRIM [12] (The TRIM input data are: $E(Ar^+) = 400$ eV, angle of incidence is 0° and total number of ions is 10000) showed a decrease of Y(Cr) and Y(N) accompanied with an increase of Y(H) when the air flow and relative humidity increase, which is in agreement with the composition evolution (Fig. **11**). For the deposition rate of chromium nitride films, we distinguish three different zones of growth that could be attributed to different stoichiometries of chromium nitride as reported in the literature (Cr-N, and CrN Cr_2N).

As mentioned elsewhere [60, 61], the LEEIXS results are completed by XPS measurements of Cr2p, O1s and N1s spectra. The simulation and deconvolution of O1s and N1s spectra clearly show the existence of satellite peak O1s at 531.5 ± 0.1 eV and N1s satellite peak at 397.4 ± 0.1 eV. The presence of the two satellite lines N1s and O1s could be attributed to a different chromium compound containing nitrogen and oxygen like $Cr_xN_yO_z$ type.

The N1s and O1s satellite peak areas were used to determine the stoichiometry of $Cr_xN_yO_z$ compound, which was found to be $(CrO_2)_3N$. It is obvious from the formation of $(CrO_2)_3$ grouping followed by nitrogen incorporation, which disturbs the O-Cr bonding energy giving a satellite peak O1s at 531.5 eV and another satellite peak N1s at 397.4 eV.

The little literature concerning chromium oxynitride studies and the total absence of thermodynamic data renders difficult all discussions of the formation mechanism. However, it is possible to suggest a reaction model (reactions 3 & 4), even though this one cannot be considered as perfect. It constitutes an attempt and a first stage of reasoning that will request some previous confirmations, especially kinematical studies and measurement of the free Gibbs energy of the proposed reactions. The proposed reactions are:

- In presence of N_2 with O_2, the following mechanism is probable:

$$3Cr + 3O_2 + \tfrac{1}{2} N_2 \rightarrow 3CrO_2 + \tfrac{1}{2} N_2 \rightarrow (CrO_2)_3\text{-}N \qquad \textbf{(3)}$$

- And in presence of N_2 with water vapour, the following mechanism is probable:

$$3Cr + 6H_2O + \tfrac{1}{2}N_2 \rightarrow 3CrO_2 + \tfrac{1}{2}N_2 + 6H_2 \rightarrow (CrO_2)_3\text{-}N + 6H_2 \qquad \textbf{(4)}$$

DECORATIVE POTENTIALITY OF $Cr_XN_YO_Z$ THIN FILMS

As mentioned before, chromium oxynitride thin films could be used for decorative applications. In fact, Fig. **12** presents a photograph of four $Cr_xN_yO_z$ thin films deposited on low carbon steel with 5 sccm of air flow at 40 % RH. Deposition times were 2, 5, 10 and 15 min and measured thicknesses 40, 100, 175 and 275 nm, respectively. We observe a large range of colours, which depend on the deposition time. Colour varies between blue and green. These differences may be attributed to various thicknesses and compositions.

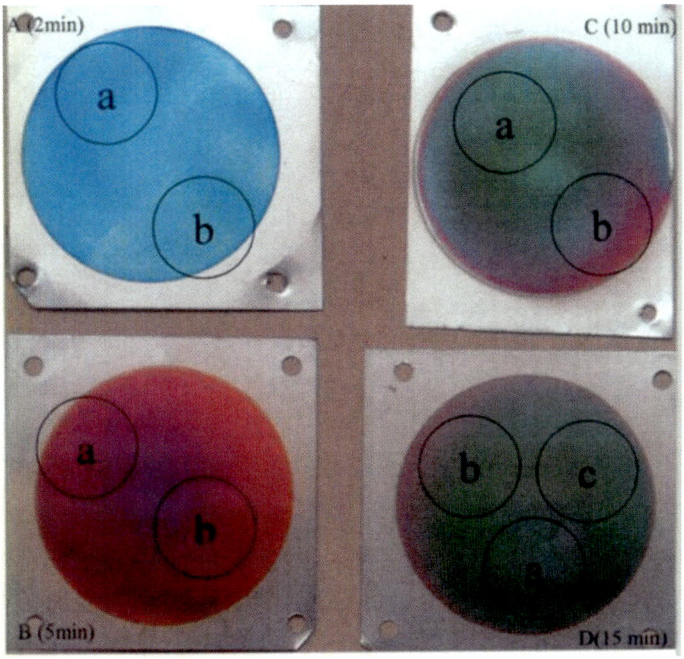

Figure 12: Photographs of four $Cr_xN_yO_z$ samples deposited on low carbon steel with 5 sccm of air flow at 40 % RH.

Resonant nuclear reaction on ^{15}N permitted the construction of N profiles. According to N profiles presented in Fig. **13**, a small difference of N content and thickness are observed between two distinct positions of the same sample. These curves show an inhomogeneity of N profiles. Close to the interface, a narrow domain is observed where the nitrogen content is higher. This may be attributed to the reaction of nitrogen with the substrate leading to the formation of an iron nitride compound.

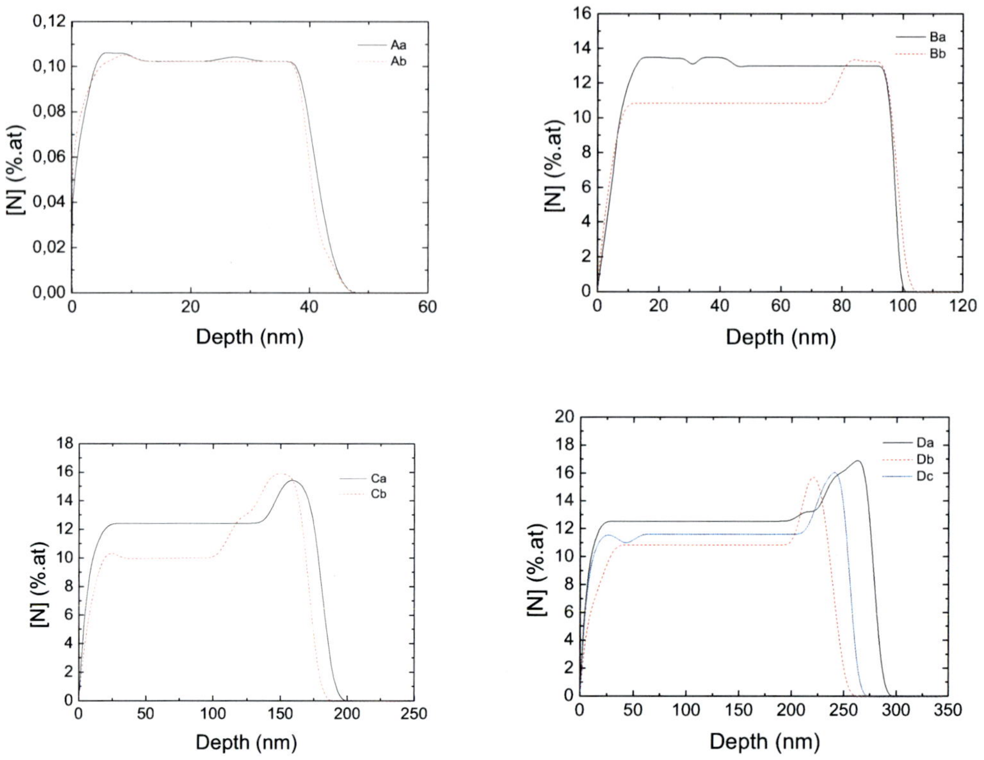

Figure 13: Nitrogen profiles measured by $^{15}N(p,\alpha\gamma)^{12}C$ for different positions (a, b and c) of $Cr_xN_yO_z$ samples (A, B, C and D samples).

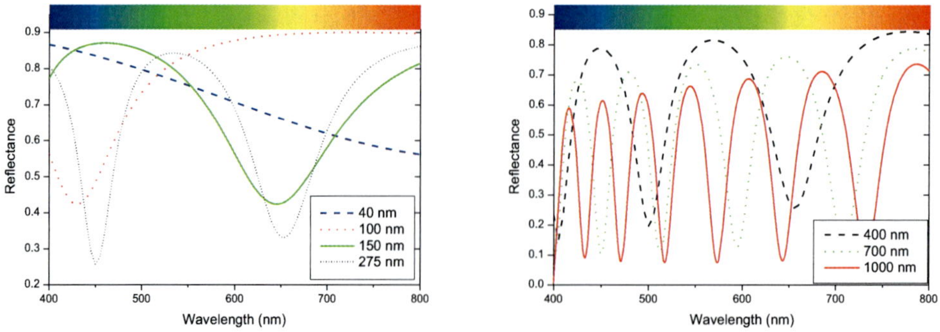

Figure 14: Reflectance calculations as a function of chromium oxynitride thickness.

During sputtering in metallic mode and at 5 sccm of air flow, the LEEIX spectroscopy results have shown that the Cr_2O_3 is in majority (80 at. %) with a small percentage of CrN (less than 20 at. %). Cr_2O_3 is a refractory material with refractive index close to 2.6. Therefore, in order to explain variations of colour according to the thickness, interference phenomena are not excluded. On the surface, light is reflective while light transmitted through the layer can be reflective repetitively between the interface and the surface. At each reflection, a fraction of light escapes producing interferences with the light reflected from the surface. Fig. **14** represents the reflectance calculations [64, 65] of chromium oxynitride layers deposited on steel.

For a thickness of 40 nm, the reflectance is maximal in the blue region and decreases for higher wavelengths, resulting in a blue colour. At a thickness of 100 nm, the reflectance is minimal in the blue domain and increases, reaching a maximum in the red domain. The appearance of the layer will exhibit a colour close to red. When the thickness is 150 nm, the reflectance is maximal in the blue-green and minimal in the red regions resulting in a colour close to blue -green. At 275 nm, the reflectance is minimal in the blue and red regions, and maximal in the green region, resulting in a green colour. We can observe how the reflectance will evolve in the case of the thick layers (400, 700 and 1000 nm). We can also note that the number of peaks of the reflectance increases with thickness. When the thickness increases from 400 nm to 1000 nm, the number of peaks passes from 3 to 7. Therefore, we will observe a mixture of different colours resulting in a colour near to the white. In conclusion, to have layers with distinct colours, it is important to limit thickness to a maximum of 300 nm.

CONCLUSION

We have shown that chromium oxynitride thin films could be grown on various substrates by using reactive sputtering in presence of air as a reactive gas. $Cr_xN_yO_z$ were characterized by RBS, RNRA, LEEIXS and XPS. Evolution of deposition rate showed a shift of transition zone between metallic and compound mode as a function of relative humidity of air. LEEIX spectroscopy measurements have shown the stoichiometry of the films. The colour of reactive unbalanced magnetron sputtered chromium oxynitride layers with air layers deposited on low carbon steel varies as a function of the thickness. The reflectance curves have permitted the explanation of the variations of colours with interference phenomena. For thin layers (40 nm), the colour of samples is blue and when the thickness reaches 300 nm the colour is green. To obtain distinct colours, reflectance calculations have demonstrated that it is interesting to limit the layer thickness to 300 nm. XPS results revealed the presence of a compound other than CrN, Cr_2O_3 and CrO_2 determined by LEEIX Spectroscopy. This compound contains nitrogen and oxygen and it has a $(CrO_2)_3N$ stoichiometry.

ACKNOWLEDGEMENT

Authors acknowledge Dr. Pierre Louette from LISE-FUNDP for XPS analysis.

CONFLICT OF INTEREST

The author(s) confirm that this chapter content has no conflict of interest.

REFERENCES

[1] S. Collard, H. Kupfer, G. Hecht, W. Hoyer, H. Moussaoui, Surf. Coat. Technol., vol. 112, pp. 181-184, 1999.
[2] W. Herr, Surf. Coat. Technol., vol. 60, pp. 428-433, 1993.
[3] E. Ertürk, H.J. Heuvel, in R-J. Peter (Ed), Beschichten mit Hartstoffen, VDI, Düsseldorf, p. 250, 1991.
[4] R. Elsing, in R-J. Peter (Ed), Beschichten mit Hartstoffen, VDI, Düsseldorf, p. 1, 1991.
[5] JCPDS-Data files, No 38-1479, International Center for Diffraction Data, Swarthmore, 1992.
[6] J. Guillot, PhD Thesis, Couches minces d'oxynitrure de titane la réactivité comme moyen original de caractérisation physico – chimique, Université de Bourgogne, France, 2002.
[7] M.J. Vasile, A.B. Emerson, F. A. Baiocchi, J. Vac. Sci. Technol., vol. A8, pp. 99-105, 1990.
[8] F.H. Lu, H.Y. Chen, Thin Solid Films, vol. 355-356, pp. 374-379 1999.
[9] T. Wierzchon, I. Ulbin-Pokorska, K. Sikorski, Surf. Coat. Technol., vol. 130, pp. 274-279, 2000.
[10] References 59-67 and 89-118, in Ref. [6].
[11] P. Sigmund, Phy. Rev., vol. 184, pp. 383-416, 1969.
[12] J.F. Ziegler and J.P. Biersack, in "The Stopping and Range of Ions in Solids", 1985.
[13] A.J. Perry, Surface Eng., vol. 2, pp. 183-190, 1986.
[14] Corrosion education Manual, Swedish Corrosion Institut, Stockholm, 1974.
[15] DIN 50 900 Blatt 1, Beuth Verlag Berlin, April 1982.
[16] H.A. Jehn, Surf. Coat. Technol., vol. 125, pp. 212-217, 2000.
[17] F. Altmayer, Choosing an accelerated corrosion test, in: M. Murphy (Ed), Metal Finishing: 65th Guidebook and Directory, Issue 97, Elsevier, New York, p. 559, 1997.
[18] H.A. Jehn, A. Zielonka, Corrosion Testing, ASM Handbook : Surface Engineering, vol.5, p. 635, 1994.

[19] D.F. Williams, ed. Definitions in biomaterials: progress in biomedical engineering. Amsterdam: Elsevier, 1987.

[20] I. Gerber, I. Gwynn, M. Alini, T. Wallimann. Eur. Cell. Mater., vol. 10, pp. 8-22, 2005.

[21] L. McClaine, C. Coppel, A.D. Little Inc. tech. Rept. AFML-TR-65-299, part I, 1965.

[22] J.M. Antunes, A. Cavaleiro, L.F. Menezes, M.I. Simões, J.V. Fernandes, Surf. Coat. Technol., vol. 149, pp. 27-35, 2002.

[23] H.S. Kim, Scripta Mater. vol. 39, pp. 1057-1061, 1998.

[24] J.A. Thornton, D.W. Hoffman, Thin Solid Films, vol. 172, pp. 5-31, 1989.

[25] P.A. Flinn, D.S. Gardner, W.D. Nix, IEEE Trans. Electron Devices ED-34, vol. 13, pp. 689-699, 1987.

[26] H. Holleck, J. Vac. Sci. Technol., vol. A4, pp. 2661-1669, 1986.

[27] S. Veprek, A.S. Argon, J. Vac. Sci. Technol., vol. B20, pp. 650-664, 2002.

[28] D. Sherman, D. Brandon, Handbook of Ceramic Hard Materials, edited by Wiley-VCH, New Jersey, 2000.

[29] W.S. Williams, Mater. Sci. Eng., vol. A105/106, pp. 1-10, 1988.

[30] B. Granier, C. Chatillon, M. Allibert, J. Am. Ceram. Soc., vol. 65, pp. 465-469, 1982.

[31] I.N Martev, Vacuum, vol. 67, pp. 261-265, 2002.

[32] Y. Li, L. Gao, Materials Letters, vol. 57, pp. 1062-1065, 2003.

[33] M.A. Centeno, I.J.A. Odriozola, Apllied Catalysis A : General, vol. 246, pp. 365-372, 2003.

[34] J.M. Chappé, N. Martin, J. Lintymer, F. Sthal, G. Terwagne, J. Takadoum, Appl. Surf. Sci., vol. 253, pp. 5312–5316, 2007.

[35] J.M. Chappé, N. Martin, G. Terwagne, J. Lintymer, J. Gavoille, J. Takadoum, Thin Solid Films, vol. 440, pp. 66-73, 2003.

[36] T. Oyama, H. Ohsaki, Y. Tachibana, Y.Hayashi, Y. Ono, N. Horie, Thin Solid Films, vol. 351, pp. 235-240, 1999.

[37] T. Wierzchon, I. Ulbin-Pokorska, K. Sikorski, Surf. Coat. Technol., vol. 130, pp. 274-279, 2000.

[38] P. Wilhartitz, S. Dreer, P. Ramminger, Thin Solid Films, vol. 447 –448, pp. 289-295, 2004.

[39] T. Suzuki, H. Saito, M. Hirai, H. Suematsu, W. Jiang, K. Yatsui, Thin solid films, vol. 407, pp. 118-121, 2002.

[40] H. Wei-Yu, H. Cheng-Hsun, H. Dung-Hau, C. Yen-Ping, L. Yen-Chun, C. Chi-Lung, Surf. Coat. Technol., vol. 188-189, pp. 129-134, 2004.

[41] T. Suzuki, J. Inoue, H. Saito, M. Hirai, H. Suematsu, W. Jiang, K. Yatsui, Thin Solid Films, vol. 515, pp. 2161-2166, 2006.

[42] L. Castaldi, D. Kurapov, A. Reiter, V. Shklover, P.Schwaller, J. Patscheider, Surf. Coat. Technol., vol. 203, pp. 545-549, 2008.

[43] Q.M. Wang, Y.N. Wu, M.H. Guo, P.L. Ke, J. Gong, C. Sun, L.S. Wen, Surf. Coat. Technol., vol. 197, pp. 68-76, 2005.

[44] C. Gautier, J. Machet, Surf. Coat. Technol., vol. 94-95, pp. 422-427, 1997.

[45] P. Wilhartitz, S. Dreer, P. Ramminger, Thin Solid Films, vol. 447-448, pp. 289-295, 2004.

[46] J. Baborowski, M. Charbonnier, M. Romand, Journal de Physique IV, Colloque C4, Supplément au Journal de Physique III, Vol. 6, pp. 429-439, 1996.

[47] J. Baborowski, M. Charbonnier, M. Romand, Surf. Coat. Technol., vol. 80, pp. 190-194, 1996.

[48] R. Mientus, R. Grötschel, K. Ellmer Surf. Coat. Technol., vol. 200, pp. 341-345, 2005.

[49] J. Shirahata, T. Ohori, H. Asami, T. Suzuki, T. Nakayama, H. Suematsu, Y. Nakajima, K. Niihara, Thin Solid Films, vol. 519, pp. 3497-3500, 2011.

[50] P. Carlsson, C. Nender, H. Barankova, S. Berg, J. Vac. Sci. Technol., vol. A11, pp. 1534-1539, 1993.

[51] S. Agouram, Mémoire de DEA, Comparison de depots pan PVD de coudhes minces Cr_xN_y et $Cr_xN_yO_t$ en presence dazote et dain. FUNDP, Namur, 2001.

[52] L.R. Doolittle, Nucl. Instrum. Meth. B, vol. 9, pp. 344-351, 1985.

[53] L.R. Doolittle, Nucl. Instrum. Meth. B, vol. 15, pp. 227-231, 1986.

[54] J.M. Meek, J.D. Craggs, « Electrical breakdown of gases », John Wiley & Sons, p. 270, 1978.

[55] G. Terwagne, H. Hody, J. Colaux, Surf. Coat. Technol., vol. 174-175, pp. 383-388, 2003.

[56] S. Kumar, M.J. Baldwin, M.P. Fewell, *et al.* Surf. Coat. Technol., vol. 123, pp. 29-35, 2000.

[57] M.P. Fewell, J.M. Priest, M.J. Baldwin, G.A. Collins, K.T. Short, Surf. Coat. Technol., vol. 131, pp. 284-290, 2000.

[58] J.M. Priests, M.J. Baldwin, M.P. Fewell, Surf. Coat. Technol., vol. 145, pp. 152-163, 2001.

[59] N. Renevier, T. Czerwiec, P. Collignon, H. Michel, Surf. Coat. Technol., vol. 98, pp. 1400-1405, 1998.

[60] S. Agouram, F. Bodart, G. Terwagne, Surf. Coat. Technol., vol. 180-181, pp. 164-168, 2004.

[61] S. Agouram, F. Bodart, G. Terwagne, J. Elect. Spect. Relat. Pheno., vol. 134, pp. 173-181, 2004.

[62] M.W. Chase, C.A. Davies *et al.* Journal of Physical and Chemical reference Data, "Janaf Thermochemical Tables", 3rd Edition, vol. 14, supplement N°1, 1985.

[63] R. Weast, Handbook of Chemistry and Physics, CRC PRESS, 1975.

[64] R. Tilley, "Colour and the Optical Properties of Materials", John Wiley & Sons, England, 2000.

[65] A. Zajac, "Optics", Hecht Optic, 2nd Ed., USA, pp. 79-427, 1987.

CHAPTER 7

A Comprehensive Study of the Properties of Sputtered NbO_xN_y Thin Films

Martin Fenker*

Forschungsinstitut Edelmetalle und Metallchemie, Katharinenstraße 17, 73525 Schwäbisch Gmünd, Germany

Abstract: Niobium oxynitride thin films have been deposited by reactive magnetron sputtering. A niobium target has been sputtered in an $Ar/N_2/O_2$ atmosphere with different gas compositions. With Rutherford backscattering spectroscopy (RBS), particle induced X-ray emission (PIXE) and energy dispersive X-ray spectroscopy (EDX) the chemical composition was acquired. Scanning electron microscopy (SEM), atomic force microscopy (AFM) and X-ray diffraction (XRD) allowed studying the microstructure of the deposited films. The electrical conductivity and the optical properties were measured by the four point probe method and spectroscopic ellipsometry. With increasing O/N ratio in the films the electrical resistance and the optical transparency increased. Mechanical properties, like indentation hardness, indentation modulus, residual stress and the tribological behaviour, have been investigated with respect to the O/N ratio. The corrosion behaviour was studied by potentiodynamic corrosion tests and neutral salt spray testing. Furthermore, the degradation behaviour of the NbON films has been tested in artificial sweat. The thermal and oxidation behaviour have been observed in vacuum and in air. Finally, the biocompatibility and the wetting behaviour of the NbON films have been examined.

Keywords: Reactive magnetron sputtering, niobium oxynitride, deposition rate, chemical composition, crystalline structure, electrical conductivity, optical transmittance, refractive index, extinction coefficient, spectroscopic ellipsometry, indentation hardness, indentation modulus, residual stress, pin-on-disk tribology, thermal stability, corrosion behaviour, potentiodynamic corrosion tests, salt spray tests, cytotoxicity tests, wetting behaviour.

INTRODUCTION

Niobium has physical and chemical properties similar to those of the element tantalum. Both are in the Va group of the periodic system of the elements. Niobium is mostly used in alloys (steel alloys) and furthermore in various superconducting materials. Other applications of niobium include its use in welding, nuclear industries, electronics, optics, numismatics and jewellery. In the

***Address correspondence to Martin Fenker:** FEM-Forschungsinstitut Edelmetalle und Metallchemie, Katharinenstraße 17, 73525 Schwäbisch Gmünd, Germany; Tel: +49-7171-1006-400; Fax: +49-7171-1006-900; E-mail: fenker@fem-online.de last two applications, niobium's low toxicity and ability to be coloured by anodisation are particular advantages [1, 2]. As oxynitride, NbON has been investigated as biocompatible material (TiNbON [3]), for optical waveguides [4], for Josephson junctions [5] or as a CrON/NbON multilayer topcoat to increase the corrosion resistance of CrN/NbN-coated high speed steel [6]. As bulk material, NbON is also used in catalysis, *e.g.* for hydrazine decomposition [7].

NbON coatings have been deposited by physical vapour deposition (PVD) methods, like sputtering [4, 6] or cathodic arc evaporation [3]. Furthermore, chemical vapour deposition (CVD) was also used to manufacture NbON coatings. Bekermann *et al.* reported on the growth of NbON films by metal organic chemical vapour deposition (MOCVD) on Si(100) and Si(111) substrates using $[Nb(NtBu)(NMe_2)\{C(NiPr)_2(NMe_2)\}_2]$ [NB; $tBu=(CH_3)_3C$; $Me=CH_3$; $iPr=(CH_3)_2CH$] and O_2 as precursors in the temperature range 400 – 600 °C [5]. It is also possible to transform Nb, NbN or Nb_2O_5 in oxidizing/nitriding gas atmospheres by rapid thermal processing (RTP [8]) or other thermochemical treatments [9].

In several studies of the system Nb-O-N [10, 11] the authors group at the research institute “fem” reported on magnetron sputtering of Nb, Nb-O and Nb-O-N thin films including the direct current (DC) and pulsed power magnetron sputtering, the use of the reactive gas pulsing process (RGPP), where the oxygen gas flow is pulsed during deposition [12-14]. The present study investigates the influence of oxygen in the system Nb-O-N on properties like thermal, mechanical, corrosion and degradation behaviour with respect to the films O/N ratio.

EXPERIMENTAL DETAILS

Thin Film Deposition

The Nb-O-N coatings were deposited by reactive magnetron sputtering in a commercial vacuum system (Leybold LH Z400). The vacuum chamber was evacuated by means of a turbomolecular pump backed by a mechanical pump leading to a base pressure of lower than 6×10^{-4} Pa. The pumping speed was 300 L s^{-1}. The depositions were performed without additional substrate heating. For adhesion improvement a thin metallic interlayer was sputtered onto the substrates prior to the Nb-O-N thin film deposition. A metallic niobium target (purity 99.8 wt.%) was sputtered in an $Ar/O_2/N_2$ atmosphere in DC and in pulsed mode with an average target current density of 100 A m^{-2}. For the pulsed magnetron sputtering a pulsed power supply (ENI, RPG-50) was used with a pulse frequency and a pulse width of 100 kHz and 1456 ns, respectively. It should be mentioned that no arcing occurred during reactive magnetron sputtering for both power modes. Arcing in reactive magnetron sputtering of insulating films occurs in consequence of charging the insulating layers formed on non-eroded areas of the sputtered target [15]. As we found no arcing – especially in DC power mode – we assume that the layers formed on the poisoned Nb cathode are not insulating (NbO is metallic, NbO_2 and sub-stoichiometric Nb_2O_5 are semiconductors). More details about the pulsed power magnetron sputtering of an Nb target in Ar/O_2 and/or $Ar/O_2/N_2$ atmosphere can be found in reference [10]. In order to deposit Nb-O-N thin films with different oxygen contents, the nitrogen flow $q(N_2)$ was kept constant while the oxygen flow $q(O_2)$ was adjusted between 0 and 7.5 sccm. The working pressure was $5x10^{-1}$ Pa, the partial pressures of oxygen and nitrogen was in the range of $0 < p(O_2) < 7.8x10^{-2}$ Pa and $0 < p(N_2) < 7.6x10^{-2}$ Pa, respectively. Mirror polished high speed steels (HSS, AISI M2, W.nr. 1.3343), stainless steels (AISI 316) and (100) Si were used as substrates.

Morphology, Microstructure and Chemical Composition

The surface morphology of the coatings was observed by high resolution scanning electron microscopy (SEM, Leo Supra 55 VP) equipped with an energy dispersive X-ray spectroscopy (EDX) system for measuring the chemical composition. The chemical composition of some of the coatings was also determined by Rutherford backscattering spectroscopy (RBS). A 2 MeV He^+ beam and a 2 MeV proton beam were used. The analysed area was 0.5×0.5 mm^2. In addition, impurities like argon were checked by particle induced X-ray emission (PIXE) using protons because of the increased sensitivity for trace or minor elements. The relative errors of RBS and PIXE are about 5 and 10%, respectively. The surface morphology of the films was additionally observed by atomic force microscopy (AFM, Nanoscope III Digital Instruments) in contact mode using a Si_3N_4 tip. The average roughness value R_a was calculated from a 1×1 μm^2 scanned surface. The crystallographic structure was investigated by X-ray diffraction (XRD, Siemens

D5000) using monochromatised Cu Kα radiation (λ = 0.15406 nm) in Bragg–Brentano configuration.

Electrical and Optical Properties

The electrical conductivity was measured in the temperature range between 293 and 473 K with the van der Pauw four point probe method, taking into account a geometry factor [15, 16]. Optical transmittance spectra of the film/glass substrate system were recorded with a Lambda 20 UV–visible Perkin-Elmer spectrophotometer in the wave wavelength range 200 – 1100 nm. The refractive index and extinction coefficient were evaluated by spectroscopic ellipsometry using a phase-modulated ellipsometer (UVISEL HR 460, Jobin-Yvon-Horiba) in the energy range 0.75 – 4.5 eV at an incidence angle of 70°.

Mechanical Properties

The instrumented indentation test (Fischerscope H100, DIN EN ISO 14577-1) provided the indentation hardness H_{IT} and the effective indentation modulus $Y_{IT} = E_{IT}/(1-\nu_f^2)$ of the Nb-O-N coatings (E_{IT}: indentation modulus, ν_f: Poisson´s ratio of the film). Y_{IT} is often called Young´s modulus. The maximum applied load during indentation was 10 mN. The residual stress of Nb-O-N coatings on stainless steel (AISI 316) disks was measured by the deflection method. The residual stress value is determined by Stoney's equation using substrate curvature radii, both before and after coating deposition [16]. The tribological investigations have been carried out with a pin-on-disk tribometer (CSM Instruments) using two different counterpart materials (100Cr6 and Al_2O_3 balls with 6 mm diameter). A load of 1 N was applied. The wear track diameter was 1 mm and the rotation speed was 500 rpm. The wear track depth was measured by profilometry at six different sites. The coating thickness was measured by using the calotest (CSM Instruments, ball diameter 40 mm).

Corrosion and Degradation Behaviour

Potentiodynamic corrosion (polarisation) tests were performed with a scan rate of 1 mV/s at a temperature of 25 °C. Furthermore, a three electrode configuration, an EG&G potentiostat and a 0.8 M NaCl solution buffered at pH 7 by acetic acid (Fixanal®, Riedel-de Haën) was used. The deaeration was performed by bubbling nitrogen gas through the solution. The reference electrode for the measurement of the sample potential was a saturated calomel electrode. The salt spray tests were performed according to DIN 50021 (ISO 9227). The test duration was 24 h. The measuring area was about 2 cm^2 for the salt spray tests and about 0.2 cm^2 for the potentiodynamic corrosion tests. Artificial sweat tests (immersion tests) according to DIN EN 1811 have been performed on Nb-O-N coatings deposited onto stainless steel disks (SS316). The test duration was 1 week at an electrolyte temperature of 30 °C. Quantitative trace analysis of the artificial sweat solution after the tests was conducted by an inductive coupled plasma spectrometer.

Thermal Behaviour

In order to study the thermal stability of the coatings, a vacuum furnace with a residual gas pressure of 10^{-4} Pa was used. The samples were heated up in 1 h to the desired temperature and then annealed for 1 h at temperatures of 400, 500, 600 and 800 ºC. The cooling was performed by free cooling, *i.e.,* the heating source was shut off. The thermal oxidation of the samples was performed in a furnace, flushed with air for temperatures ranging from 400 to 600 °C and an annealing time of 1 h.

Biocompatible Behaviour

Cytotoxicity tests have been performed on three selected Nb-O-N coatings according to the ISO standard 10993-5 (at Institute of Cell Biology, ETH, Zürich). The Nb-O-N coatings (three different compositions, eight identical samples of the same composition) were deposited on glass substrates. The samples were sterilized in 70% ethanol before use and subsequently washed 3 times in phosphate buffered saline (pH = 7.4). To assess the cytotoxicity of the coatings, 3T3 cells (embryonic mouse fibroblast cell line) have been directly grown on the samples to be tested. Polystyrene (culture dish surface) is used as a negative control and uncoated glass substrates as a positive control. The parameters investigated are viability based on the physical uptake of neutral red (NR) and metabolic activity by MTT (3-(4,5-dimethylthiazol-2yl-)-2,5-diphenyltetrazoliumbromide) assay, which depends on the activity of intracellular enzymes.

Wetting Behaviour

The wetting behaviour of Nb-O-N coatings was acquired by the water contact angle measurements with the sessile drop-method (contact angle system OCA 20 from Data Physics).

TUNEABLE PROPERTIES

Deposition Rate

The deposition rate in dependence of the reactive gas flux in DC and pulsed power magnetron sputtering is shown in Fig. **1**.

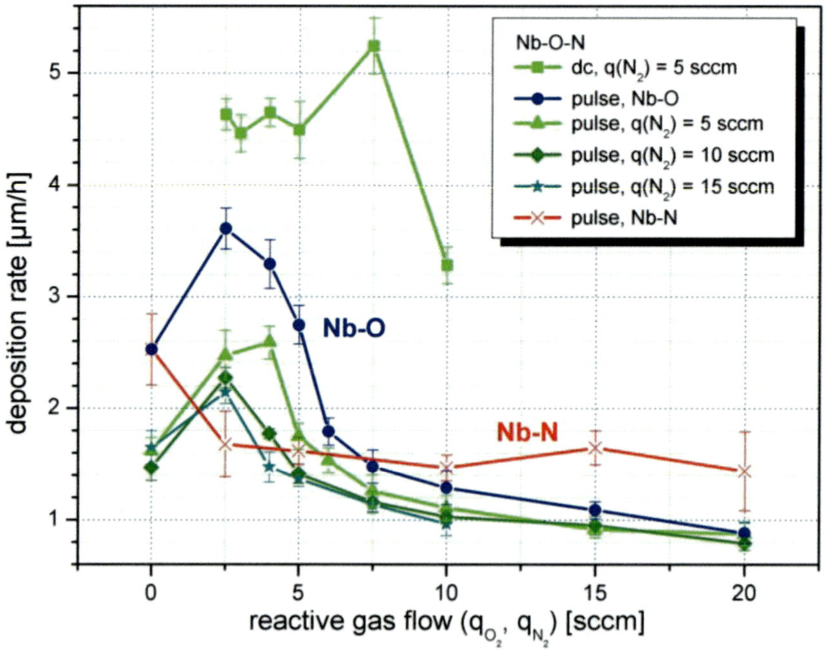

Figure 1: Deposition rates *vs.* reactive gas flow rate in DC and pulsed power magnetron sputtering of NbO, NbN and NbON coatings (pulse frequency 100 kHz, pulse width 1456 ns). One curve for DC magnetron sputtering is shown (green curve with full squares, $q(N_2) = 5$ sccm).

Obviously, the deposition rate is much higher for DC compared to pulsed power magnetron sputtering with $q(N_2) = 5$ sccm (DC: green curves with full squares, pulsed: green curve with full triangles). The deposition rates for pulsed power magnetron sputtering were in the range of 0.8 – 3.6 µm/h. The highest rates were found for low oxygen flows in the Nb–O system. The slight increase of the deposition rate at low oxygen gas flows (Nb–O, Nb–O–N) can be explained by the additional integration of oxygen in the Nb lattice leading to the NbO phase which has a lower density than pure Nb. With increasing oxygen gas flow more complex Nb–O phases like NbO_2 or Nb_2O_5 with lower densities will be generated, but now the effect of target poisoning overcompensates this effect of volume gain and causes a strong decrease in the deposition rates. For the deposition of Nb–N coatings such a maximum does not occur. The target poisoning effect leads to a decrease of the deposition rate from 2.5 µm/h (metallic niobium) to 1.5 – 1.7 µm/h (niobium nitride).

Morphology, Microstructure and Chemical Composition

Chemical Composition

EDX analysis of Nb–O coatings deposited by pulsed power magnetron sputtering revealed that already at low oxygen flow rates (ca. 2.5 sccm) high oxygen contents in the coatings are present. The addition of nitrogen to the reactive gas atmosphere has only a minor effect on this behaviour as can be seen in Fig. **2**. The chemical composition of stoichiometric Nb_2O_5 is obtained at about 5 sccm of oxygen flow rate. With the addition of 5 sccm of N_2 the chemical composition of stoichiometric Nb_2O_5 is obtained at about 6–7 sccm of O_2. As for other Me–O–N systems (Me = metal), the oxygen has a strong influence on the chemical composition of the NbON coatings due to the higher reactivity compared with nitrogen. In fact, the standard heat of formation is -239 kJ/mol for NbN and about -400, -800 and -1900 kJ/mol for NbO, NbO_2 and Nb_2O_5, respectively [17].

Nb–O coatings deposited at oxygen gas flows >5 sccm look transparent, at lower oxygen gas flows the coatings appear metallic. The addition of nitrogen shifts the transition “metallic–transparent” to lower oxygen gas flows, despite the fact that the oxygen content is lower in these Nb–O–N films at this transition zone as for stoichiometric Nb_2O_5.

It has to be noted that the oxygen content is much higher for Nb–O–N coatings deposited in pulsed mode compared to the coatings deposited in DC mode (at the same oxygen gas flow) [11]. The reason for this is the lower deposition rate in pulsed mode (about a factor of 2 for low O/N ratios compared to DC mode).

During the positive pulse no sputtering takes place and the average cathode voltage is about 100 V lower compared to DC mode, leading to a lower power density on the target. Therefore, much higher concentrations of oxygen can be incorporated into the as-deposited film due to the higher reactivity of oxygen as compared to nitrogen with niobium.

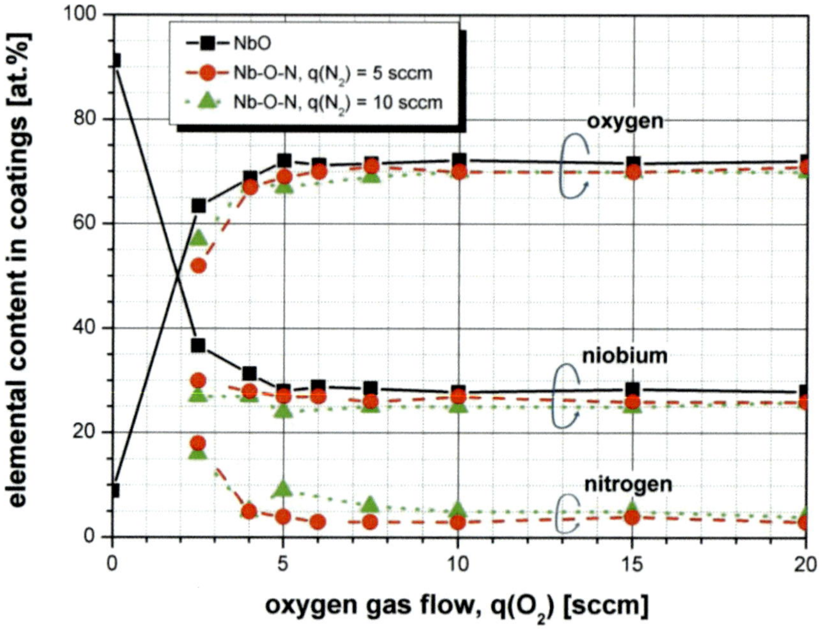

Figure 2: Elemental content in the coatings *vs.* oxygen flow rate for different nitrogen flow rates in pulsed power magnetron sputtering (pulse frequency 100 kHz, pulse width 1456 ns).

Structural and Morphological Properties

SEM micrographs of the fracture cross sections of two Nb–O–N coatings are shown in Fig. **3**. Only for the DC sputtered Nb–O–N coating deposited with the lowest oxygen flow (2.5 sccm, coating with 33.7 at.% of oxygen) a columnar microstructure can be observed (Fig. **3a**).

All other coatings possess a nearly featureless microstructure, which is usually characteristic for amorphous coatings (as an example, a Nb–O–N coating with 59.2 at.% of oxygen is shown in Fig. **3b**). This non-columnar microstructure is also responsible for the smooth surface, which can be noticed by AFM measurements (Fig. **4**).

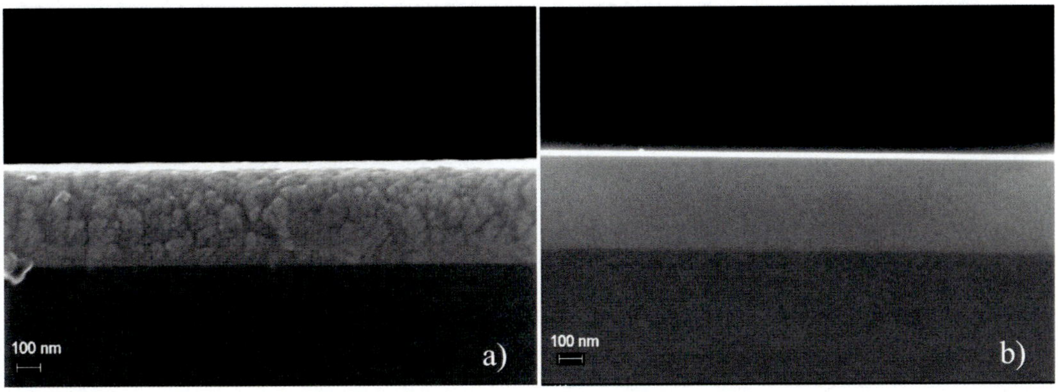

Figure 3: SEM micrographs of fracture cross sections of two DC sputtered Nb–O–N coatings ($q(N_2)$ = 5 sccm) with a) $q(O_2)$ = 2.5 sccm (33.7 at.% of oxygen, O/N = 1.2) and b) $q(O_2)$ = 7.5 sccm (59.2 at.% of oxygen, O/N = 6.0) [11].

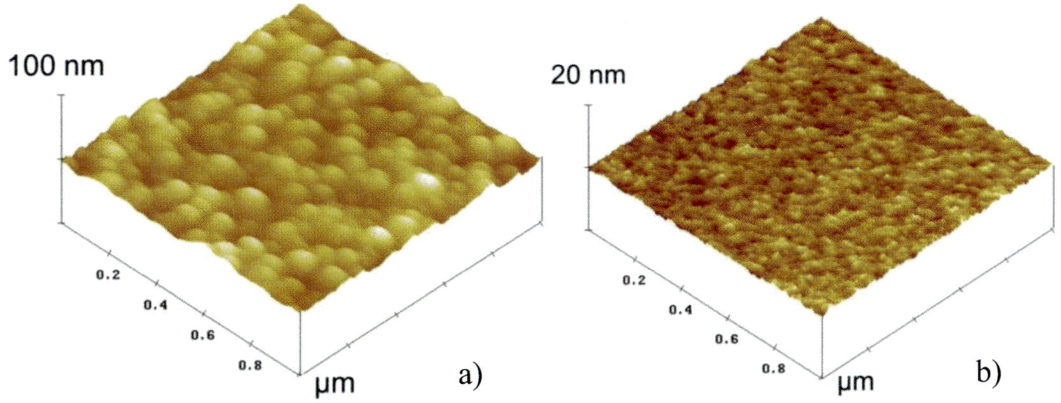

Figure 4: AFM images of the surface morphology of two DC sputtered Nb–O–N coatings ($q(N_2)$ = 5 sccm) with a) $q(O_2)$ = 2.5 sccm (33.7 at.% of oxygen, O/N = 1.2, Ra = 1.65 nm) and b) $q(O_2)$ = 7.5 sccm (59.2 at.% of oxygen, O/N = 6.0, Ra = 0.32 nm) [11].

Again, only the DC sputtered coating with the lowest oxygen flow rate showed a coarse-grained surface structure with a nodular aspect more or less bumped (Fig. **4a**). All other coatings exhibit a fine-grained surface topography (note that the z scale is different in Fig. **4a** and **b**). As an example, the Nb–O–N coating with 59.2 at.% of oxygen is displayed in Fig. **4b**. More roughness values measured by AFM can be found in reference [11].

X-ray diffraction patterns of Nb–O–N coatings deposited on glass substrates at room temperature (*i.e.*, without additional heating) are represented in Fig. **5**. Only for the coating with an oxygen concentration of 33.7 at.% some reflections can be observed. These reflections are located at 2θ positions of about 36°, 61°, 72° and 76°. There exist three phases in the Nb–O–N system, which show reflections at these positions: a) a face-centred cubic NbN phase (PDF 71-162), b) a body-centred tetragonal Nb–O–N phase with a sum formula of $Nb_{3.49}N_{4.56}O_{0.44}$ (PDF 50-320) and c) a body-centred cubic $NbN_{0.6}O_{0.3}$ phase (PDF 00-013-0467). As a result, no clear phase identification can be made. For the Nb–O–N coatings with higher oxygen contents (O > 40 at.%) only a diffuse peak centred at about 22 – 38° is visible.

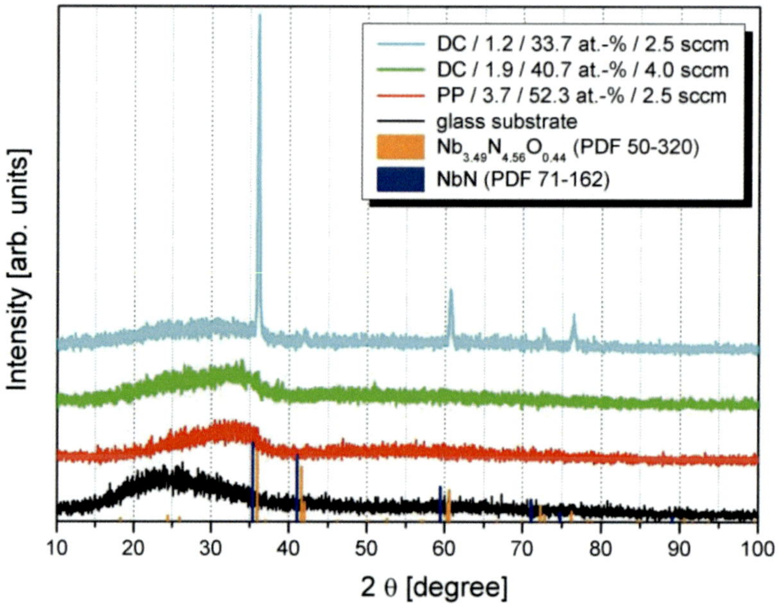

Figure 5: XRD patterns of Nb–O–N coatings deposited on glass substrates ($q(N_2)$ = 5 sccm). The description in the legend indicates the power mode (DC or pulsed power = PP), the O/N ratio, oxygen content and $q(O_2)$.

The influence of additional heating during Nb-O-N deposition on the Nb-O-N phase formation can be recognized in Fig. **6**. An Nb-O-N coating ($q(N_2)$ = 5 sccm, $q(O_2)$ = 3 sccm) which shows no clearly visible reflections in the XRD pattern without additional heating during deposition has been chosen (denoted with RT = room temperature in Fig. **6**). But when using additional heating (curves with 150 or 250 °C) reflections from the $Nb_{3.49}N_{4.56}O_{0.44}$ phase (PDF 50-320) occur. The intensity of the reflections of this Nb-O-N phase increases with the deposition temperature.

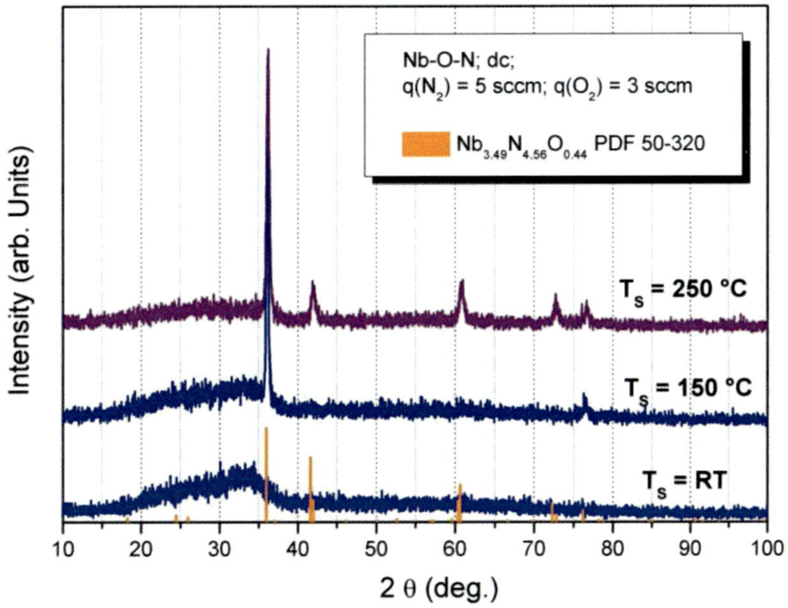

Figure 6: XRD patterns of an Nb-O-N coating deposited with the same reactive gas flows but with different substrate temperatures (RT = room temperature, *i.e.,* without additional heating).

Electrical and Optical Properties

Electrical Properties

The DC electrical conductivity *vs.* temperature of Nb–O–N coatings on glass substrate is shown in Fig. **7** [11]. As a reference, an NbN coating deposited by DC reactive magnetron sputtering was also prepared and measured. This coating exhibits a typical metallic behaviour (*i.e.,* linear variation of the electrical conductivity *vs.* temperature) with $\sigma_{300\ K} = 4.6\times10^4$ S m^{-1} and with a temperature coefficient of resistance TCR $\approx -10^{-5}$ K^{-1}. This value for electrical conductivity is much lower as usually reported for NbN, which is about 50 – 170×10^4 S m^{-1} [18, 19], due to an oxygen contamination of the NbN coating, which originates from the residual gas in the vacuum deposition chamber (3.6 at.% of oxygen in the film as measured by EDX). Similarly to NbN, the electrical conductivity of the Nb–O–N coatings increases with temperature.

Such behaviour has been reported also for other metal oxynitride systems such as Ti–O–N [20], Zr–O–N [21] or Al–O–N [22]. In addition, an increase of the oxygen content in the film leads to a systematic decrease of the electrical conductivity. For Nb–O–N coatings, conductivity is largely reduced from $\sigma_{300\ K} = 10^4\ Sm^{-1}$ down to 7.5 S m^{-1} with increasing O content from 33.7 to 52.3 at.%. The activation energy for thermally induced conduction mechanism, E_a, calculated from an Arrhenius plot, is in the range 61 – 284 meV and also increases with oxygen content. A further increase of the oxygen content in the film (≥ 59.2 at.%) leads to insulating oxynitride compounds (not measurable with the 4 point-probe method) with an electrical conductivity tending to that of niobium oxide (E_a > 500 meV). Consequently, a gradual change of the electrical behaviour of Nb–O–N films from metallic to semi-conducting and finally insulating-like as a function of oxygen content is evidenced. In [11] the progressive decrease of conductivity of Nb–O–N films is attributed to the increase of ionicity due to the formation of Nb–O bonds as oxygen concentration increases.

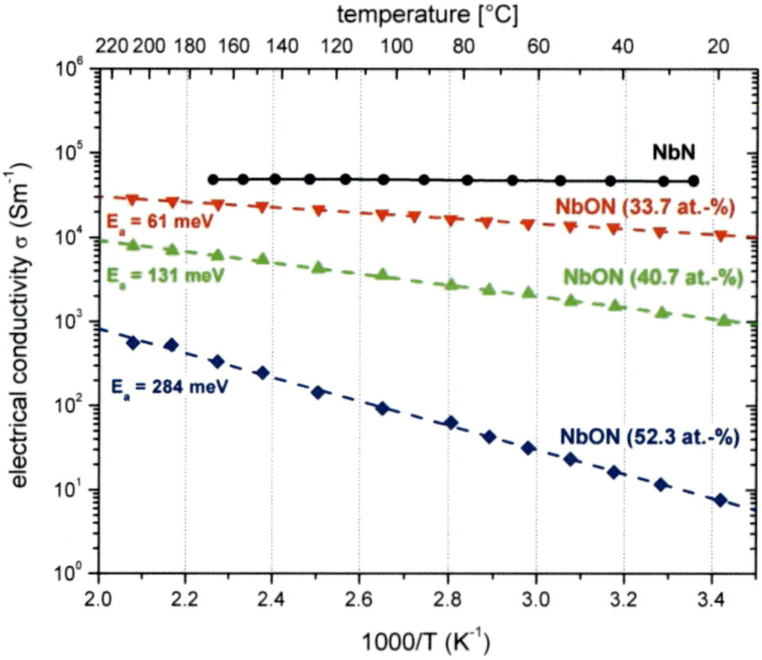

Figure 7: Electrical conductivity *vs.* temperature of NbN and NbON coatings with different $q(O_2)$ at $q(N_2)$ = 5 sccm (oxygen content in brackets) [11].

Dielectric Characteristic of Nb-O-N

In order to study the dielectric characteristics of the films, Al//Nb–O–N//SiO_2//Si metal-oxide-semiconductor (MOS) structures were fabricated and the film capacitance–voltage (C–V) characteristics were measured. An NbO_x and an Nb-O-N film were deposited onto p-type (100) and 525 µm thick Si wafers (diameter 50 mm) with a 100 nm thick SiO_2 top layer. The metallization was done by evaporation of 300 nm thick Al dots through a shadow mask with an area of 0.78 cm^2 on the surface of the films. To improve the quality of the MOS structures, a two-step post-metallization annealing (PMA) was performed. PMA is usually used in microelectronics in MOS capacitor fabrication in order to densify the Al dots and to reduce the number of charges present at the interfaces and in the film [23]. The first annealing step was executed in N_2 at 723 K for 30 min and the second one in forming gas (H_2:N_2 = 10:90) at 723 K for 30 min. The C–V characteristics of the MOS capacitor were measured using a HP 4275A multifrequency LCR METER at 1 MHz. The temperature was kept constant at 299 K by a temperature controller. Details on the C–V measurements can be found in [24].

C_{ox} [pF]	N_{sub}	V_{fb} [V]	N_{ss}
276	1.97 x 10^{14}	-0.811	9.84 x 10^{9}
271	1.43 x 10^{15}	-1.64	1.70 x 10^{11}
263	1.95 x 10^{15}	-1.61	1.57 x 10^{11}
267.00	1.69 x 10^{15}	-1.63	1.64 x 10^{11}

C_{ox}: capacity of the dielectric films
N_{sub}: electric charges at the film-substrate interface
V_{fb} : flat band voltage
N_{ss}: electric charge concentration
ε_s: permittivity

ε_s
28

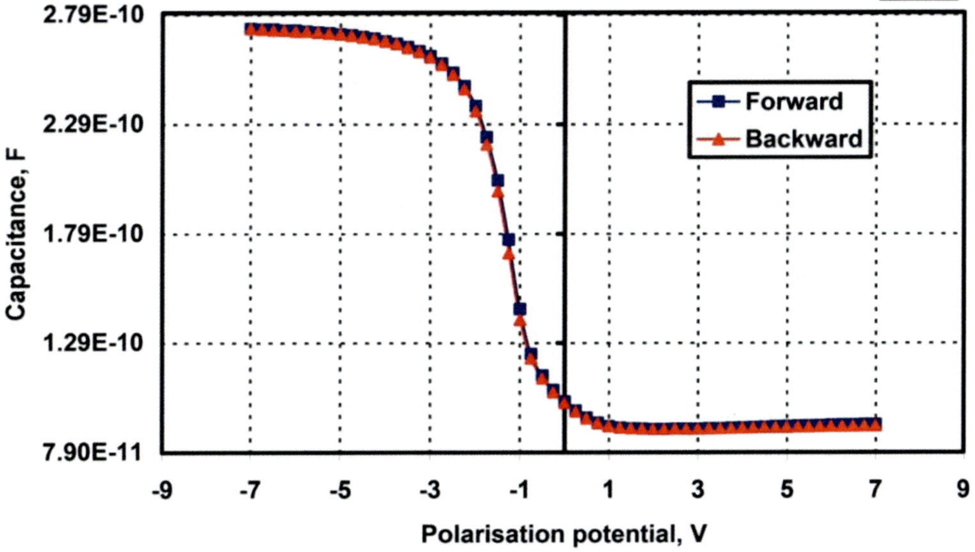

a)

Fig. 8: contd….

C_{ox} [pF]	N_{sub}	V_{fb} [V]	N_{ss}
245	2.23×10^{15}	-1.76	1.74×10^{11}
244	1.91×10^{15}	-1.65	1.52×10^{11}
243	2.19×10^{15}	-1.38	1.03×10^{11}
244.00	2.11×10^{15}	-1.60	1.43×10^{11}

ε_s
28

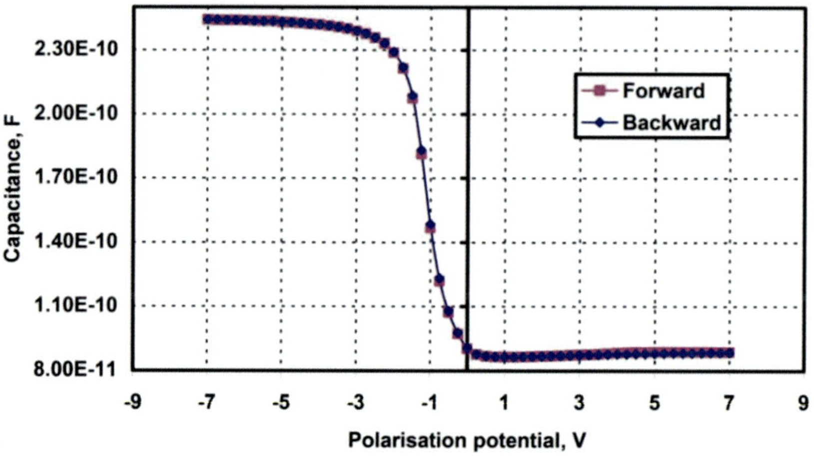

b)

Figure 8: C-V characteristics of a) NbO_xN_y film and b) of the NbO_x film.

Both, Nb-O-N (about 60 at.% oxygen content, O/N ≈ 5.6) and NbO_x (presumably a-Nb_2O_5) exhibit a very good dielectric behaviour thanks to stability in accumulation and inversion regimes (Fig. **8a** and **b**). The electric charge concentration (N_{ss}) and the flat-band voltage (V_{fb}) are almost equivalent but the NbO_x shows a small advantage due to the lower value of electric charges at the film-substrate interface (N_{sub}). A promising result is that the permittivity ε_s of the Nb-O-N is higher (28) in comparison with the NbO_x (22), which is widely-known and attracted great interest for usage as microelectronic dielectric layers.

Optical Properties

The transmittance of the Nb–O–N coatings deposited on glass substrates is shown in Fig. **9**. Only the insulating (oxidic) Nb–O–N coatings show a high transmittance at wavelengths longer than 350 nm (with typical interference fringes). The highest transmittance of 60 – 90% is found for the coatings deposited in pulsed mode with oxygen flows $q(O_2)$ of 4.0 and 7.5 sccm (63.4 and 67.3 at.% of oxygen) at $q(N_2) = 5$ sccm, respectively. The optical band gap has been determined from transmission measurements from the Tauc plot as described in reference [11]. The optical band gap of the insulating Nb–O–N coatings is strongly dependent on the oxygen concentration and increases from 2.2 to 3.3 eV when the oxygen content increases from 59.2 to 67.3 at.%, respectively.

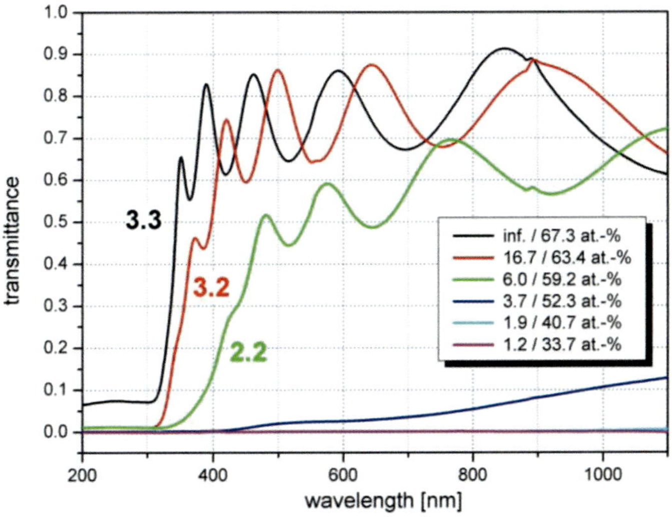

Figure 9: UV–visible transmission spectra of Nb–O–N coatings on glass (legend: O/N ratio / oxygen concentration at $q(N_2)$ = 5 sccm). The numbers at the curves with high transmittance are the optical band gap values [11].

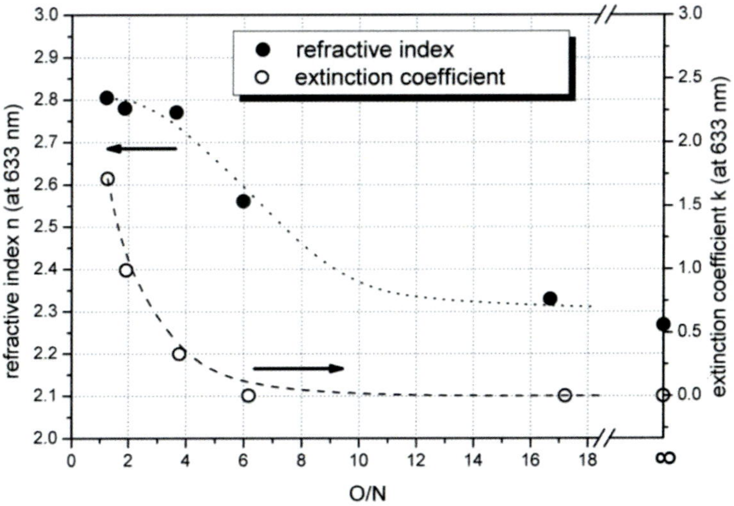

Figure 10: Refractive index and extinction coefficient of Nb–O–N coatings *vs.* O/N ratio (open symbols: extinction coefficient, full symbols: refractive index, the lines are a guide for the eye) [11]. The values for Nb_2O_5 are located at an infinity O/N ratio.

Refractive index, n and extinction coefficient, k (both taken at 633 nm) of the Nb–O–N coatings deposited on (100) silicon are plotted *vs.* the O/N ratio in Fig. **10**. The refractive index values are the highest for "metallic"-like coatings and decrease with increasing O/N ratio. The refractive index n of transparent and semi-transparent Nb–O–N coatings is in the range of 2.3 and 2.6. The extinction coefficient k is strongly decreasing with increasing O/N ratio.

Mechanical Properties

Hardness and Residual Stress

The microhardness H_{IT} (indentation hardness) and the residual stress σ values are represented in Fig. **11** *vs.* O/N ratio. H_{IT} and the compressive stress σ decreases drastically with increasing O/N ratio. Hardness and stress starts at high values for the NbN coating (H_{IT} = 30 GPa, σ = −4.5 GPa) and end at low hardness and stress values for the Nb_xO_y coating (H_{IT} = 5.6 GPa, σ = −0.1 GPa). Niobium nitride is known as a hard or superhard coating with very high hardness in the range of 34 to 53 GPa [25]. For Nb_2O_5 thin films a hardness of 10 GPa was reported by Szymanowksi *et al.* (they measured the plastic hardness which is a little bit higher as the indentation hardness H_{IT}) [26]. The increase of the non-metal content in the film leads to a lower fraction of the strong covalent TM–N bonds, which results in the decrease of H_{IT} values. Furthermore, a lower compressive stress level induces a lower hardness value [32].

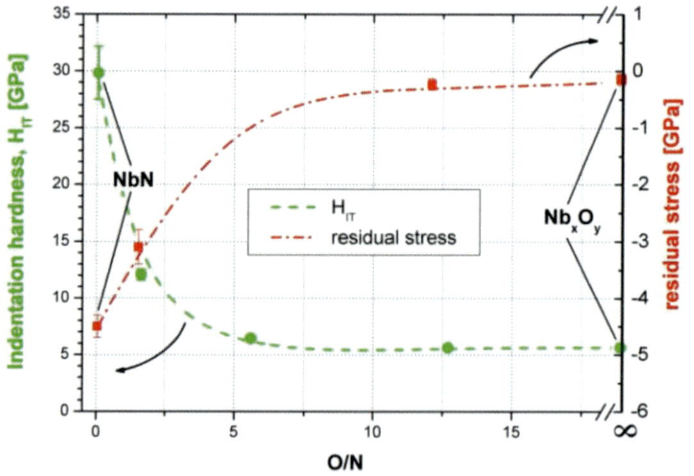

Figure 11: Indentation hardness H_{IT} and residual stress σ *vs.* O/N ratio of Nb–O–N coatings on high speed steel (lines are a guide to the eye) [32].

Tribological Properties

The friction coefficients (μ) of Nb–O–N coatings against a 100Cr6 ball are presented in Fig. **12**.

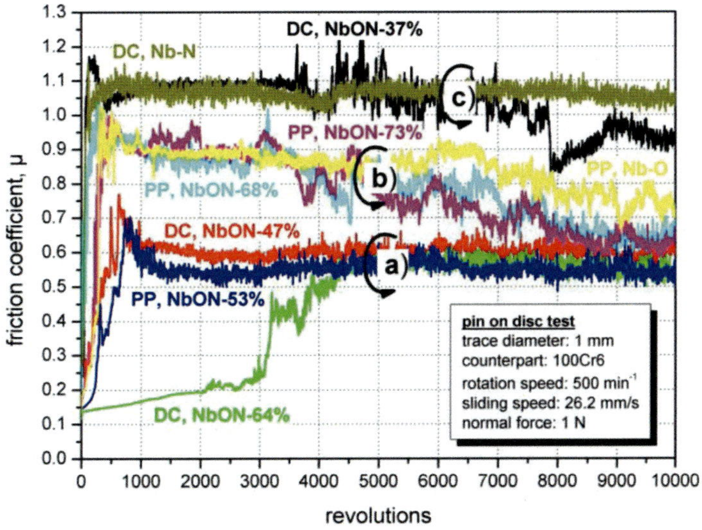

Figure 12: Friction coefficient of Nb–O–N coatings on HSS against 100Cr6 (curves smoothed, numbers at the curves represent the oxygen content; a), b) and c): classification with respect to μ) [32].

For a comprehensive understanding of the tribotests all wear tracks have been investigated by profilometry, light optical microscope, SEM and EDX. From light optical microscope and SEM images it can be seen that there has been a transfer of ball material to the sample surface for coatings with an O/N ratio of ≥ 2.2. The EDX measurements showed that the transferred material is iron oxide. Wear track depth measurements by profilometry revealed that the Nb–O–N coatings with an O/N ratio of 1.2 and ≥ 12.1 were worn through after the tribotests (see Fig. **13**). In SEM micrographs the carbides of the HSS can be observed in the wear track [32]. Additionally Iron-based material transferred to the surface of the wear track, where the Nb–O–N coating has been worn is visible. It is assumed that this material originates from the worn ball and not from the HSS.

The results of these pin-on-disk tests can be classified in the following three categories with respect to the friction coefficient (the classification is indicated in Fig. **12**):

a) Low μ of about 0.15–0.6: coatings with medium oxygen content (O/N = 2.2 – 7.1), no coating wear visible (see Fig. **13**), transfer of ball material to the coating, low stress (σ = 0.3 – 0.5 GPa [32]), featureless/amorphous microstructure, smooth coating surface,

b) Medium μ of about 0.6–0.9: coatings with high oxygen content (O/N = 12.1 – infinity), high coating and ball wear (adhesive wear), transfer of ball material to the coating, negligible stress, amorphous microstructure, smooth coating surface,

c) High μ of about 0.9 – 1.1: coatings with low oxygen content (O/N $\leq$ 1.2), high coating wear (mainly abrasive wear), high stress values, columnar microstructure, rough coating surface.

From Fig. **12** and the experimental details given in this classification it can be seen that the highest friction coefficient is found for the coatings with a columnar microstructure exhibiting the highest surface roughness (category c). Higher friction coefficients for a higher surface roughness were also reported by Harlin *et al.* for TiN coatings [27]. A distinct reduction of μ is observed when there is a transition from a columnar (category c) to an amorphous microstructure (category a). After running-in, the friction coefficient of 100Cr6 against M2 steel – which was measured to be 0.6 – is obtained. This can be ascribed to the transfer of ball material to the coating surface. A similar behaviour was reported by Savisalo *et al.* for CrON/NbON multilayer coatings [6]. A further increase of the oxygen content (transition from category a) to b) in the Nb–O–N coatings (O/N $\geq$ 12.1) results in an increase of μ, due to tribochemical reactions caused by the increasing affinity of the coating to the counterpart material (increasing adhesive forces).

Similar results are found for the wear of the coatings (see Fig. **13**). Abrasive wear prevails for the coatings with a columnar microstructure (category c). It is assumed that the high friction energy caused by the high friction coefficients in combination with the high residual stresses results in the high wear of the coatings of category c). A quarrying out of the columns of the coatings during the tribotest could be also a reason for the abrasive wear mechanism of these hard coatings against the softer 100Cr6 ball. With the transition to an amorphous microstructure (category a), the wear is practically approaching zero, caused by the well-adhering iron oxide transfer layer on the Nb–O–N coating. The 100Cr6 ball is rubbing against the transfer layer. For increasing oxygen content (transition from category a) to b) of the coatings the effect of the adhesion to the counterpart prevails. This causes a pronounced adhesive wear. It is assumed that the chemical composition plays here a decisive role for the higher μ of samples from category b) (high O content) with respect to samples from category a) (medium O content). A plausible explanation would be a high affinity of the iron or iron oxide from the 100Cr6 ball to the oxide surface, depending on the O content of the Nb–O–N coatings.

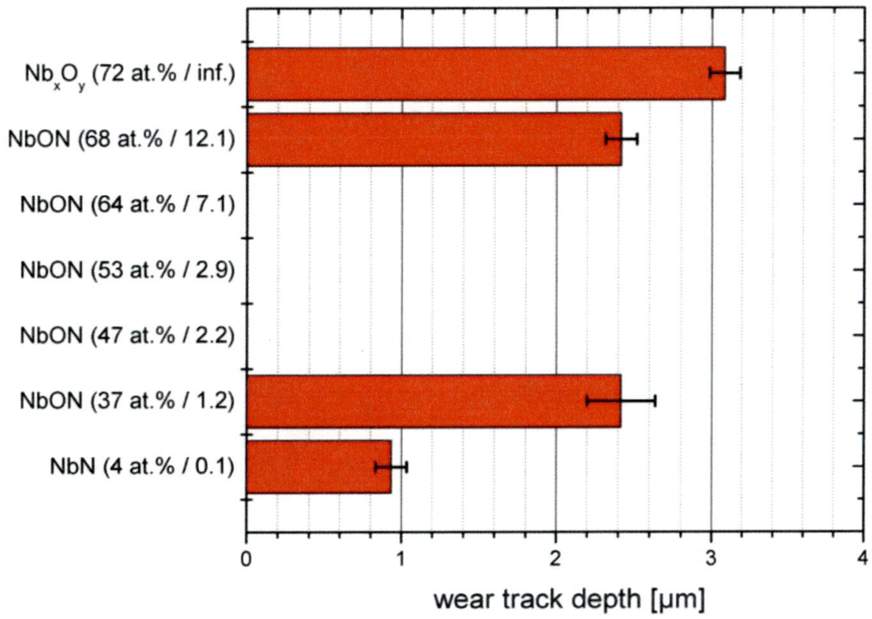

Figure 13: Wear track depths on Nb–O–N coated HSS against a 100Cr6 ball after pin-on-disk tests extracted from [32] (in brackets behind coating material: oxygen content/O/N ratio).

We also studied the tribological behaviour of the Nb-O-N coated HSS substrates against an Al_2O_3 ball and correlated these results with the hardness, H_{IT}/Y_{IT} and H_{IT}^3/Y_{IT}^2 values, which are represented in Fig. **14**. The ratio H_{IT}/Y_{IT} is a measure for the elastic strain to failure and a high H_{IT}/Y_{IT} value is often a reliable indicator for good wear resistance [28]. The ratio H_{IT}^3/Y_{IT}^2 is a measure for the resistance to plastic deformation [29]. The averaged friction coefficients of Nb–O–N coatings against an Al_2O_3 ball are in the range of 0.8–0.9 (Fig. **14**; the friction values are average values from revolutions 1,000 – 10,000 in the pin-on-disc test). The lowest friction coefficient is obtained for the NbN coating. However, no tendency can be observed with respect to the O/N ratio. Profilometry measurements revealed that only the Nb–O–N coatings with high oxygen contents ($\geq$ 68 at.%, O/N $\geq$ 12.1) were worn through after the tribotests (Fig. **14**). SEM and EDX measurements showed that the wear mechanism of the pairing Nb–O–N/Al_2O_3 ball is abrasive wear, due to the high hardness of the counterpart material [32]. To some extent a correlation with the hardness of the Nb–O–N coatings can be made: the higher the hardness of the Nb–O–N coatings, the lower is the wear rate in pin-on-disk tests (Fig. **14**). PalDey *et al.* reported that high hardness is beneficial in resisting abrasive wear [30]. However, a high hardness is only found for the NbN coating. All Nb-O-N coatings with an oxygen content $\geq$ 37 at.% possess a hardness < 10 GPa. Whereas, the wear depth is low for all Nb-O-N coatings with an oxygen content < 64 at.%. Despite the fact, that the friction coefficient is higher and the hardness and the H_{IT}/Y_{IT} is lower for the Nb-O-N coatings with an oxygen content from 37-53 at.% compared to the NbN coating. Both ratios, the H_{IT}/Y_{IT} and the H_{IT}^3/Y_{IT}^2, do not help to explain the wear behaviour of the Nb-O-N coatings in detail.

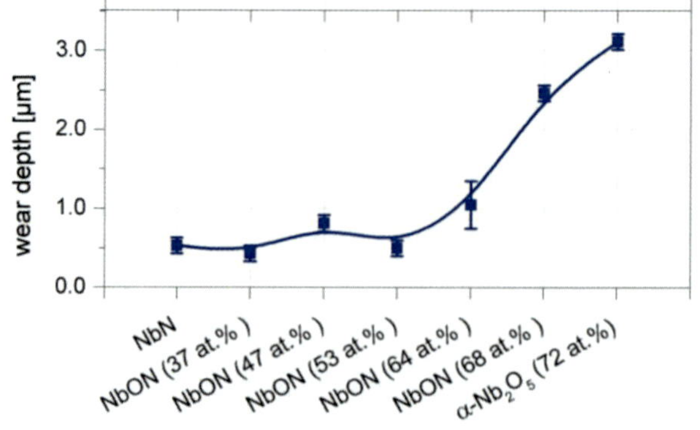

Fig. 14: contd....

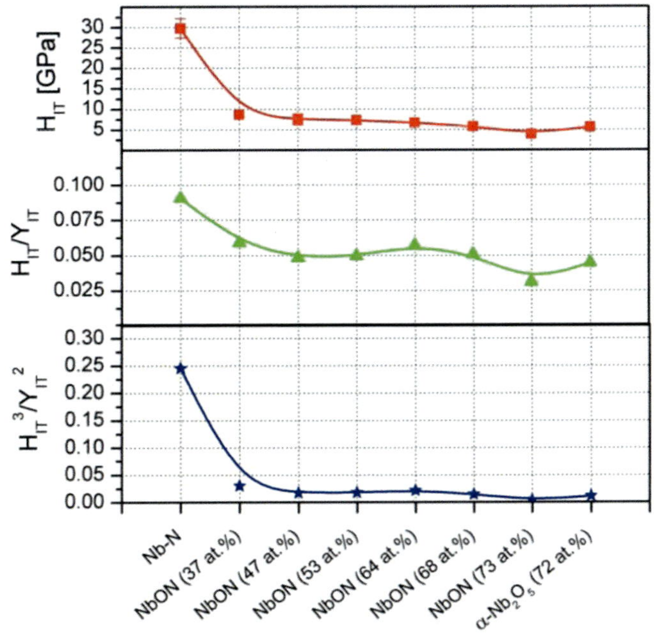

Figure 14: Friction and wear depth (pin-on-disc, counterpart Al_2O_3), hardness, H_{IT}/Y_{IT} and H_{IT}^3/Y_{IT}^2 for NbN, NbON and Nb_2O_5 coatings on HSS.

Corrosion and Degradation Behaviour

Potentiodynamic Corrosion Tests

Potentiodynamic corrosion tests have been performed in 0.8 M NaCl solution (pH 7) on samples deposited in the same batch as the samples for salt spray tests. The voltage-current density curves of these samples are represented in Fig. **15**. All Nb-O-N-coated HSS substrates show a lower corrosion current density as the bare steel substrate. Additionally, it can be observed that the addition of oxygen to NbN dramatically increases the corrosion resistance of the coatings deposited in DC sputtering mode [31]. Already the NbN-coated HSS shows a very low corrosion current density of about $4x10^{-6}$ A/cm^2 in the passivation range up to a pitting potential of 1000 mV. But the current density for the NbON-coated HSS with an O/N ratio of 12.8 (68 at. % of oxygen) is even more than 2 orders of magnitude lower as compared to the NbN/HSS sample. Pitting corrosion only occurs at a potential of > 1400 mV. Two factors can be responsible for the improved corrosion resistance of Nb-O-N coatings:

1) The Nb-O-N coatings are getting denser and they lose their columnar microstructure with increasing oxygen content. This reduces the number of defects.

2) The electrical conductivity is strongly decreasing with increasing oxygen content. This impedes the transport of electrons through the coating, *i.e.,* the dissolving of iron is hindered or even stopped due to the increase of the electrical resistance in the circuit of the anode and the cathode.

The behaviour of the coatings deposited in pulsed power (PP) sputtering mode as represented in the E – i curves is not yet understood at the moment, as these samples do not show any pitting corrosion after the corrosion test, despite the high measured current densities during potentiodynamic corrosion tests. Obviously, the insulating behaviour of the coating influences the measurement. A more detailed discussion of the corrosion mechanism can be found in reference [32].

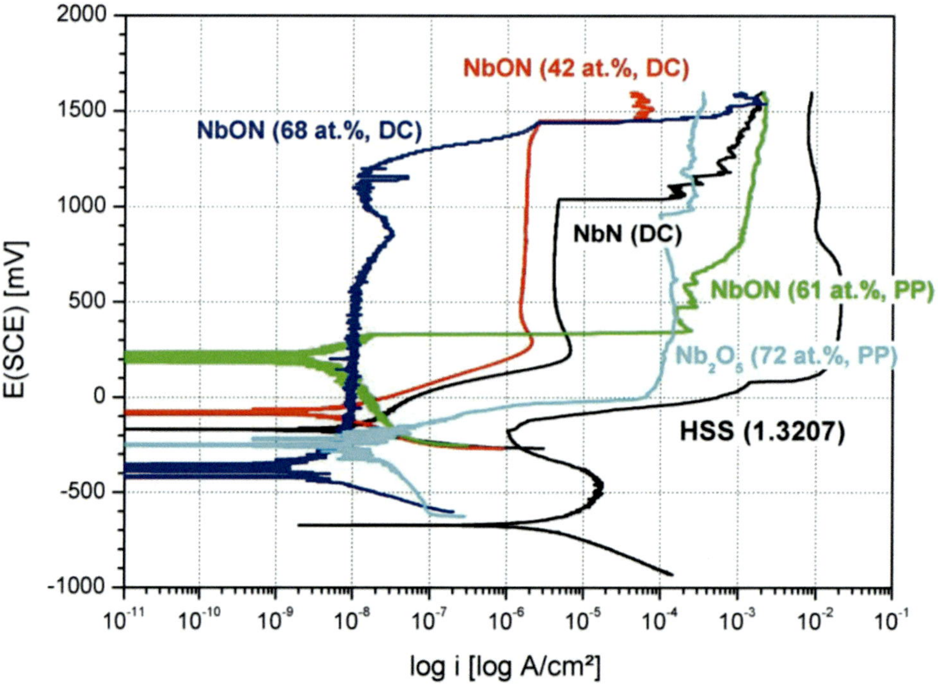

Figure 15: Potentiodynamic corrosion tests of Nb–O–N coatings on HSS in 0.8 M NaCl solution (pH 7) [32].

Neutral Salt Spray Tests

The photographs and the corrosion pit density of Nb-O-N-coated HSS samples after 24 h of salt spray test can be found in Fig. **16**.

Coating O/N O content	NbN, DC 0.1 3.7 at.%	NbON, DC 1.6 41.5 at.%	NbON, PP 5.6 61.0 at.%	NbON, DC 12.8 67.8 at.%	Nb_2O_5, PP infinity 72.1 at.%
NSS test after 24 h					
Pit density [pit/cm^2]	6-8	14-15	8-9	< 1	< 1

Figure 16: Photographs of different Nb-O-N coatings after 24 h of neutral salt spray testing. The pit density was calculated by counting the visible rust sites with the unaided eye.

It can be observed that the two samples with the highest oxygen content show the best corrosion performance, *i.e.,* the lowest pit density. Additionally, a much lower bleeding of the steel at the pits of the Nb-O-N coatings is visible compared to the NbN coating. As observed in potentiodynamic corrosion tests, the anodic current density is decreasing for (conducting) Nb-O-N coatings with increasing oxygen content, *i.e.,* a lower amount of iron is dissolved which leads to less bleeding at the pits. Furthermore, for the NbN-coated steel, the number of the corrosion sites is underestimated, due to the coalescence of corrosion pits during the salt spray test (this can be concluded from the irregular shape of the pits). If we take this into account, we can draw the conclusion that with increasing oxygen content in the coatings the corrosion resistance increases dramatically. It must be emphasized that the excellent corrosion behaviour of the Nb-O-N-coated steel is even better than for state-of-the-art nitride coatings, like *e.g.* CrN, Nb_2N, TiN and TiBN [31]. An improvement of the corrosion behaviour of CrN/NbN nano-scale multilayers with a CrON/NbON topcoat was also reported in reference [6]. Likewise, Wierzchoń *et al.* found an improvement of the corrosion behaviour in a 0.5 M NaCl solution of oxidized or oxynitrided electroplated chromium coatings on AISI 1045 steel [33].

Artificial Sweat Tests

After one week of artificial sweat test on Nb-O-N coatings deposited on glass or stainless steel (SS316), niobium could be detected in the electrolyte (0.06-0.3 mg/l) [32]. It was observed that with increasing O/N ratio the amount of dissolved Nb decreases. It was concluded that the dissolution of Nb in artificial sweat depends on the chemical composition of the coating and seems to be independent on the substrate material (stainless steel, glass) [32]. No information about the dissolution of Nb from Nb-O-N in artificial sweat solution could be found in literature. The mechanism of Nb dissolution from Nb-O-N coatings in artificial solution is not clear at present.

Thermal Behaviour

Thermal Stability

The crystallographic structure of the vacuum-annealed Nb-O-N coatings was investigated by XRD using a Philips PW 1710 apparatus in normal θ-2θ geometry (Bragg-Brentano) and monochromatic Cu Kα radiation (λ = 0.15406 nm) [32].

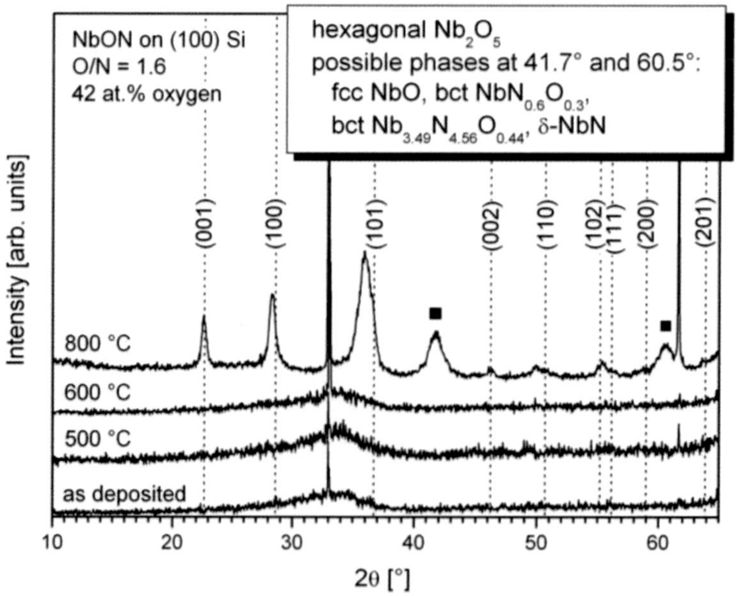

Figure 17: XRD patterns of a thermally treated NbON-coated (100) Si sample with O/N=1.6 (dotted vertical lines represent the hexagonal Nb_2O_5 phase) [32].

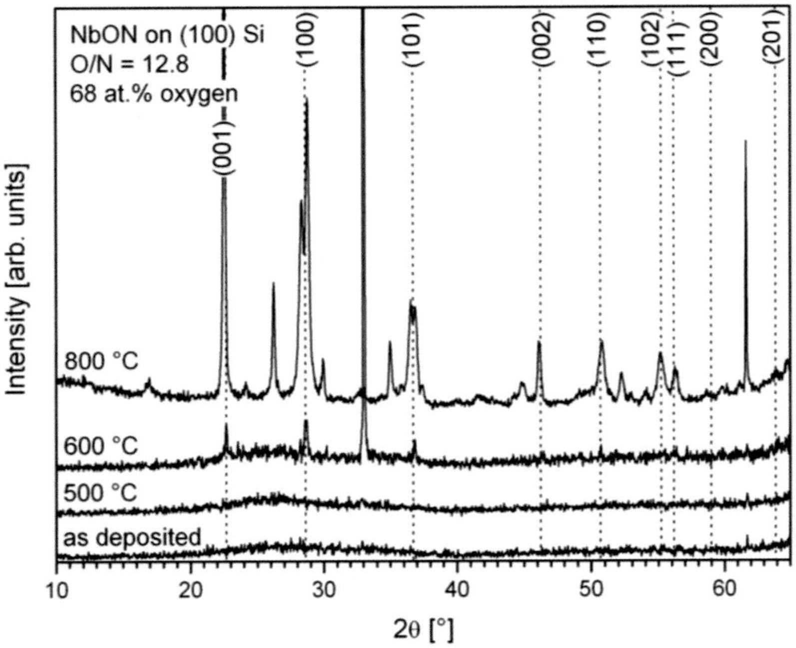

Figure 18: XRD patterns of a thermally treated NbON-coated (100) Si sample with O/N=12.8 (68 at.% oxygen) [32].

All coatings – except the NbN coating – of the system Nb-O-N were X-ray amorphous in the as-deposited state. Nb-O-N coatings with O/N atomic ratios of 1.6 and 5.6 show crystalline phases at temperatures of 800 °C [32]. However, the Nb-O-N coating with an O/N atomic ratio of 12.8 and the Nb_xO_y coating crystallize already at 600 °C. Obviously, there is a change in the crystallization temperature between an O content of 61 and 68 at.% in the Nb-O-N coatings. Maybe there is still enough nitrogen in the Nb-O-N coating with 61 at.% to stabilize the coating at these temperatures. As an example, XRD patterns of vacuum annealed samples with an O/N ratio of 1.6 and of 12.8 are shown in Figs. **17** and **18**, respectively. The reflections at 33° and 62° originate from the (100) Si substrate. For both Nb-O-N coatings, a hexagonal Nb_2O_5 phase develops with lattice parameters of a = 0.3607 nm and c = 0.3925 nm (PDF 07-0061). Additional peaks are visible at 41.7° and 60.5° which can originate from four different phases: fcc NbO (PDF 42-1125), bct $NbN_{0.6}O_{0.3}$ (PDF 13-467), bct $Nb_{3.49}N_{4.56}O_{0.44}$ (PDF 50-320) or δ-NbN (PDF 38-1125). These phases have also a peak at about 36°. At 800 °C also for the Nb-O-N coating with O/N = 12.8

additional peaks occur at about 24°, 26°, 30°, 35°, 45°, 52° and 54°. Monoclinic Nb_2O_5 and $NbO_{2.4}$ have some reflections at these locations. A mixture of the hexagonal Nb_2O_5 phase plus other O-containing phases was observed at 800 °C for Nb-O-N coatings with an oxygen content ≥ 42 at. %. From a thermodynamic point of view, Nb_2O_5 is the most favourable niobium compound [32], which is experimentally confirmed by our annealing tests.

Oxidation Behaviour

The oxidation behaviour of Nb–O–N coatings was studied on coatings with a multilayer structure (the reader is referred to reference [12] for the complete investigation). For the coating preparation, DC reactive magnetron sputtering was used and the multilayer structure was obtained by sputtering a metallic niobium target at a constant nitrogen flow $q(N_2)$ = 2.9 sccm and pulsing the oxygen flow between a minimum and a maximum flow of 0.5 sccm and 4.2 sccm, respectively. This technique is called reactive gas pulsing process. Films with different oxygen contents were deposited on (100) silicon substrates. A mounting triangle was used as the pulse shape for the oxygen gas flow. Pulse durations (T) of 10 s and duty cycles $\alpha = t_{ON}/T$ of 0.3, 0.6 and 0.9 were chosen (t_{ON} = injection time of oxygen).

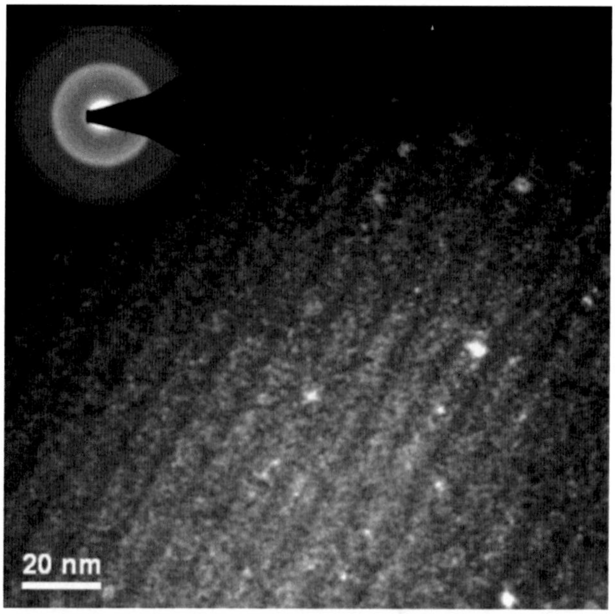

Figure 19: TEM dark field image of cross sections of Nb–O–N coatings with T=10 s (α=0.6) [14].

TEM investigations [14] revealed a multi-layered structure (see Fig. **19**). The film was deposited with a duty cycle of 0.6 and has a total thickness of 520±10 nm and a period λ (amorphous layer + crystalline layer) of about 10 nm. Some small crystallites (< 7 nm) have been observed in the layers.

The chemical composition analysis [12] showed that at 400 °C all films develop a very thin oxide layer at the surface with enhanced O concentration. At 500 °C, this oxide layer becomes thicker, encompassing about half the film thickness. At 600 °C, the process is finished and a single oxide layer is formed with reduced Nb and increased O concentration.

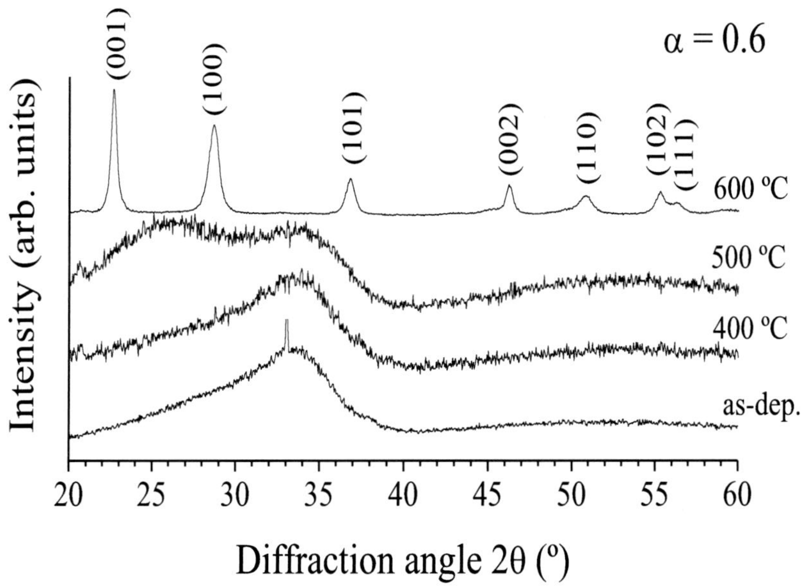

Figure 20: XRD analysis for the as-deposited and annealed samples at 400, 500 and 600 °C, prepared with a duty cycle α = 0.6 [12].

XRD patterns of in air annealed Nb-O-N coatings deposited with a duty cycle of 0.6 are shown in Fig. **20**. The as-deposited film is X-ray amorphous and crystallizes in the form of Nb_2O_5 after annealing at 600 °C. At this temperature all the films became completely oxidized and the nitrogen is depleted from the oxide layer. The oxidation kinetics of the films was determined to be diffusion-limited [12], indicating that it is the oxygen diffusion through the oxide scale which limits the process.

Biocompatibility

Cytotoxicity tests have been performed on three selected Nb-O-N coatings: One a-Nb_2O_5 and two Nb-O-N coatings (O/N ratio of 1.6 and 5.6). All coatings deposited onto glass slides were well tolerated by 3T3 cells directly grown on the coatings or beside the glass slides. The NR uptake was 0.55-0.63 NR/cm^2 (control: 0.23 NR/cm^2) and the MTT uptake was 0.58-0.67 MTT/cm^2 (control: 0.33 MTT/cm^2). There were no statistically significant differences observed between the tested coatings and the control for all parameters analysed.

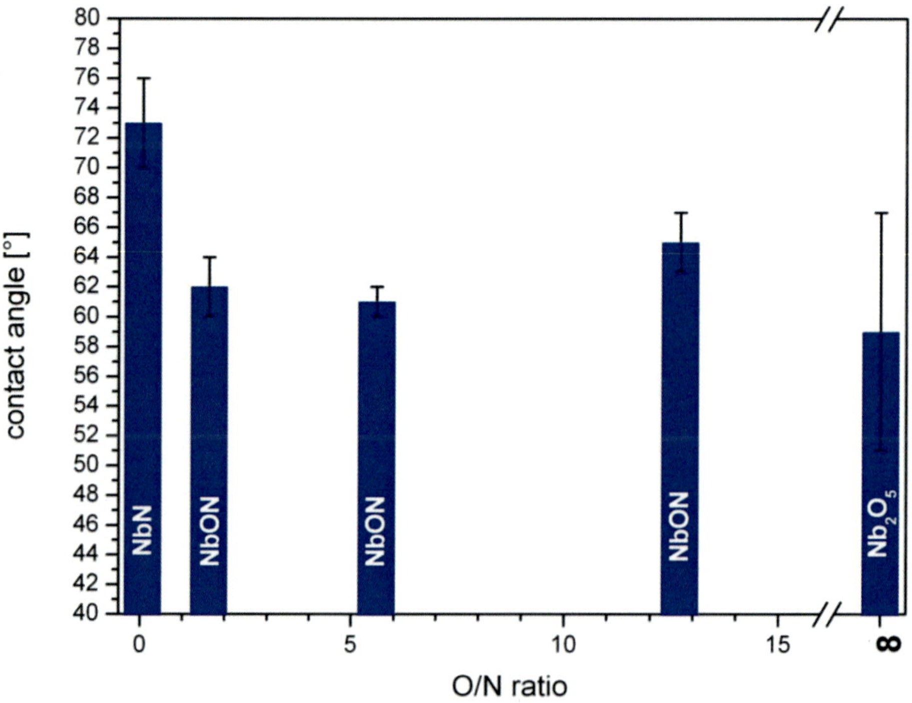

Figure 21: Water contact angle *vs.* O/N ratio of NbN, NbON and Nb_2O_5 coatings ($q(N_2)$ = 2.9 sccm).

Wetting Behaviour

The wettability tests showed that the water contact angle of Nb-O-N coatings is in the range of 59° – 73° (Fig. **21**). Ramírez *et al.* have published a water contact angle of 78.7 ± 1.9 and 80.7 ± 2.1 for NbN and Nb_2O_5, respectively [34]. The value for NbN fits quite well with our result. Only for Nb_2O_5 we measured a lower value. From the results presented in Fig. **21**, it can be concluded that the chemistry of the Nb-O-N coatings, *i.e.,* the O/N ratio, does not have a strong influence on the water contact angle. It is assumed that the roughness plays a more significant role with respect to the water contact angle for coatings of the system Nb-O-N. The different roughnesses could be the reason for the deviation in the water contact angle for Nb_2O_5 in the paper of Ramírez *et al.* compared to our result. They deposited the coatings on sandblasted stainless steel surfaces with an average roughness Ra of 2 µm.

The wettability can be an important property in the corrosion behaviour. The lower the wettability, the lower will be the contact of the electrolyte with the coating surface and therefore the penetration of the electrolyte into pores or cracks within the coating.

CONCLUSION

A comprehensive study of the Nb-O-N coating system is presented in this chapter. The coatings have been deposited by reactive magnetron sputtering process using a niobium target and a gas mixture of $Ar/N_2/O_2$. Starting with NbN and adding more and more oxygen to the gas mixture, the properties of NbON coatings could be tuned by the oxygen content in the coatings. However, oxygen contents lower than about 34 at.% could not be achieved due to the limitation of the oxygen mass flow controller for low flows and the strong reaction of the oxygen with niobium.

The addition of oxygen to the sputter gas atmosphere results in smooth coatings with featureless fracture cross sections. For oxygen contents > 40 at.%, Nb–O and Nb-O-N coatings seem to be X-ray amorphous. Vacuum annealing of the Nb-O-N coatings shows that the crystallisation depends on the O content and starts at 600 °C for coatings with higher O content (≥ 68 at.%). The crystallisation temperature is higher (about 800 °C) for Nb-O-N coatings with lower O content (≤ 61 at.%). The hardness decreases with increasing oxygen content from 30 GPa for NbN down to 6 to 8 GPa for Nb-O-N coatings with oxygen contents > 50 at.%. The tribological investigations on a pin-on-disc machine showed that the results strongly depend on the counterpart material (100Cr6, Al_2O_3). For 100Cr6 as counterpart material, an oxygen content in the range of about 47 – 64 at.% seems to be beneficial for a low friction coefficient of about 0.6 and the lowest wear rates.

The electrical conductivity at room temperature is decreasing with increasing oxygen content from 4.6×10^{4} S m^{-1} for NbN to values of 7.5 S m^{-1} for Nb-O-N coatings with an oxygen content of 53 at.%. Insulating Nb–O–N coatings (≥ 59 at.% of oxygen) show a high transmittance at wavelength longer than 350 nm. Transparent and semi-transparent Nb-O-N coatings have a refractive index in the range of 2.3 – 2.6 (at 633 nm).

The results of the salt spray test revealed an excellent improvement of the corrosion resistance of NbN-coated steel by the insertion of oxygen into the coating by a factor of > 6-8 with respect to the number of corrosion sites. Additionally, potentiodynamic corrosion tests showed a dramatically improved corrosion resistance with increasing oxygen content. In artificial sweat tests a low amount of Nb is dissolved, decreasing with increasing O/N ratio. Finally the Nb-O-N coatings passed the cytotoxicity tests.

ACKNOWLEDGEMENTS

This work was supported by the European Community under grant no. NMP3-CT-2003-505948 (project acronym "HARDECOAT"). The author would like to thank H. Kappl for doing most of the experimental work and R. Bretzler and K. Petrikowski (both **fem**) for performing SEM/EDX and XRD analyses. Furthermore the AFM and 4-point probe measurements in the group of N. Martin, the thermal stability tests in the group of F. Vaz, the UV-Vis spectrometry investigations in the group of C. Rousselot and the ellipsometry measurements by O. Banakh (HESSO) are gratefully acknowledged.

CONFLICT OF INTEREST

The author(s) confirm that this chapter content has no conflict of interest.

DISCLOSURE

Part of information included in this chapter has been previously published in Thin Solid Films Volume 519, Issue 8, 1 February 2011, Pages 2457–2463.

REFERENCES

[1] J.F. Papp, "Niobium (Columbium) and Tantalum", USGS 2006 Minerals Yearbook. Retrieved 2008-09-03.

[2] J.J. Olaya, S.E. Rodil, S. Muhl, “Comparative study of niobium nitride coatings deposited by unbalanced and balanced magnetron sputtering”, Thin Solid Films, vol. 516, pp. 8319–8326, April 2008.

[3] J. Probst, U. Gbureck, R. Thull, “Binary nitride and oxynitride PVD coatings on titanium for biomedical applications”, Surf. Coat. Technol., vol. 148, pp. 226-233, July 2001.

[4] F. Richard, F. Hickernell, F. Cho, “Optical waveguide including superstrate of niobium or silicon oxynitride and method of making same”, U.S. Patent 4,701,008, October 20, 1987.

[5] D. Bekermann, D. Barreca, A. Gasparotto, H. W. Becker, R. A. Fischer, A. Devi, “Investigation of niobium nitride and oxy-nitride films grown by MOCVD”, Surf. Coat. Technol., vol. 204, pp. 404–409, July 2009.

[6] T. Savisalo, D.B. Lewis, P.Eh. Hovsepian, “Microstructure and properties of novel wear and corrosion resistant CrON/NbON nano-scale multilayer coatings”, Surf. Coat. Technol., vol. 200, pp. 2731-2737, March 2005.

[7] R. Brayner, G. Djéga-Mariadassou, G.M. Da Cruz, J.A.J. Rodrigues, “Hydrazine decomposition over niobium oxynitride with macropores generation”, Catal. Today, vol. 57, iss. 3-4, pp. 225-229, March 2000.

[8] V.A. Matylitskaya, W. Bock, K. Thoma, B.O. Kolbesen, “Formation of niobium oxynitrides by rapid thermal processing (RTP)”, Appl. Surf. Sci., vol. 252, pp. 205–210, February 2005.

[9] H.S. Kim, C.H. Shin, G. Bugli, M. Bureau-Tardy, G. Djega-Mariadassou, “Catalytic activity of niobium oxynitride and carbide I. Preparation, characterization and thermal stability of finely divided niobium oxynitrides”, Appl. Catal., A, vo. 119, iss. 2, pp. 223-240, 1994.

[10] M. Fenker, H. Kappl, K. Petrikowski, R. Bretzler, “Pulsed power magnetron sputtering of a niobium target in reactive oxygen and/or nitrogen atmosphere”, Surf. Coat. Technol., vol. 200, pp. 1356-1360, October 2005.

[11] M. Fenker, H. Kappl, O. Banakh, N. Martin, J.F. Pierson, “Investigation of Niobium oxynitride thin films deposited by reactive magnetron sputtering”, Surf. Coat. Technol., vol. 201, pp. 4152-4157, September 2006.

[12] J.M. Chappé, P. Carvalho, A.V. Machado, F. Vaz, E. Alves, N.P. Barradas, M. Fenker, H. Kappl, “Structural study of the oxidation process and stability of NbOxNy coatings “, Nucl. Instrum. Methods Phys. Res., Sect. B, vol. 266, pp. 4927-4932, August 2008.

[13] J.M. Chappé, P. Carvalho, F. Vaz, M. Fenker, H. Kappl, N.M.G. Parreira, A. Cavaleiro, E. Alves, “Influence of air oxidation on the properties of decorative NbOxNy coatings prepared by reactive gas pulsing”, Surf. Coat. Technol., vol. 202, pp. 2363-2367, September 2007.

[14] M. Fenker, H. Kappl, C.S. Sandu, “Precise control of multilayered structures of Nb–O–N thin films by the use of reactive gas pulsing process in DC magnetron sputtering”, Surf. Coat. Technol., vol. 202, pp. 2358-2362, August 2007.

[15] J. Musil, P. Baroch, J. Vlcek, K.H. Nam, J.G. Han, “Reactive magnetron sputtering of thin films: present status and trends”, Thin Solid Films, vol. 475, pp. 208-218, September 2004.

[16] G.G. Stoney, "The tension of metallic films deposited by electrolysis", Proc. R. Soc. Lond. A, vol. 82, pp. 172-175, May 1909.

[17] Gmelins Handbuch der anorganischen Chemie, 8. Auflage, Niob – Teil B1, Weinheim: Verlag Chemie GmbH, 1970.

[18] CRC Handbook of Chemistry and Physics, 67th ed., Boca Raton, FL: CRC Press, 1986.

[19] H. Holleck, "Möglichkeiten und Grenzen einer gezielten Stoffauswahl für verschleißfeste Hartstoffschichten", Z. Werkst.Tech., vol. 17, pp. 334-341, September 1986.
[20] E. Vogelzang, J. Sjollema, H.J. Boer, J.T.M. de Hosson, "Optical absorption in TiNxOy-compounds", J. Appl. Phys., vol. 61, pp. 4606-4611, 1987.
[21] P. Carvalho, F. Vaz, L. Rebouta, L. Cunha, C.J. Tavares, C. Moura, E. Alves, A. Cavaleiro, Ph. Goudeau, E. Le Bourhis, J.P. Rivière, J.F. Pierson, O. Banakh, "Structural, electrical, optical, and mechanical characterizations of decorative ZrOxNy thin films", J. Appl. Phys., vol. 98, pp. 23715, July 2005.
[22] N.J. Ianno, H. Enshashy, O. Dillon, "Aluminum oxynitride coatings for oxidation resistance of epoxy films", Surf. Coat. Technol., vol. 155, pp. 130-135, January 2002.
[23] E.H. Nicollian, J.R. Brews, "MOS (metal oxide semiconductor) physics and technology", New York: Wiley, pp. 920, 1982.
[24] D. Briand, G. Mondin, S. Jenny, P.D. van der Wal, S. Jeanneret, N.F. de Rooij, O. Banakh, H. Keppner, "Metallo-organic low-pressure chemical vapor deposition of Ta2O5 using TaC12H30O5N as precursor for batch fabrication of microsystems", Thin Solid Films, vol. 493, pp. 6-12, June 2005.
[25] M. Fenker, M. Balzer, R.V. Büchi, H.A. Jehn, H. Kappl, J.-J. Lee, "Deposition of NbN thin films onto high-speed steel using reactive magnetron sputtering for corrosion protective applications", Surf. Coat. Technol., vol. 163– 164, pp. 169-175, 2003
[26] H. Szymanowski, O. Zabeida, J.E. Klemberg-Sapieha, L. Martinu, "Optical properties and microstructure of plasma deposited Ta_2O_5 and Nb_2O_5 films", J. Vac. Sci. Technol., A, vol. 23, iss.2, pp. 241-247, Mar/Apr 2005.
[27] P. Harlin, P. Carlsson, U. Bexell, M. Olsson, "Influence of surface roughness of PVD coatings on tribological performance in sliding contacts", Surf. Coat. Technol. vol. 201, pp. 4253- 4259, September 2006.
[28] A. Leyland, A. Matthews, "On the significance of the H/E ratio in wear control: A nanocomposite coating approach to optimized tribological behavior", Wear, vol. 246, pp. 1-11, 2000.
[29] T.Y. Tsui, G.M. Pharr, W.C. Oliver, C.S. Bhatia, R.L. White, S. Anders, A. Anders, I.G. Brown, "Nanoindentation and nanoscratching of hard carbon coatings for magnetic disks", Mater. Res. Soc. Symp. Proc., vol. 383, pp. 447-452, 1995.
[30] S. PalDey, S.C. Deevi, "Single layer and multilayer wear resistant coatings of (Ti,Al)N: a review", Mater. Sci. Eng. A, vol. 342, pp. 58-79, April 2002.
[31] M. Fenker, M. Balzer, H. Kappl, "Corrosion behavior of decorative and wear resistant coatings on steel deposited by reactive magnetron sputtering–Tests and improvements", Thin Solid Films, vol. 515, pp. 27-32, January 2006.
[32] M. Fenker, H. Kappl, P. Carvalho, F. Vaz, "Thermal stability, mechanical and corrosion behaviour of niobium-based coatings in the ternary system Nb-O-N", Thin Solid Films, vol. 519, pp. 2457–2463, December 2010.
[33] T. Wierzchoń, I. Ulbin-Pokorska, K. Sikorski, "Corrosion resistance of chromium nitride and oxynitride layers produced under glow discharge conditions", Surf. Coat. Technol., vol. 130, pp. 274-279, April 2000.
[34] G. Ramírez, S.E. Rodil, H. Arzate, S. Muhl, J.J. Olaya, "Niobium based coatings for dental implants", Appl. Surf. Sci., vol. 257, pp. 2555–2559, October 2010.

CHAPTER 8

Tuneable Properties of Aluminium Oxynitride Thin Films

Joel Borges[1,*], Nuno P. Barradas[2], Eduardo Alves[2], Nicolas Martin[3], Marie-France Beaufort[4], Sophie Camelio[4], Dominique Eyidi[4], Thierry Girardeau[4], Fabien Paumier[4], Jean-Paul Rivière[4], Filipe Vaz[1] and Luis Marques[1]

[1]Centro de Física, Universidade do Minho, 4710-057 Braga, Portugal; [2]Instituto Superior Técnico/ITN, Universidade Técnica de Lisboa, E.N. 10, 2686-953 Sacavém, Portugal; [3]Institut FEMTO-ST, 32, avenue de l'observatoire, 25044 Besançon, France and [4]Institut PRIME - UPR 3346 CNRS-Université de Poitiers-ENSMA, Département de Physique et Mécanique des Matériaux Bât. SP2MI - Téléport 2, BP 30179 F86962 Futuroscope Chasseneuil Cedex – France

Abstract: In this subchapter is discussed some characteristics and properties of AlO_xN_y thin films produced by reactive DC magnetron sputtering. The films were deposited using Ar as working gas and a reactive gas mixture of N_2+O_2 (17:3). The reactive gas flow was varied in order to produce a wide range of chemical compositions. Sub-stoichiometric AlO_xN_y films, with C_{O+N}/C_{Al} atomic ratios up to 0.85 were produced, with Al-type crystalline structure. Transmission electron microscopy (TEM) analysis and X-ray photoelectron spectroscopy (XPS) spectra suggests that the films are a percolating network, composed by aluminium nanocrystals with different shapes and sizes embedded in an oxide/nitride matrix. The particular composition, structure and morphology of the films results in very different electrical properties, which can be explained using a tunnel barrier conduction mechanism for the electric charge transport, as well as distinct optical responses, such as an unusual large broadband absorption for some films.

Keywords: Reactive DC magnetron sputtering, thin films, AlO_xN_y, target potential, deposition rate, composition, structure, grain size, morphology, percolating network, XPS, TEM, electrical and optical properties, reflectance, refractive index, extinction coefficient, electrical resistivity, temperature coefficient of resistance (TCR), broadband absorption.

INTRODUCTION

Aluminium oxynitride (AlON) is known to be a polycrystalline material with a cubic spinel structure, resulting from alumina (Al_2O_3) stabilised in a cubic

***Address correspondence to Joel Borges:** Centro de Física, Universidade do Minho, 4710-057 Braga, Portugal; E-mail: joelborges@fisica.uminho.pt

Filipe Vaz, Nicolas Martin and Martin Fenker (Eds)

structure with the incorporation of nitrogen, at very high temperature [1]. It is a ceramic material with high strength and hardness, revealing also excellent optical properties, offering some advantages compared to glass and sapphire [2, 3]. To produce this material, some special conditions are required, such as high temperatures and time to reach thermodynamic equilibrium. This structure, in principle, cannot be synthesised using DC magnetron sputtering at low temperatures, but obviously this is not the major concern of this work. Indeed the important task of this investigation is to establish limits for practical applicability of devices coated with AlO_xN_y, by correlating the discharge parameters with the chemical composition of the films and studying its influence in the electrical and optical responses.

The interest of studying the Al-O-N system arises from the importance of the three base materials: aluminium (Al), aluminium nitride (AlN) and aluminium oxide (Al_2O_3), which are used in different technological fields. The wide difference between the properties of these three materials opens the door to combine some of their advantages by varying the chemical concentration of aluminium, oxygen and nitrogen in the film, and study the physical and chemical properties of the resulting ternary aluminium oxynitride system (AlO_xN_y), according to the desired application [4].

Aluminium is the third most abundant element on the Earth crust [5], being a metallic material used in many areas of engineering and industry. It is one of the most important electrical conductor in the domain of microelectronic applications [6], being used as electrode in many types of devices such as in diodes [7, 8], metal-insulator-metal (MIM) capacitors [9], and, in the last years, it is being studied as a potential candidate as buffer layer for very/ultra large scale integration (VLSI/ULSI) technology [10, 11]. It is also used as optical coating due to its optical properties [12], such as in surface plasmon-coupled emission (SPCE) devices [13], and it is a good candidate to be used as nonresonant plasmonic nanoparticle in thin-film silicon solar cells [14, 15].

Aluminium nitride is an excellent thermal conductor (thermal conductivity as high as 320 $W.m^{-1}.K^{-1}$) [16, 17], with high stability and resistance to caustic chemical etching [18], possessing a low thermal expansion coefficient (10^{-6} K^{-1}) [19, 20], being an important ceramic material used in applications such as substrate in microelectronic devices. Normally it is mentioned as being a semiconductor with a large band gap (~ 6.2 eV) [16], in its more stable and common hexagonal (wurtzite) crystalline structure (lattice constants: a= 0.311 nm and c= 0.498 nm) [21, 22]. It also exhibits high electrical resistivity (~10^{12} Ω.m) [20] and high hardness (~25GPa) [23]. Besides this structure, AlN has also two cubic structures (with different lattice parameters, a = 0.412 nm and a = 0.791 nm) [22], where the thermal conductivity and electrical resistivity are even higher than that of the hexagonal one. Polycrystalline aluminium nitride has also a high dielectric strength (4 - 5.5 MV.cm^{-1}), which can be improved if the produced material is amorphous [16]; it reveals a moderated dielectric constant (~9 at 1 MHz) [20]; and it can be used as gate dielectric in high voltage and high power electronic devices. It is also an excellent piezoelectric material [24], with a high surface acoustic wave velocity (above 600 m.s^{-1}) [21] and it was considered as a good alternative to zinc oxide (ZnO) in the last decade [25, 26]. All these physical/chemical properties allow the use of AlN material in the fabrication of optical sensors in the ultraviolet-visible region; light emitting diodes (LEDs) with one of the shortest emission wavelength reported (210 nm) [27]; high power and high temperature electronic devices [28] and surface acoustic wave (SAW) devices [28]. It is also widely used in resonators and band-pass filters in communication systems, high-frequency (GHz) bulk acoustic wave (BAW) devices [29], film bulk acoustic resonator (FBAR) [30], electronic packages, among several other examples.

Aluminium oxide, or simply alumina, (Al_2O_3) is one of the best corrosion protector materials, even at high temperatures [31], providing a protective barrier against both chemical corrosion and mechanical wear at temperatures up to 1250 K [32]. It is an electrical insulator, which is commonly prepared in the form of a polymorphous material, since it can exist in many metastable structures that are divided in two broad categories, according to the arrangement of aluminium cations. The first is a face-centered-cubic (fcc) packing of the oxygen anions, while the second is based on a hexagonal close packed (hcp) packing of oxygen. The first case includes γ- Al_2O_3 and η- Al_2O_3 (cubic arrangement), θ- Al_2O_3 (monoclinic), and δ- Al_2O_3 (either tetragonal or orthorhombic); while in the second case we have α- Al_2O_3 (trigonal), κ- Al_2O_3 (orthorhombic), and χ- Al_2O_3 (hexagonal) phases [33]. The dominant and stable phase of alumina, α- Al_2O_3, possesses trigonal symmetry with a rhombohedral Bravais lattice. An important fact about α- Al_2O_3 (or corundum phase of alumina) deposited by Physical Vapour Deposition (PVD) techniques is that it can only be produced at temperatures above ~1000 K [31]. Alumina also reveals a high electrical breakdown field (5 – 30 $MV.cm^{-1}$), large band gap (> 9 eV), high permittivity (8.6 – 10) [34], and possesses a low refractive index [35]. Its dielectric properties allows the use of this coating in gates instead of SiO_2, such as in flash memory circuits [36, 37], in organic thin films transistors (OTFTs) [38], in metal–oxide–semiconductor field-effect transistor (MOSFET) [39], among others [40]. It is also a protective film for metal reflectors, and it is used in metal-oxide-semiconductor (MOS) devices [41]. The sub-stoichiometric aluminium oxide is also important in solar selective coatings because it exhibits very high solar selectivity [42].

Tailoring the properties of AlO_xN_y thin films from those of Al, Al_2O_3, and AlN to achieve a material with different physical responses is a challenge with an enormous technical interest. Using DC reactive magnetron sputtering, with different discharge parameters, the control of the non-metallic (oxygen, nitrogen)/metallic (aluminium) ratio in the film is possible. This fact can create a new microstructure with a large gradient in different properties [43-45], according to the industrial application required.

The use of AlO_xN_y films is not yet very common, despite some very few examples that are known in the field of protective coatings against wear, diffusion and corrosion and optical coatings, taking the advantage of the mentioned spinel structure properties [46]. Amorphous aluminium oxynitride can also be used as dielectric in multilayer capacitors with high energy density and wide temperature properties [1] and in cermet solar coatings since it exhibits a very high solar absorbance [47].

MATERIALS AND METHODS

The coatings were deposited by reactive DC magnetron sputtering, in a laboratory-sized deposition system. The films were prepared with the substrate holder positioned at 70 mm from the target in a rotation mode-type (9 r.p.m.). A DC current density of 75 $A.m^{-2}$ was used on the aluminium target (99.6% purity).

The substrates (glass and single crystalline silicon wafers with <100> orientation) were kept at a constant temperature of approximately 373 K before discharge ignition (using a Joule effect resistor). Before the deposition, they were subjected to an etching process, using pure argon with a partial pressure of 0.3 Pa (70 sccm), a pulsed current of 0.6 A (T_{on}= 1536 ns and f= 200 kHz) for 900 s. The substrates were grounded during etching and deposition.

The aluminium target, with dimensions 200x100x6 mm^3, was sputtered using a gas atmosphere composed of argon (working gas), with a partial pressure fixed at 0.3 Pa (70 sccm), and a reactive gas mixture composed of nitrogen and oxygen with a constant N_2:O_2 ratio of 17:3. Before each deposition, a target cleaning process was carried out in pure argon with a partial pressure of 0.3 Pa until the target voltage remained constant.

The discharge parameters: target potential and current, gas pressure, argon flow and reactive gas flow, were monitored using a Data Acquisition/Switch Unit Agilent 34970A, with a multifunction module (334907A). This unit uses a RS-232 interface and the data is acquired with Benchlink Data Logger III software. The power supply, the pressure sensors and the flow controllers have analog outputs, which allowed the connections to the acquisition system. The partial pressure of reactive gas was also measured prior to discharge ignition. These values, which are directly proportional to gas flow, will be used is results discussion.

The chemical composition of the films was investigated by Rutherford Backscattering Spectrometry (RBS) with scattering angles of 140º (standard detector) and 180º (annular detector) and incidence angles of 0º and 20º. Measurements were made at 2 MeV with He^+ and, for some samples, at 1.25 MeV with protons. The data were analyzed with the NDF code [48].

The structure and the phase distribution of the coatings were analyzed by X-ray diffraction (XRD), using a Philips PW 1710 diffractometer (Cu-Kα radiation) operating in a Bragg–Brentano configuration and also a PANalytical X'Pert PRO – MPD. The XRD patterns were deconvoluted, assuming to be Pearson VII functions to yield the peak position, peak intensity and integral breadth, using Winfit software [49]. From these parameters the interplanar distance, preferential orientation and grain size were calculated.

Scanning electron microscopy (SEM) observations were carried out in order to determine the thickness of the films and to take images from the cross section of the coatings, using a Leica Cambridge S360 apparatus. Atomic Force Microscopy (AFM) was carried out with a Nanoscope III, in tapping mode. The AFM images were processed by WSxM software [50]. Transmission Electron Microscopy (TEM) observations were performed on a JEOL 2200-FS operating at 200 kV and equipped with a field-emission gun and an energy-filter (omega). Cross-sectional specimens were investigated in bright-field, dark-field, high-resolution and selected area diffraction modes. The electrical resistivity at room temperature of the conducting films was measured using the four-point probe method (in linear geometry) [51], and the Van der Pauw geometry was used for the measurements in temperature. For the high resistivity films, aluminium contacts (1x6 mm^2) were vapour deposited on the top of the coating and the electrical resistivity of the films was obtained from the I-V characteristics of the metal-insulator-semiconductor (MIS) capacitor structure.

The colour coordinates of the films were represented in the CIELab 1976 colour space and computed using a Minolta CM-2600d portable spectrophotometer with a wavelength range of 300 – 700 nm under primary illuminant D65 (specular component included).

The total reflectance and transmittance were measured between 290 and 2500 nm using a Shimadzu UV-3101 PC UV-Vis-NIR with an attached integrating sphere of 60 mm (inner diameter). To eliminate experimental artefacts caused by the integrating sphere, two standards were used: a STAN-SSH High-Reflectivity Specular Reflectance Standard and a WS-1-SL White Reflectance Standard with Spectralon. Also an UV-Vis-NIR spectrophotometer Cary5000 from Agilent Technologies was used in a configuration (VW) that allows to measure simultaneously the beam reference and the sample absolute specular reflectance, between 250 and 3000 nm.

XPS experiments were carried out using an Isa-Riber setup equipped with a Mac 2 detector of 0.5 eV resolution. The pressure in the analytical chamber was kept below 10^{-8} Pa. For XPS spectra, a Mg anode was used as operating source (hv = 1253.6 nm). For each sample, a survey scan was performed, followed by a core level spectrum of the Al-2p, O-1s and N-1s region. The binding energies were charged shifted with respect to the C1s peak at 284.6 eV.

TUNEABLE PROPERTIES OF ALUMINIUM OXYNITRIDE

Target Potential, Deposition Rate and Morphology of the Films

The evolution of the target potential, -V, and deposition rate of the different coatings as a function of the partial pressure of the gas mixture, $p(N_2+O_2)$, is plotted in Fig. **1**. As it can be observed, the equilibrium target potential varies almost linearly from a maximum (absolute) value of 452 V, for an atmosphere without reactive gas, to a value of about 320 V when the reactive gas partial pressure is approximately 5.4×10^{-2} Pa, corresponding to a significant variation of ~30%. Then a sudden jump can be observed in the plot corresponding to a decrease towards a value of about 267 V, which corresponds to the minimum value that was measured. This occurs for a pressure of 5.6×10^{-2} Pa and, for pressures above this value, no significant changes in the target potential can be observed. According to this brief analysis of the data, there are clearly two different regimes of the target. In the first one, denoted in Fig. **1** as regime I, the values of the cathode voltage gradually decrease as the pressure of reactive gas increases. In this case it is expected that the films reveal a wide range of different chemical compositions, which in turn may induce a gradient in the electrical and optical responses. The second regime, denoted as regime II, corresponds to pressures of 5.6×10^{-2} Pa and above, where the target potential is approximately constant, and thus one can expect to obtain films with similar characteristics.

Regarding the deposition rate, also plotted in Fig. **1**, one can observe approximately constant values close to 35 nm.min^{-1} within the range of up to 2.0×10^{-2} Pa. A significant increase of ~31%, to a value of about 46 nm.min^{-1}, is observed when the partial pressure of reactive gas is 2.2×10^{-2} Pa. Increasing further the reactive gas pressure, the deposition rate continues to increase reaching a maximum value of 63 nm.min^{-1} for a pressure of 3.0×10^{-2} Pa. Above this pressure, the deposition rate decreases steeply until a value of about 24 nm.min^{-1} is reached for a pressure of 5.4×10^{-2} Pa. Furthermore, the drop of the target potential to the second regime implies constant deposition rates with values around 5 nm.min^{-1}. This group of samples, with p(N_2+O_2) ≥ 5.6×10^{-2} Pa, was indexed as zone II, Fig. **1**. Concerning the regime I of the target, one can ascribe some films to zone Ia, corresponding to approximately constant deposition rates. One can also report the existence of a transition zone where the deposition rate evolution is quite peculiar, aggregating a group of films ascribed to zone Ib, with increasing deposition rates, followed by a systematic decrease of those values for other group of samples, ascribed to zone Ic, as showed in Fig. **1**.

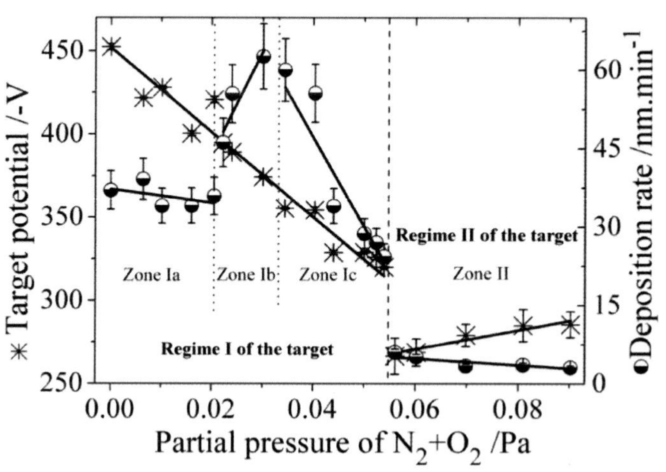

Figure 1: Evolution of the target potential and deposition rate as a function of the partial pressure of reactive gas (N_2+O_2). The error bars of the target potential were estimated by the mean absolute deviation of the target potential values and the error bars of the deposition rate from the maximum deviation to the average value of thickness from SEM observations.

In order to clarify the differences in target potential and deposition rate for different pressures of reactive gas, one must take into account that reactive magnetron sputtering is a complex process, depending on many parameters that are commonly strongly correlated [4, 43, 52]. The chemical species coming from the reactive gas used in the deposition of the films (a mixture of N_2 and O_2) do not only bond with the sputtered aluminium during compound material formation but also interact with the cathode surface [53], leading both to the formation of an oxide/nitride layer by chemisorption at its surface [54] and to ion implantation [55, 56]. These processes modify the sputtering yield and the ion induced secondary electron emission (γ_{ISEE}) coefficient of the target [54], which, in turn, is closely related to the minimum sustaining discharge voltage, according to the Thornton equation [53]. Moreover, for magnetron sputtering the cathode voltage essentially depends on this γ_{ISEE} coefficient, which is 0.091 for aluminium, according to Depla *et al.* [57], representing the number of electrons emitted per incoming ion on the target.

Concerning the reactive magnetron sputtering process, it is commonly accepted that there are three general working regimes of the sputtered target, as the flow of reactive gas is increased: (i) a metallic mode where the target remains almost in a clean state, since the available reactive gas is consumed by the getter effect [56] at the condensation sites [58]; (ii) a transition regime where the target gradually becomes "poisoned" since the flow of reactive gas is becoming more important than the gettering rate of the sputtered material; and (iii) a compound mode where the target can be completely covered with a compound layer [4, 58]. In the case of the target used to prepare the films in the framework of this study, these three regimes are not well defined, as can be observed in Fig. **1**. In fact, only two different tendencies concerning the aluminium cathode voltage can be observed, and there are two major reasons that can explain this behaviour. First, the rise of the gas pressure in the chamber leads to a higher ionization probability which should decrease the minimum target potential to sustain the discharge, since the two parameters are inversely proportional, according to the Thornton equation [53]. On the other hand, the γ_{ISEE} coefficient of the target material is changing as the pressure of the reactive gas in the chamber rises.

Indeed, the degree of oxidization/nitridation of a metallic target strongly influences its ion induced secondary electron emission coefficient, which is different for each target material and chosen reactive gas [57].

In the case of an aluminium target, the increment of reactive gas partial pressure (oxygen and nitrogen) favours the formation of the compound layer on the target surface, and a reduction of the target potential can be reported, Fig. **1**, in agreement with other works related with magnetron sputtering of aluminium target using oxygen or nitrogen as reactive gases [53, 59, 60]. This behaviour is mainly due to the

Figure 2: SEM images of representative films of each zone. Some images were reprinted from [4] with permission from Elsevier.

increase of the γ_{ISEE} coefficient, which is inversely proportional to the target voltage [61], as the target becomes poisoned [53, 54]. When the pressure of reactive gas is higher (values of 5.6×10^{-2} Pa and above) the sputtered target should be completely "poisoned" (compound mode) and this would be the case of the films prepared for this study ascribed to regime II of the target, or zone II.

The amount of material that is deposited in the substrate per unit of time, the so-called deposition rate, is correlated with the amount of atoms sputtered from the target. Knowing that the sputtering yield of a poisoned target is lower than that from a metallic one [58], it is expectable that the deposition rate decreases as the target is becoming "poisoned". Considering the range of pressures within the zone Ia, it can be observed that the values of the deposition rate are not changing significantly. Within this range of pressures, the slight decrease of the sputtering yield of the target might be compensated by the increase in the incorporation of reactive gas atoms in the films, thus explaining the approximately constant deposition rates observed. Still in regime I of the target, a transition zone was identified where the deposition rate evolution may look unusual. Indeed, a sudden increase is reported and the values obtained are actually higher than those obtained for the films of the previous zone (zone Ia). This fact can be understood by the analysis of the scanning electron microscopy (SEM) images shown in Fig. **2**(a-h).

It is clear that the type of growth of the films prepared within the transition zone (which encloses zones Ib and Ic) is significantly different from those indexed to zone Ia. Whereas the films deposited with pressure up to 2.0×10^{-2} Pa, have a typical columnar growth [62, 63], the films deposited with high reactive gas pressure up to 5.4×10^{-2} Pa reveal a cauliflower growth and appear very porous. Therefore, the high deposition rates within zone Ib do not imply that a higher quantity of material was deposited, but films with lower density were produced. Further increasing the partial pressure, within the transition zone, namely zone Ic, the deposition rate is reduced as a direct consequence of the reduction of the target sputtering yield, although the cauliflower growth is maintained.

Finally, in regime II the aluminium target is expected to be completely poisoned and thus the sputtering yield is very low, explaining the roughly constant and low deposition rates of the films ascribed to this zone.

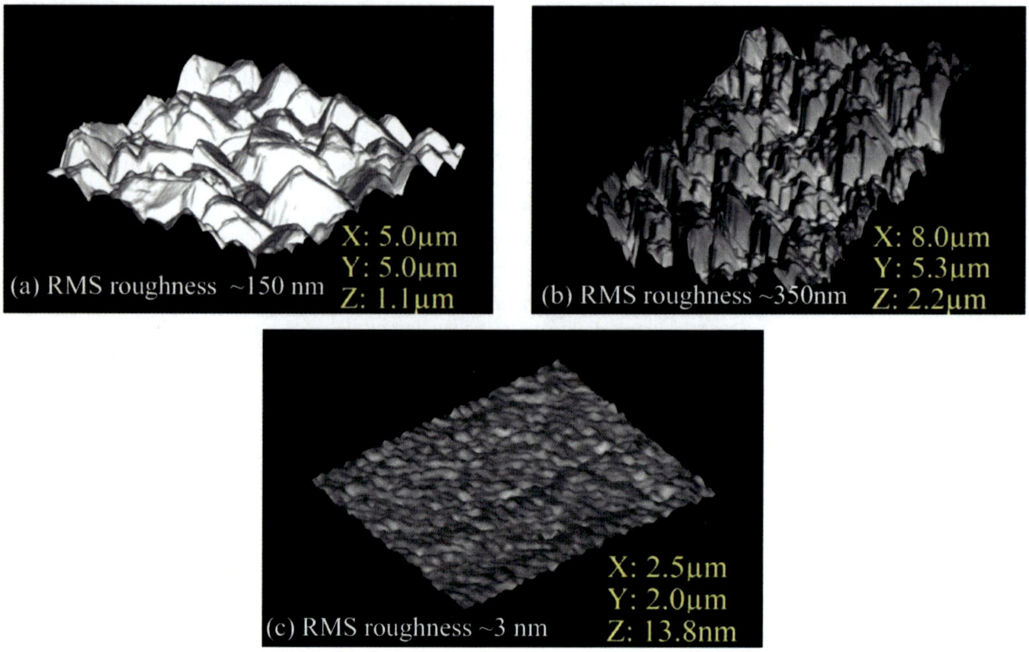

Figure 3: Atomic force microscopy images for (a) zone Ia; (b) transition zone and (c) zone II.

Fig. **2** shows that the films indexed to zone Ia, as already stated, have a typical columnar growth as can be observed in Fig. **2**(a-b). Regarding the SEM image corresponding to the film deposited with a pressure of 2.2×10^{-2} Pa, Fig. **2**(c), one can claim that it is somewhat different, compared to the one deposited with a pressure of 2.0×10^{-2} Pa, Fig. **2**(b), but it is also true that the transition between the two zones is very smooth. Thereby, the type of growth of the film represented in Fig. **2**(c), where the columns are not so well defined as those from previous zone Ia, can be considered a transition from the columnar growth towards cauliflower. For pressures within the range between 2.4×10^{-2} Pa and 5.4×10^{-2} Pa, the films revealed a cauliflower growth and the images displayed in Fig. **2**(d-g) are representative. Another interesting feature of the transition zone is the size of the microscopic round shape grains, which is gradually increasing until better defined columns are again starting to form at the end of the zone Ic, Fig. **2**(g). The SEM image displayed in Fig. **2**(h) reveal a dense, featureless-type of growth consistent with films deposited with the target in a compound mode (regime II).

Regarding the surface morphology of the films, atomic force microscopy (AFM) was performed on three samples, one of each type of growth. The images, displayed in Fig. **3**(a-c), reinforce the existence of at least three different behaviours. The columnar growth of a representative film indexed to zone Ia induces a relatively high RMS roughness, ~150 nm, Fig. **3**(a). The films with a cauliflower growth are even rougher, with RMS roughness of ~350 nm, as it can be seen in Fig. **3**(b), which, taking into account the SEM images (Fig. **2**) is the expected behaviour. Finally, the film deposited with the target in regime II reveals a smooth surface, with a corresponding low value of RMS roughness of about 3 nm, Fig. **3**(c).

Composition

Fig. **4**(a-b) shows the chemical composition results of the deposited coatings obtained from RBS spectra analysis. While in Fig. **4**(a) the variation of the atomic concentration (at.%) of the different chemical elements is represented, in Fig. **4**(b) the concentration ratios of non-metallic over metallic elements of the films is plotted, in both cases as a function of the partial pressure of reactive gas mixture.

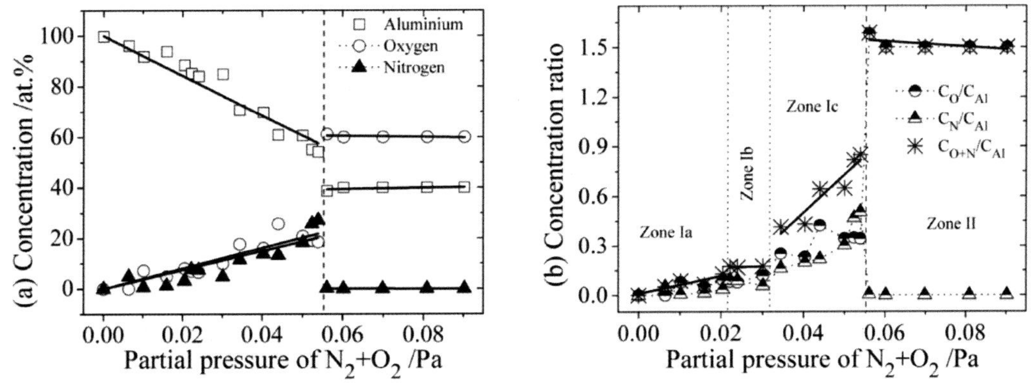

Figure 4: Evolution of the (a) chemical composition (at.%) and (b) concentration ratio, as a function of the partial pressure of reactive gas (N_2+O_2). The chemical composition of all samples was determined within an error of about 5 at.%.

According to the results shown in Fig. **4**(a), it is reasonable to argue that there are two different tendencies depending on the atomic fraction of each chemical element. The first one corresponds to the films prepared with partial pressures of reactive gas up to 5.4×10^{-2} Pa, and the second one for pressures of 5.6×10^{-2} Pa and above. It can be observed that the atomic concentration of aluminium is gradually decreasing with the increase of the reactive gas pressure, from 100 at.%

(roughly pure aluminium coating) to a value close to 54 at.% for a pressure of 5.4×10^{-2} Pa. Concerning the non-metallic elements, it is clear from Fig. **4**(a) that the content of oxygen and nitrogen is quite similar for pressures up to 5.4×10^{-2} Pa, where a gradual increase of these two chemical elements occurs. While the atomic concentration of oxygen varies from 0 at.% to approximately 19 at.%, the nitrogen content reaches a maximum value of barely 27 at.%. Although the reactive gas is composed of a mixture of N_2+O_2 with a fixed 17:3 ratio, the incorporation of oxygen in the film that is being deposited is similar to that of nitrogen due to the well-known higher reactivity of oxygen compared to nitrogen [64-66].

An interesting feature about the chemical composition results is that the samples that were prepared with the highest partial pressures, $p(N_2+O_2) \geq 5.6\times10^{-2}$ Pa, revealed very similar compositions, where the oxygen amount is very close to 60 at.%, and the aluminium concentration is around 40 at.%, while the nitrogen concentration drops to a residual value (non-detectable within the resolution of the experimental setup, meaning that it should be below 5 at.%). This is the result of the higher affinity of oxygen to bond aluminium, as demonstrated by the values of the Gibbs free energy of formation of aluminium oxide (alumina) at room temperature, ΔG^0 (Al_2O_3,) = -1.58 $MJ.mol^{-1}$ in comparison with that of aluminium nitride, ΔG^0 (AlN) = -0.29 $MJ.mol^{-1}$ [67]. Of course this comparison is just qualitative, since the film growth is a non-equilibrium kinetic process. This set of results induces a roughly stoichiometric composition of the films in the form of Al_2O_3 (alumina), with a concentration ratio (C_O/C_{Al}) of about 1.5 as displayed in Fig. **4**(b). In accordance with this concentration ratio, the films have interference colours consistent with their semi-transparency.

Although the identification of the previously defined four zones (displayed in Fig. **1**) is somewhat difficult only from the observation of the atomic concentration values, Fig. **4**(a), the values of the concentration ratio of non-metallic over aluminium elements display clearer tendencies. In fact, the non-metallic over aluminium atomic ratio slightly increases within the films indexed to zone Ia with concentration ratios (C_{O+N}/C_{Al}) rising up to 0.13. In the case of the films ascribed to zone Ib, where the deposition rate is rapidly increasing, one can observe very similar concentration ratios (C_{O+N}/C_{Al}), around 0.17-0.18. This is evidence that the density of the films must be decreasing while the non-metallic/metallic atomic ratios are approximately constant and thus the fraction of voids is becoming important within the correspondent range of partial pressures, as suggested by the type of growth, Fig. **2**(c-d). Entering in zone Ic, where the deposition rate is gradually decreasing, one can observe a stronger increase of the concentration ratio of non-metallic elements over aluminium (C_{O+N}/C_{Al}) ranging from a value of 0.41 towards a value of 0.85. In fact, the difference in chemical ratio from zone Ib towards the beginning of zone Ic is very high (~128%), considering the fact that the pressure of reactive gas has a very low increase (~14%). All these results confirm that it is possible to divide the deposited films into four distinct compositional regions, in accordance with zones identified in Fig. **1** and the films characteristics shown in Fig. **2**.

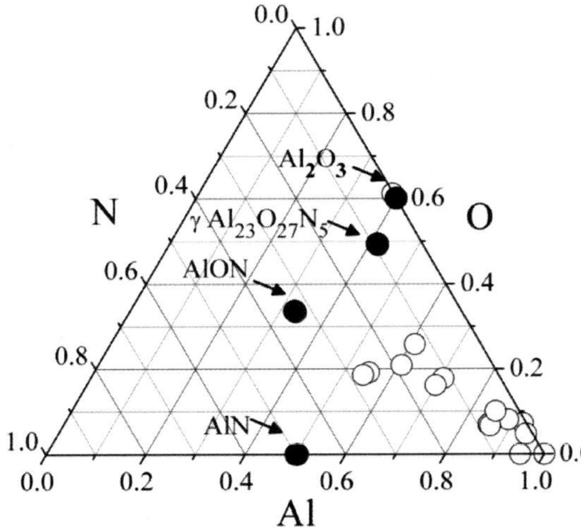

Figure 5: Ternary diagram of the deposited AlO_xN_y films. Open circles represent the samples produced, and closed circles represent some well known stoichiometric compounds.

Displayed in Fig. **5** is a ternary diagram where the location of the AlO_xN_y samples deposited is represented, together with the well known compounds AlN and Al_2O_3, the stoichiometric aluminium oxynitride compound, AlON, and the nominal composition of spinel aluminium oxynitride, γ-AlON [46]. An interesting feature that can be observed from the analysis of this diagram is that the samples have stoichiometries somewhere between a pure aluminium nitride and a pure aluminium oxide compound and it seems that the composition tends to a stoichiometric AlON compound just before the transition to the close-stoichiometric Al_2O_3 films.

Crystalline Structure

One of the most important properties of any given thin films system is its crystalline structure, which depends significantly on its particular composition. In order to study the phase composition of the thin films obtained, structural characterization was carried out. Fig. **6**(a-d) shows X-ray diffraction (XRD) diagrams of the AlO_xN_y films prepared within the scope of the present work.

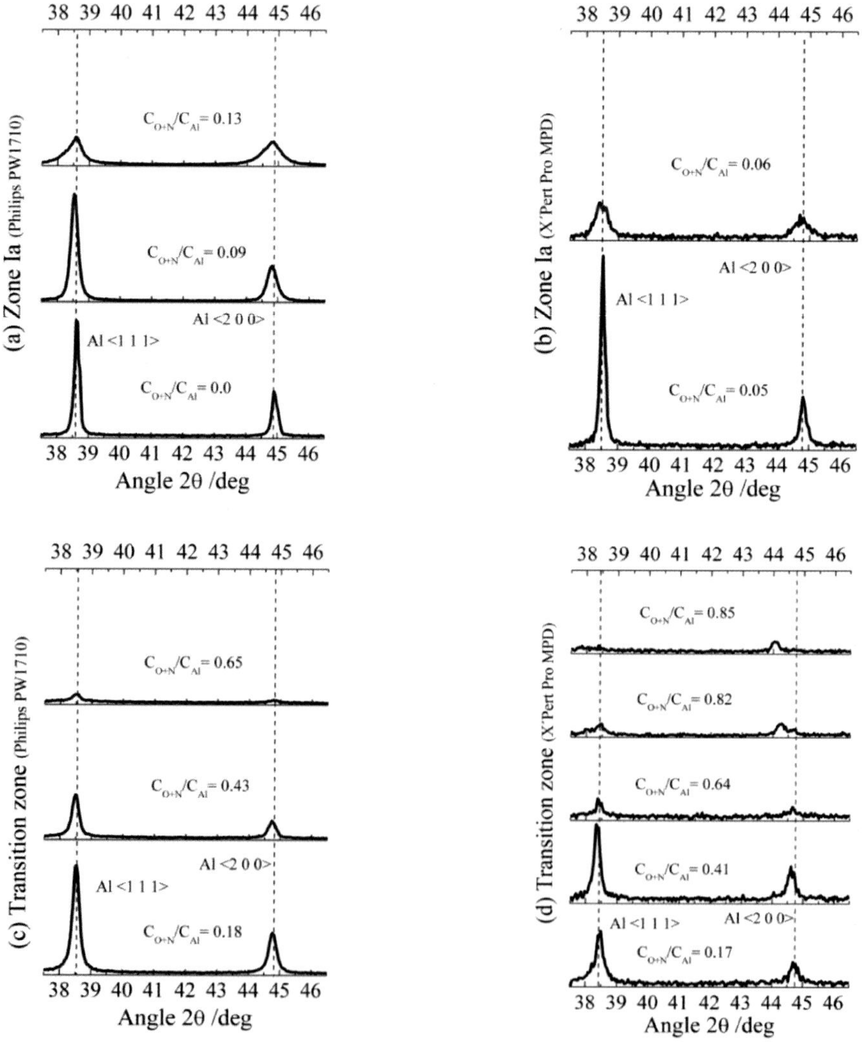

Figure 6: X-ray diffraction patterns for (a-b) zone Ia and (c-d) transition zone films, using two different equipments – a Philips PW1710 Reflection Diffractometer and a X'Pert Pro MPD – Multi Purpose X-ray Diffractometer (PANalytical). The films indexed to zone II (Al_2O_3) are amorphous.

The results obtained revealed crystalline thin films deposited with the aluminium target in regime I, and amorphous ones in the case of the films deposited with the target in the regime II. The film deposited with the reactive gas turned off ("pure" aluminium thin film) exhibits, as expected, the characteristic face-centered cubic (fcc) structure of (bulk) aluminium. The two diffraction peaks represented in the XRD patterns, Fig. **6**, correspond to the <111> and <200> planes of such structure, with a clear <111> preferential growth. The bulk aluminium diffraction angles are, respectively, $2\theta = 38.50^{o}$ and $2\theta = 44.76^{o}$ (ICSD collection code: 52255).

A second important feature is that in spite of the differences in the chemical composition, the structure of the films is initially maintained; becoming completely amorphous only when near-stoichiometric Al_2O_3 films are formed. It is also important to note the evolution of the crystalline samples. In fact, these coatings are gradually becoming amorphous in the transition zone as the non-metallic content increases. As shown in Fig. **6**(c-d), the samples with concentration ratios between 0.64 and 0.85 are quasi-amorphous, and a high disorder is expected to occur in these films. Furthermore, with the increase of the non-metallic over aluminium atomic ratio, there is no significant shift of the diffraction peaks, meaning that probably there is no significant incorporation of non-metallic elements in the Al-type crystalline structure as the content of oxygen and nitrogen in the film increases.

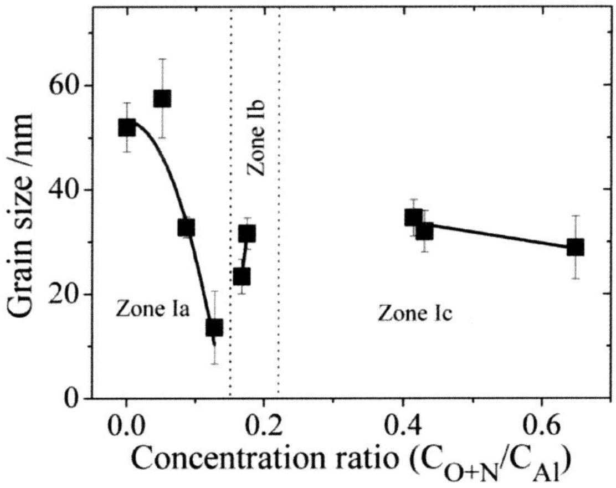

Figure 7: Grain size of the face-centered cubic (Al-type) crystals as a function of the non-metallic over aluminium atomic ratio. The error bars were estimated from the reliability of the fit (Winfit software).

The values of the lattice parameter and grain size (estimated by XRD peak fitting with a Pearson VII function, using the integral breadth) were also evaluated. The deposited aluminium coating reveals a lattice parameter of a= 0.403 nm, very close to the reference value of this material found in the available literature [68] (a_{Al}= 0.40496 nm). The grain sizes of the aluminium crystals are displayed in Fig. **7**, where a sharp decrease as a function of the concentration ratio, C_{O+N}/C_{Al}, can be observed in the zone Ia. In fact, the estimated average size of the grains varies from 52 nm ($C_{O+N}/C_{Al} = 0.0$) to 14 nm ($C_{O+N}/C_{Al} = 0.13$), within the zone of films with columnar growth. This decrease could be related with the incorporation of oxygen in the films that is segregated to the surface and grain boundaries, which inhibits the grain coarsening during film growth due to the reduced mobility of aluminium atoms on oxide layers [63]. When the type of growth evolves from columnar to cauliflower, the grain size suffers an inflection and increases again, in zone Ib, towards a value of about 32 nm. This behaviour is consistent with the formation of round grains whose growth is completely blocked by a surrounding oxide layer in an early stage of grain formation [63]. Still in the transition zone, namely zone Ic, the grain size of the crystals is approximately the same, around 30 nm for ratios between 0.41 and 0.65, contrarily to the case of the films with ratios of 0.82 and 0.85 where any measurable crystals could be detected using this technique, in agreement with their quasi-amorphous structure.

The previous analysis of XRD patterns, as well as the grain size estimation, strongly suggests the presence of pure aluminium nano-crystals in the sub-stoichiometric AlO_xN_y films.

Finally, in the case of the Al_2O_3 films, as expected, no diffraction peaks were detected since the deposition temperature is too low to allow the crystallization of alumina [63]. These results follow the different tendencies observed in morphology and deposition rate and are also consistent with the changes observed in chemical composition. The existence of a transition zone between an Al-rich group of films towards Al_2O_3 films is again confirmed by X-ray diffraction.

Electrical Properties

The electrical resistivity (ρ) of the AlO_xN_y films was measured at room temperature using the four point probe method (in line) for the high conductive samples and the I-V curve method (in bulk) in the case of Al_2O_3 films.

The polycrystalline aluminium film has a measured electrical resistivity of about 3×10^{-8} Ω.m, which is very close to the literature value for bulk aluminium: $\rho_{Al,bulk} = 2.78\times10^{-8}$ Ω.m [69]. For low values of the concentration ratio ($C_{O+N}/C_{Al} \leq 0.13$) the electrical resistivity of the films increases about one order of magnitude towards 2×10^{-7} Ω.m, as can be seen in the Fig. **8**. This fact allows indexing these nearly metallic films to the zone Ia where the increase of non-metallic elements in the films gradually removes the electrons of the aluminium conduction band, and thus an increase of the electrical resistivity is expected. Concerning the zone Ib, the content of non-metallic elements is more important than in the previous zone, and an increase of the electrical resistivity was found. Furthermore, one can also observe a substantial increase of the electrical resistivity within this zone where the chemical composition is almost constant and thus the hypothesis of an increase of the void fraction within this zone is again reinforced. Still in the transition zone, namely zone Ic, a resistivity increase of about one order of magnitude from about 1×10^{-5} Ω.m to 1×10^{-4} Ω.m was measured. This behaviour is again related with the reduction of the amount of free electrons in the lattice, due to an increase of the non-metallic concentration in the films. Finally, the electrical resistivity of zone II films is typical of an insulator material. The Al_2O_3 films formed exhibits values of electrical resistivity between 10^{10} and 10^{11} Ω.m, in agreement with the values found in the literature for alumina coatings [36, 70].

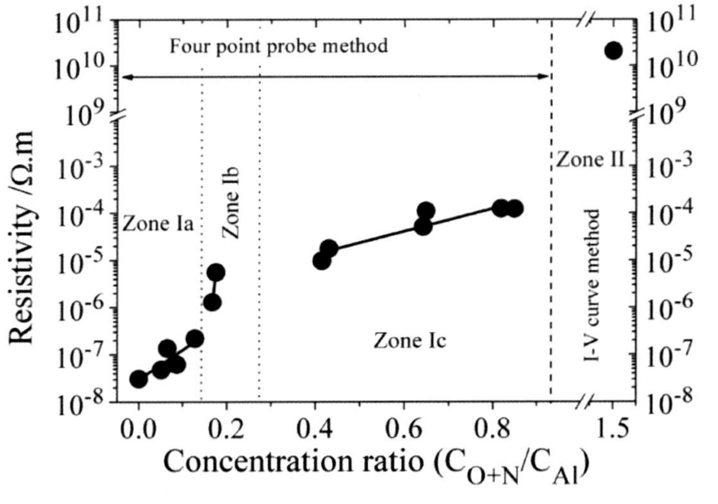

Figure 8: Electrical resistivity, at room temperature, of the AlO_xN_y films as a function of the concentration ratio.

In order to better understand the conduction mechanisms of the low resistivity films, the results obtained are insufficient [43]. Therefore, the electrical resistivity was measured as a function of the temperature, using the four point probe technique in a Van der Pauw geometry [71, 72].

The temperature coefficient of resistance (TCR_{300K}) values, calculated from the measurements of the electrical resistivity as a function of the temperature (between ~300 K and ~500 K), are displayed in Fig. **9**. The aluminium coating has a TCR_{300K} value of 4×10^{-3} K^{-1}, which is very close to the TCR of bulk aluminium found in literature (4.4×10^{-3} K^{-1}) [73]. A zone where the TCR sharply decreases as a function of the concentration ratio to a value around 1×10^{-3} K^{-1} is clearly observed in Fig. **9**. These films are indexed to the zone Ia. For ratios between 0.17 and 0.85 the TCR values undergo a slower decrease and become negative for higher ratios, until a minimum of -3.6×10^{-4} K^{-1} is reached for C_{O+N}/C_{Al} = 0.85. This means that the resistivity of some samples, ascribed to the transition zone, actually decreases as the sample is heated, namely for concentration ratios between 0.64 and 0.85.

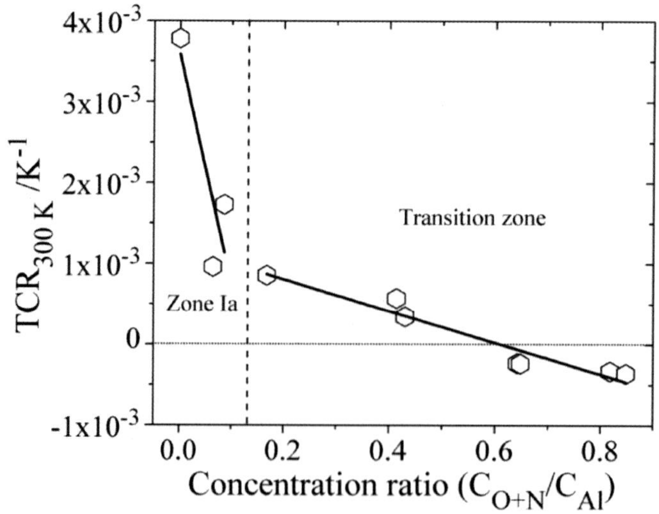

Figure 9: Temperature coefficient of resistance ($TCR_{300\ K}$) of the AlO_xN_y films as a function of the concentration ratio.

The variation of the electrical properties with non-metallic over metallic atomic ratios can be explained assuming that the electrical transport in the films takes place through a series of channels formed by aluminium grains (nanoparticles), that can be in contact with each other or separated by insulator/semiconductor aluminium oxide/nitride layers or even by empty space (voids), as suggested by the SEM images, Fig. **2**, and by the structural analysis. When the Al nanoparticles are in contact, the current conduction is governed by the constrictions between them, however other grains can be separated by the semiconductor/insulating barrier, and thus the electrical transport is controlled by tunnelling processes, such as thermally activated tunnelling and/or tunnelling through localized states in the insulating layer [74, 75]. In fact, transmission electron microscopy (TEM) analysis strongly supports the hypothesis of the growth of Al nanoparticles randomly embedded in an oxide/nitride or oxynitride matrix, as can be observed in Fig. **10**.

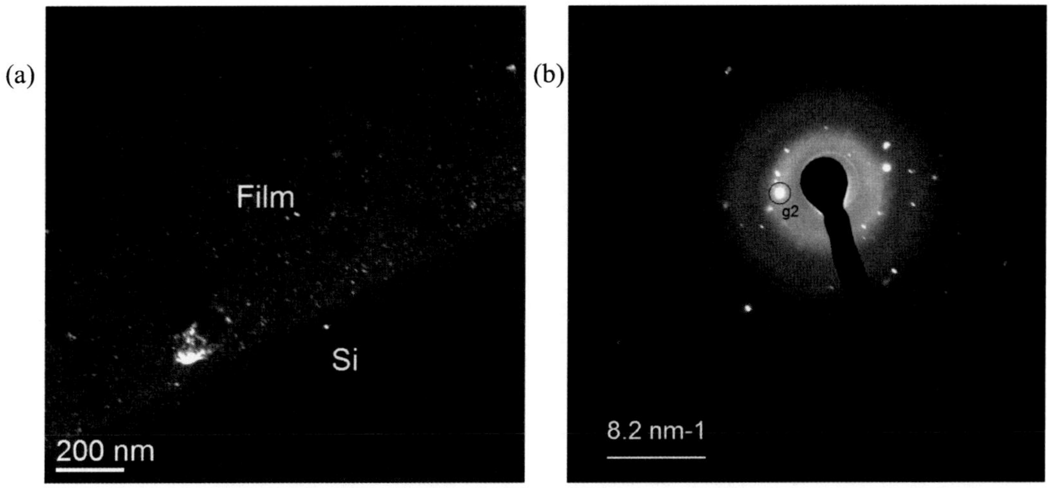

Figure 10: (a) Energy-filtered dark-field image obtained by selecting a spot given by one set of <200> aluminium reflecting planes in Al, having interplanar distance of 0.215 ± 0.005 nm, showing the distributions of nanocrystals (bright spots in DF image). The diffraction pattern, (b), was obtained by probing a circular area of 1 μm diameter (size of the selected area aperture) in the film. Intense spots, indicating crystalline phases, and faint large rings, revealing amorphous phases, were observed.

Also the X-ray photoelectron spectra (XPS) of the 2s and 2p orbitals of Al, Fig. **11**, clearly suggest the existence of pure aluminium with binding energies of, respectively, 72.6 eV and 117.8 eV, in accordance with the binding energy values found in literature [76]. The existence of pure aluminium can be detected in zone Ia and in the transition zone, along with the lines of Al in Al_2O_3 at 75.2 eV and

120 eV. The pure Al disappears in zone II, as expected, since the films are composed of quasi-pure Al_2O_3. Fig. **11** also shows an Al plasmon-loss peak [76] at ~88 eV accompanying the 2p core level of Al (binding energy of ~73 eV), which gives a plasmon energy of ~15 eV, in agreement with Al-plasmon energy found in the available literature [12]. This peak can be seen in a sample of zone Ia as well as in the transition zone, reinforcing the idea of the existence of pure aluminium in the matrix of the samples ascribed to these zones.

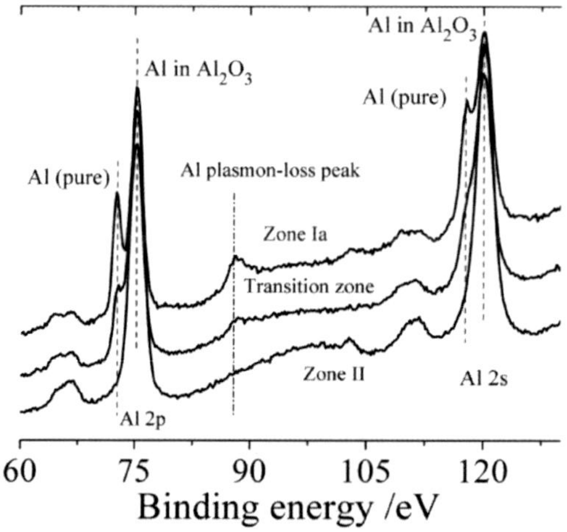

Figure 11: XPS spectra of representative samples of each zone.

The above discussion of composition, structure and morphology of the deposited films clearly justifies the electrical behaviour observed. In fact, the increase of film resistivity observed as a function of the non-metallic over aluminium atomic ratios, Fig. **8**, can be explained by growth of the oxide/nitride (or oxynitride) matrix leading to a larger electron scattering at grain boundaries [77-79], due to a limitation of the aluminium grain size, and to an increase of the barrier thickness between grains. Simultaneously, the barrier component of the resistance, which has a negative dependence on temperature [74, 75], becomes dominant, thus explaining the negative TCR of the films with large C_{O+N}/C_{Al}.

Optical Properties

The colour coordinates L* (brilliance), a* (redness) and b* (yellowness) of the films with chemical ratios, C_{O+N}/C_{Al}, up to 0.85 were determined and represented in the CIELab (1976) colour space [80] and are displayed in Fig. **12**. The Al_2O_3

films revealed, as expected, interference colorations, consistent with their semi-transparency, and are not represented in the plot. Low values of a* and b* can be observed in Fig. **12**, which indicates intrinsic grey colours in accordance with the visual aspect of the coatings. Concerning the brilliance (L*), one can observe a very slight decrease of its magnitude in films with high aluminium content (low C_{O+N}/C_{Al}), indexed to zone Ia. As the non-metallic over aluminium atomic ratio increases, and the type of growth changes from columnar towards cauliflower, a very sharp decrease of the brilliance can be observed, dropping to a value below 30 in the zone Ib. The relatively high roughness of films ascribed to the transition zone can be one of the factors that can explain the reported low values of L*, which are approximately constant in zone Ic. These results indicate a weak influence of oxygen and nitrogen content on the films colour, contrarily to other metal oxynitride systems such as zirconium [80] and titanium [65]. In fact, one can only report a gradual transition from a metallic-white intrinsic colour towards dark-grey tones in the AlO_xN_y system.

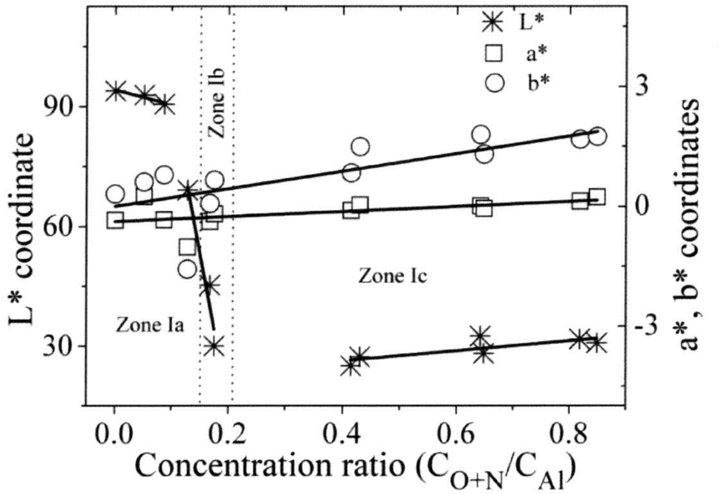

Figure 12: Average colour coordinates of the AlO_xN_y films, represented in the CIELab colour space under primary illuminant D65 (specular component included), as a function of the concentration ratio of non-metallic over metallic content.

The total reflectance and transmittance of the AlO_xN_y films were also measured using a spectrophotometer with an attached integrating sphere. According to the results obtained, the sub-stoichiometric AlO_xN_y films, with atomic ratios C_{O+N}/C_{Al} up to 0.85, are opaque with reflectance spectra strongly correlated with the chemical composition and morphology of the films. One can observe a typical profile of polycrystalline aluminium [81] in the sample deposited without reactive gas, as can be confirmed by the results plotted in Fig. **13**. It can also be observed that the reflectance has a pronounced decrease as the ratio of non-metallic over aluminium elements increases. An interesting feature about the films indexed to the transition zone is that the reflectance is approximately constant and with values below 10% in all spectrum measured (250 to 2500 nm).

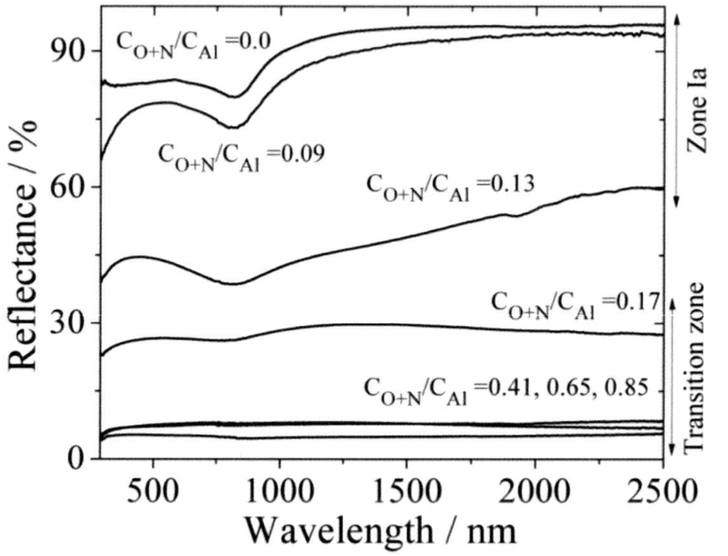

Figure 13: Reflectance (total) spectra of the AlO_xN_y films indexed to zone Ia ($C_{O+N}/C_{Al} \leq 0.13$) and transition zone ($0.17 \leq C_{O+N}/C_{Al} \leq 0.85$).

Fig. **14** shows the absolute specular reflectance, measured in the range of 250 to 3000 nm using the VW configuration, which is a technique that can evaluate whether the surfaces of the films are mirror-like, especially the ones with high content of aluminium, zone Ia. Although the total reflectance of the polycrystalline aluminium is very high in all wavelengths, as showed in Fig. **13**, Fig. **14** suggests that the diffuse reflectance plays an important role in the total reflectance. In fact, the specular reflectance of the aluminium coating drops in the visible range, meaning that the surface of the film is not mirror smooth. This behaviour is not a surprise since the surface is covered by a thin oxide layer which gives a metallic-white colour to the films indexed to zone Ia. Also, the columnar growth induces roughness in the film and thus increases the diffuse reflectance.

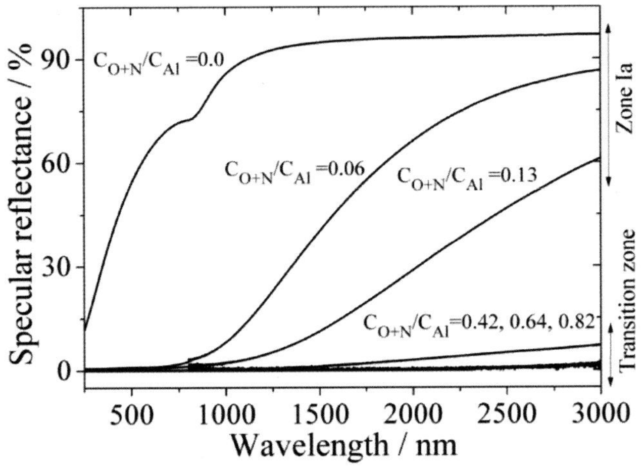

Figure 14: Absolute specular reflectance of some representative samples ascribed to zone Ia and to the transition zone.

It was possible to simulate the optical properties of the films indexed to zone Ia [44]. Due to the high content of aluminium, the optical standard model of aluminium [12] was used. The model includes an intraband component (Drude model) and also interband transitions described by one OJL oscillator [82] and two Kim oscillators [83]. The reflectance and transmittance of semi-transparent Al_2O_3 coating (not represented in Fig. **13**) were also used to simulate the optical properties of the material [44]. The Cauchy equations were used in this case [84].

The plasma frequency extracted from the fit is about 10.4 eV and the damping coefficient of 210 meV, for the polycrystalline Al coating. The plasma frequency is lower than the value calculated for the bulk material (~12.5 eV) found in literature [12], and the damping coefficient is in the same order of magnitude as the results of other authors [81]. The existence of defects in the lattice, residual elements like oxygen, and a contaminated and rough surface can explain the value of plasma frequency for Al coating, which is lower than the one predicted for a crystalline and uncontaminated material. According to the model used to adjust the reflectance spectrum, the typical interband absorption at 1.5 eV [85], which results from the so-called "parallel-band" effect [86, 87], is located at 1.4 eV for the Al coating, according to the results of the model. At this photon energy a significant drop in the reflectance can be observed, Fig. **13**. This interband absorption, as well as other at 0.5 eV that is hidden by the intraband absorption, occurs between the occupied and unoccupied (parallel) conduction bands of aluminium. There are also two further absorptions, positioned at 2.97 eV and 1.73 eV, determined by the model, possibly due to weak absorptions also predicted by the aluminium band-structure calculations [12].

As the incorporation of oxygen and nitrogen in the film increases, with a consequent increase of the concentration ratio (C_{O+N}/C_{Al}), a pronounced decrease of the reflectance in all wavelengths measured can be observed, and a drop in the plasma frequency is verified in zone Ia, reaching a value of 2.33 eV for $C_{O+N}/C_{Al} = 0.13$. This is again a consequence of the decrease of the conduction electrons due to incorporation of oxygen and nitrogen, in accordance with the results obtained for the electrical resistivity.

The reflectance of the transition zone films is more interesting than that of the films with Al-type behaviour (zone Ia), since it is clear that the transition zone films have a large broadband absorption, independent of the wavelength, in the range of 250 – 2500 nm. This behaviour suggests a percolating network of aluminium nanoparticles embedded in the oxide/nitride matrix, as discussed in the electrical properties section. The coalescence between aluminium nanoparticles can result on the formation of irregular clusters with different shapes and sizes, randomly distributed through the matrix. This creates, not a strong absorption maximum at a particular wavelength as observed in other materials such as TiO_2 with Au nanoparticles [88, 89], but a flat absorption from UV to IR wavelengths. This is an interesting behaviour for possible applications in concentrated solar power systems (CSP) [90, 91], solar collectors [92], sensors, and others [93]. As demonstrated in other materials, such as Ag nanoparticles in polymer matrices [93], this morphology is suitable for surface plasmons excitation, with resonances at a longer wavelength, over the aluminium network thus opening the range of absorption wavelengths.

The results of simulated complex refractive index (n and k), as well as the energy loss function of representative films of zones Ia and II, are plotted in Fig. **15**. The influence of chemical composition on those quantities is evident. The results of n and k obtained for the Al coating agree with other works [12, 94]. Furthermore, the n, k and energy loss of the sample deposited with a concentration ratio of 0.09 has approximately the same shape as the aluminium coating, although a small difference can be observed. Indeed, the refractive index is somewhat high for some wavelengths in the NIR, the extinction coefficient is lower in all wavelengths and the energy loss function is almost the same, except in the UV region. As the content of non-metallic over aluminium elements is more important, the shape of the plots, *i.e.*, n, k and energy loss *vs.* wavelength, changes and a high refractive index can be observed. Displayed in Fig. **15** is also the sample with a concentration ratio of 0.13 with a higher refractive index than the other two mentioned samples. The transparent Al_2O_3 film reveals a refractive index of 1.7-1.8 which is in agreement with values found in the literature [12], as well as the extinction coefficient and energy loss function which are very close to zero.

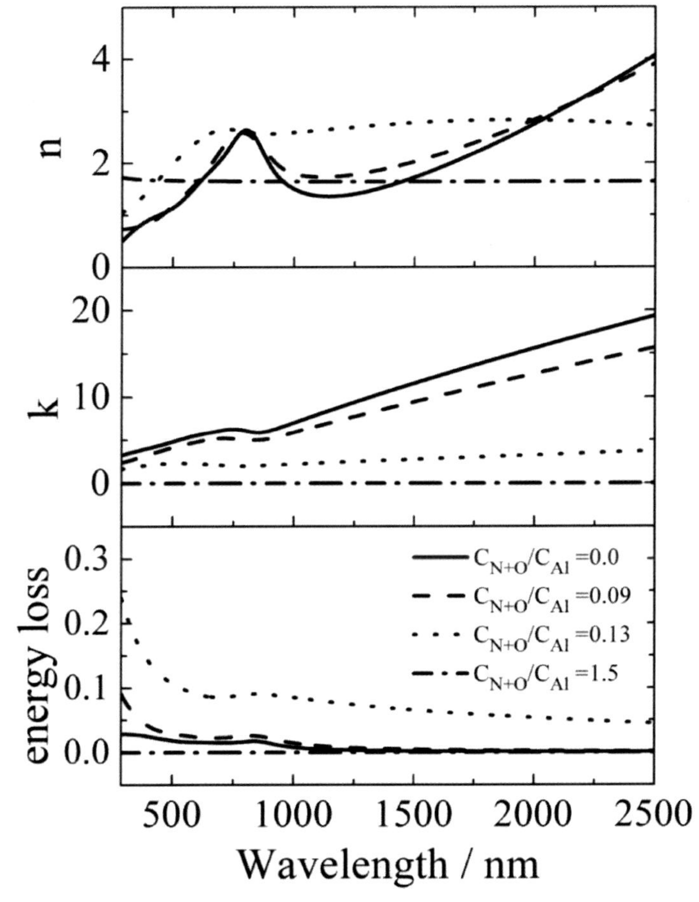

Figure 15: Simulated refractive index, extinction coefficient and energy loss function for the films indexed to zone Ia ($C_{O+N}/C_{Al} \leq 0.13$) and zone II (Al_2O_3).

CONCLUSION

AlO_xN_y thin films were deposited by DC reactive magnetron sputtering, using an aluminium target, in an atmosphere composed of Ar and a mixture of nitrogen and oxygen as reactive gases (17:3 ratio). The discharge parameters were correlated with the morphology, deposition rate, chemical composition and structure of the films. Increasing the partial pressure of reactive gas allowed the production of films with different features concerning their characteristics and physical properties.

One can report that the target potential has two different regimes: a regime where the target potential gradually decreases as it is becoming more poisoned, regime I; and a regime II where the target potential values are approximately constant as a result of a complete poisoning. Furthermore, it was possible to categorize the different coatings in three different zones, in respect to their type of growth. With the target in regime I, a first zone (zone Ia) can be identified where the films revealed a typical columnar growth, followed by a transition zone where a cauliflower growth can be observed. The regime II of the target induced the deposition of dense and featureless-type films. The deposition rate has a quite interesting evolution since it is approximately constant in zone Ia, while within the transition zone it has two different tendencies: it increases in a short range of partial pressures (zone Ib) and again decreases (zone Ic). The zone II is characterized by low deposition rate values. The RBS experiments revealed that the films produced with the target in regime I are sub-stoichiometric, AlO_xN_y, meaning that the aluminium is always in excess, contrarily to those prepared with the target in regime II, where quasi-stoichiometric Al_2O_3 coatings were formed. The concentration ratio of oxygen and nitrogen over aluminium, C_{O+N}/C_{Al}, increases slightly in zone Ia and this group of films is characterized by high content of Al. In the case of zone Ib the concentration ratio is approximately constant although the partial pressure of the reactive gas is increasing and in zone Ic there is a sharper increase of C_{O+N}/C_{Al} atomic ratios. The structure of the sub-stoichiometric films ($C_{O+N}/C_{Al} \leq 0.85$) is Al-type, although they are becoming amorphous in zone Ic. The Al_2O_3 films (zone II) are amorphous according to their XRD diffractograms.

The electrical resistivity, measured at room temperature, is very well correlated with the chemical composition and morphology. Four different zones were found and indexed to the previous mentioned zones Ia, Ib, Ic and II. Even in zone Ib, where the chemical composition is roughly the same for all samples, there is an increase of electrical resistivity resulting from the slight differences in the morphology, which by itself can enlarge the voids area. The stoichiometric alumina films, ascribed to zone II, are electrical insulators. The temperature coefficient of resistance sharply decreases in zone Ia, decreasing slower in the transition zone, and finally it becomes negative for samples with atomic ratios, C_{O+N}/C_{Al}, between 0.64 and 0.85. The colour of the films gradually changes from metallic-like towards dark grey tones. The optical reflectance revealed that a small increase of non-metallic elements over aluminium in the films is sufficient to induce a transition from a typical polycrystalline aluminium profile towards a marked decrease of the reflectance in the NIR range, as observed in zone Ia. More peculiar is the reflectance of the samples indexed to the transition zone, where the values can be as low as 5% in all spectral range (from 250 to 2500 nm).

The electrical and optical properties observed in the transition zone can be explained assuming that the films are in fact a percolation network of aluminium nanoparticles (crystalline grains) embedded in an oxide/nitride matrix, as TEM and XPS results suggests. The nanoparticles can coalescence in irregularly formed clusters with different shapes and sizes through the matrix, inducing a broadband absorption nearly independent of the wavelength (250 to 2500 nm). The conductivity of these films is governed by the constrictions between grains that can be in contact or separated by insulating layers. This barrier component of the resistance, which has a negative dependence on the temperature, becomes dominant for atomic ratios, C_{O+N}/C_{Al}, between 0.64 and 0.85 and thus explains the negative TCR of these films.

ACKNOWLEDGEMENTS

This research was supported by FEDER through the COMPETE program and by the Portuguese Foundation for Science and Technology (FCT) in the framework of the Strategic Project PEST-C/FIS/UI607/2011, as well as the project PTDC/CTM-NAN/112574/2009 and by Programa Pessoa 2010/2011, Cooperação

Portugal/França, Proc.°441.00, Project *"COLOURCLUSTER"*. J. Borges also acknowledges FCT financial support under PhD grant N° SFRH/BD/47118/2008 (*financiado por POPH – QREN – Tipologia 4.1 – Formação Avançada, comparticipado pelo Fundo Social Europeu e por fundos nacionais do MCTES*).

CONFLICT OF INTEREST

The author(s) confirm that this chapter content has no conflict of interest.

REFERENCES

[1] N. D. Corbin, "Aluminum oxynitride spinel: A review", *Journal of the European Ceramic Society,* vol. 5, pp. 143-154, 1989.

[2] T. M. Hartnett, S. D. Bernstein, E. A. Maguire, and R. W. Tustison, "Optical properties of ALON (aluminum oxynitride)", *Infrared Physics & Technology,* vol. 39, pp. 203-211, 1998.

[3] A. Pallone, J. Demaree, and J. Adams, "Application of nondestructive ion beam analysis to measure variations in the elemental composition of armor materials", *Nuclear Instruments and Methods in Physics Research Section B: Beam Interactions with Materials and Atoms,* vol. 219-220, pp. 755-758, 2004.

[4] J. Borges, F. Vaz, and L. Marques, "AlN_xO_y thin films deposited by DC reactive magnetron sputtering", *Applied Surface Science,* vol. 257, pp. 1478-1483, 2010.

[5] J. Emsley, *Nature's Building Blocks*: Oxford University Press, 2001.

[6] J. M. Camacho and A. I. Oliva, "Morphology and electrical resistivity of metallic nanostructures", *Microelectronics Journal,* vol. 36, pp. 555-558, 2005.

[7] Z. Wei, T. P. Chen, L. Yang, Y. Ming, Z. Sam, W. L. Zhang, and S. Fung, "Charging-Induced Changes in Reverse Current-Voltage Characteristics of Al/Al-Rich Al_2O_3/p-Si Diodes", *IEEE Transactions on Electron Devices,* vol. 56, pp. 2060-2064, 2009.

[8] D. E. Yildiz and S. Altindal, "On the temperature dependence of series resistance and interface states in Al/SiO_2/p-Si (MIS) Schottky diodes", *Microelectronic Engineering,* vol. 85, pp. 289-294, 2008.

[9] G. Yip, J. Qiu, W. T. Ng, and Z. H. Lu, "Effect of metal contacts on the electrical characteristics of Al_2O_3 dielectric thin films", *Applied Physics Letters,* vol. 92, p. 122911, 2008.

[10] N. Joshi, A. K. Debnath, D. K. Aswal, K. P. Muthe, M. Senthil Kumar, S. K. Gupta, and J. V. Yakhmi, "Morphology and resistivity of Al thin films grown on Si(111) by molecular beam epitaxy", *Vacuum,* vol. 79, pp. 178-185, 2005.

[11] D. K. Aswal, J. Niraj, A. K. Debnath, K. P. Muthe, S. K. Gupta, J. V. Yakhmi, and V. Dominique, "Morphology-dependent electric transport in textured ultrathin Al films grown on Si", *Journal of Applied Physics,* vol. 98, p. 026103, 2005.

[12] E. D. Palik, *Handbook of Optical Constants of Solids*: Elsevier, 1997.

[13] I. Gryczynski, J. Malicka, Z. Gryczynski, K. Nowaczyk, and J. R. Lakowicz, "Ultraviolet Surface Plasmon-Coupled Emission Using Thin Aluminum Films", *Analytical Chemistry,* vol. 76, pp. 4076-4081, 2004.

[14] Y. Akimov and W. S. Koh, "Design of Plasmonic Nanoparticles for Efficient Subwavelength Light Trapping in Thin-Film Solar Cells", *Plasmonics,* vol. 6, pp. 155-161, 2011.

[15] Y. A. Akimov and W. S. Koh, "Resonant and nonresonant plasmonic nanoparticle enhancement for thin-film silicon solar cells", *Nanotechnology,* vol. 21, p. 235201, 2010.

[16] K. R. Bray, R. L. C. Wu, S. Fries-Carr, and J. Weimer, "Aluminum oxynitride dielectrics for multilayer capacitors with higher energy density and wide temperature properties", *Thin Solid Films,* vol. 518, pp. 366-371, 2009.

[17] Q. Xia, H. Xia, and A. L. Ruoff, "Pressure-induced rocksalt phase of aluminum nitride: A metastable structure at ambient condition", *Journal of Applied Physics,* vol. 73, pp. 8198-8200, 1993.

[18] N. Rakov, A. Mahmood, and M. Xiao, "Surface metallic state of aluminum-nitride (AlN) thin films prepared by direct current (DC)-reactive magnetron sputtering: optical spectroscopic analysis with incoherent light", *Scripta Materialia,* vol. 50, pp. 589-592, 2004.

[19] Z. An, C. Men, Z. Xu, P. K. Chu, and C. Lin, "Electrical properties of AlN thin films prepared by ion beam enhanced deposition", *Surface and Coatings Technology,* vol. 196, pp. 130-134, 2005.

[20] H. C. Barshilia, B. Deepthi, and K. S. Rajam, "Growth and characterization of aluminum nitride coatings prepared by pulsed-direct current reactive unbalanced magnetron sputtering", *Thin Solid Films,* vol. 516, pp. 4168-4174, 2008.

[21] R. D. Vispute, J. Narayan, and J. D. Budai, "High quality optoelectronic grade epitaxial AlN films on [alpha]-Al_2O_3, Si and 6H-SiC by pulsed laser deposition", *Thin Solid Films,* vol. 299, pp. 94-103, 1997.

[22] J. Wang, W. L. Wang, P. D. Ding, Y. X. Yang, L. Fang, J. Esteve, M. C. Polo, and G. Sanchez, "Synthesis of cubic aluminum nitride by carbothermal nitridation reaction", *Diamond and Related Materials,* vol. 8, pp. 1342-1344, 1999.

[23] L. Yate, J. C. Caicedo, A. H. Macias, F. J. Espinoza-Beltrán, G. Zambrano, J. Muñoz-Saldaña, and P. Prieto, "Composition and mechanical properties of AlC, AlN and AlCN thin films obtained by r.f. magnetron sputtering", *Surface and Coatings Technology,* vol. 203, pp. 1904-1907, 2009.

[24] V. Mortet, M. Nesladek, K. Haenen, A. Morel, M. D'Olieslaeger, and M. Vanecek, "Physical properties of polycrystalline aluminium nitride films deposited by magnetron sputtering", *Diamond and Related Materials,* vol. 13, pp. 1120-1124, 2004/8// 2004.

[25] D. Marc-Alexandre and M. Paul, "Properties of aluminum nitride thin films for piezoelectric transducers and microwave filter applications", *Applied Physics Letters,* vol. 74, pp. 3032-3034, 1999.

[26] F. Martin, P. Muralt, M. A. Dubois, and A. Pezous, "Thickness dependence of the properties of highly c-axis textured AlN thin films", *Journal of Vaccum Science & Technology A,* vol. 22, pp. 361-365, 2004.

[27] Y. Taniyasu, M. Kasu, and T. Makimoto, "An aluminium nitride light-emitting diode with a wavelength of 210 nanometres", *Nature,* vol. 441, pp. 325-328, 2006.

[28] Q. X. Guo, M. Yoshitugu, T. Tanaka, M. Nishio, and H. Ogawa, "Microscopic investigations of aluminum nitride thin films grown by low-temperature reactive sputtering", *Thin Solid Films,* vol. 483, pp. 16-20, 2005.

[29] J. Xiong, H.-s. Gu, K. Hu, and M.-z. Hu, "Influence of substrate metals on the crystal growth of AlN films", *International Journal of Minerals, Metallurgy, and Materials,* vol. 17, pp. 98-103, 2010.

[30] W.-J. Liu, S.-J. Wu, C.-M. Chen, Y.-C. Lai, and C.-H. Chuang, "Microstructural evolution and formation of highly c-axis-oriented aluminum nitride films by reactively magnetron sputtering deposition", *Journal of Crystal Growth,* vol. 276, pp. 525-533, 2005.

[31] A. Aryasomayajula, S. Canovic, D. Bhat, M. H. Gordon, and M. Halvarsson, "Transmission electron microscopy and X-ray diffraction analysis of alumina coating by alternate-current inverted magnetron-sputtering technique", *Thin Solid Films,* vol. 516, pp. 397-401, 2007.

[32] J. M. Schneider, W. D. Sproul, and A. Matthews, "Reactive ionized magnetron sputtering of crystalline alumina coatings", *Surface and Coatings Technology,* vol. 98, pp. 1473-1476, 1998.

[33] L. Igor and B. David, "Metastable Alumina Polymorphs: Crystal Structures and Transition Sequences", *Journal of the American Ceramic Society,* vol. 81, pp. 1995-2012, 1998.

[34] R. K. Bhan, A. Jain, D. Kumar, and S. K. Mehta, "Conduction, dielectric and interface properties of Al_2O_3 films on GaAs deposited by the e-beam evaporation technique", *Semiconductor Science and Technology,* vol. 24, p. 095017, 2009.

[35] H. Demiryont, L. R. Thompson, and G. J. Collins, "Optical and electrical characterizations of laser-chemical-vapor-deposited aluminum oxynitride films", *Journal of Applied Physics,* vol. 59, pp. 3235-3240, 1986.

[36] J. Kolodzey, E. A. Chowdhury, T. N. Adam, Q. Guohua, I. Rau, J. O. Olowolafe, J. S. Suehle, and C. Yuan, "Electrical conduction and dielectric breakdown in aluminum oxide insulators on silicon", *IEEE Transactions on Electron Devices,* vol. 47, pp. 121-128, 2000.

[37] Y. B. Li, S. Zhang, Y. Liu, T. P. Chen, T. Sritharan, and C. Xu, "Magnetron Sputtered nc-Al/alpha-Al_2O_3 Nanocomposite Thin Films for Nonvolatile Memory Application", *Journal of Nanoscience and Nanotechnology,* vol. 9, pp. 4116-4120, Jul 2009.

[38] W. H. Ha, M. H. Choo, and S. Im, "Electrical properties of Al_2O_3 film deposited at low temperatures", *Journal of Non-Crystalline Solids,* vol. 303, pp. 78-82, 2002.

[39] C. Lin, J. Kang, D. Han, D. Tian, W. Wang, J. Zhang, M. Liu, X. Liu, and R. Han, "Electrical properties of Al_2O_3 gate dielectrics", *Microelectronic Engineering,* vol. 66, pp. 830-834, 2003.

[40] H. Bartzsch, D. Glöß, B. Böcher, P. Frach, and K. Goedicke, "Properties of SiO_2 and Al_2O_3 films for electrical insulation applications deposited by reactive pulse magnetron sputtering", *Surface and Coatings Technology,* vol. 174-175, pp. 774-778, 2003.

[41] B. G. Segda, M. Jacquet, and J. P. Besse, "Elaboration, characterization and dielectric properties study of amorphous alumina thin films deposited by r.f. magnetron sputtering", *Vacuum,* vol. 62, pp. 27-38, 2001.

[42] H. C. Barshilia, N. Selvakumar, G. Vignesh, K. S. Rajam, and A. Biswas, "Optical properties and thermal stability of pulsed-sputter-deposited Al_xO_y/Al/Al_xO_y multilayer absorber coatings", *Solar Energy Materials and Solar Cells,* vol. 93, pp. 315-323, 2009.

[43] J. Borges, N. Martin, N. P. Barradas, E. Alves, D. Eyidi, M. F. Beaufort, J. P. Riviere, F. Vaz, and L. Marques, "Electrical properties of AlN_xO_y thin films prepared by reactive magnetron sputtering", *Thin Solid Films,* vol. 520, pp. 6709-6717, 2012.

[44] J. Borges, N. P. Barradas, E. Alves, M. F. Beaufort, D. Eyidi, F. Vaz, and L. Marques, "Influence of stoichiometry and structure on the optical properties of AlN_xO_y films", *Journal of Physics D: Applied Physics,* vol. 46, p. 015305, 2013.

[45] J. Borges, E. Alves, F. Vaz, and L. Marques, "Optical properties of AlN_xO_y thin films deposited by DC magnetron sputtering," in *International Conference on Applications of Optis and Photonics*, Braga, 2011, p. 80010F.

[46] J. J. Guo, K. Wang, T. Fujita, J. W. McCauley, J. P. Singh, and M. W. Chen, "Nanoindentation characterization of deformation and failure of aluminum oxynitride", *Acta Materialia,* vol. 59, pp. 1671-1679, 2011.

[47] Z. Qi-Chu, K. Zhao, B. C. Zhang, L. F. Wang, Z. L. Shen, D. Q. Lu, D. L. Xie, and B. F. Li, "High performance Al--N cermet solar coatings deposited by a cylindrical direct current magnetron sputter coater", *Journal of Vacuum Science & Technology A: Vacuum, Surfaces, and Films,* vol. 17, pp. 2885-2890, 1999.

[48] N. P. Barradas, C. Jeynes, and R. P. Webb, "Simulated annealing analysis of Rutherford backscattering data", *Applied Physics Letters,* vol. 71, pp. 291-293, 1997.

[49] A. Ruhm, B. P. Topeverg, and H. Dosch, "Supermatrix approach to polarized neutron reflectivity from arbitrary spin structures", *Physical Review B,* vol. 60, p. 16073, 1999.

[50] I. Horcas, R. Fernandez, J. M. Gomez-Rodriguez, J. Colchero, J. Gomez-Herrero, and A. M. Baro, "WSXM: A software for scanning probe microscopy and a tool for nanotechnology", *Review of Scientific Instruments,* vol. 78, p. 013705, 2007.

[51] A. W. Robert, "An algorithm for computing linear four-point probe thickness correction factors", *Review of Scientific Instruments,* vol. 72, pp. 3580-3586, 2001.

[52] R. Arvinte, J. Borges, R. E. Sousa, D. Munteanu, N. P. Barradas, E. Alves, F. Vaz, and L. Marques, "Preparation and characterization of CrN_xO_y thin films: The effect of composition and structural features on the electrical behavior", *Applied Surface Science,* vol. 257, pp. 9120-9124, 2011.

[53] D. Depla, J. Haemers, and R. De Gryse, "Discharge voltage measurements during reactive sputtering of oxides", *Thin Solid Films,* vol. 515, pp. 468-471, 2006.

[54] D. Depla and R. D. Gryse, "Influence of oxygen addition on the target voltage during reactive sputtering of aluminium", *Plasma Sources Science and Technology,* vol. 10, p. 547, 2001.

[55] D. Depla and R. De Gryse, "Target poisoning during reactive magnetron sputtering: Part I: the influence of ion implantation", *Surface and Coatings Technology,* vol. 183, pp. 184-189, 2004.

[56] D. Depla and R. De Gryse, "Target poisoning during reactive magnetron sputtering: Part II: the influence of chemisorption and gettering", *Surface and Coatings Technology,* vol. 183, pp. 190-195, 2004.

[57] D. Depla, G. Buyle, J. Haemers, and R. De Gryse, "Discharge voltage measurements during magnetron sputtering", *Surface and Coatings Technology,* vol. 200, pp. 4329-4338, 2006.

[58] I. Safi, "Recent aspects concerning DC reactive magnetron sputtering of thin films: a review", *Surface and Coatings Technology,* vol. 127, pp. 203-219, 2000.

[59] J. Schulte and G. Sobe, "Magnetron sputtering of aluminium using oxygen or nitrogen as reactive gas", *Thin Solid Films,* vol. 324, pp. 19-24, 1998.

[60] S. Venkataraj, D. Severin, R. Drese, F. Koerfer, and M. Wuttig, "Structural, optical and mechanical properties of aluminium nitride films prepared by reactive DC magnetron sputtering", *Thin Solid Films,* vol. 502, pp. 235-239, 2006.

[61] D. Depla, H. Tomaszewski, G. Buyle, and R. De Gryse, "Influence of the target composition on the discharge voltage during magnetron sputtering", *Surface and Coatings Technology,* vol. 201, pp. 848-854, 2006.

[62] J. A. Thornton, "The microstructure of sputter-deposited coatings", *Journal of Vacuum Science & Technology A: Vacuum, Surfaces, and Films,* vol. 4, pp. 3059-3065, 1986.

[63] I. Petrov, P. B. Barna, L. Hultman, and J. E. Greene, "Microstructural evolution during film growth", *Journal of Vacuum Science & Technology A: Vacuum, Surfaces, and Films,* vol. 21, pp. S117-S128, 2003.

[64] N. M. G. Parreira, N. J. M. Carvalho, F. Vaz, and A. Cavaleiro, "Mechanical evaluation of unbiased W-O-N coatings deposited by d.c. reactive magnetron sputtering", *Surface and Coatings Technology,* vol. 200, pp. 6511-6516, 2006.

[65] F. Vaz, P. Cerqueira, L. Rebouta, S. M. C. Nascimento, E. Alves, P. Goudeau, and J. P. Riviere, "Preparation of magnetron sputtered TiNxOy thin films," in *Eight International Conference on Plasma Surface Engineering*, 2003, pp. 197-203.

[66] P. Carvalho, J. M. Chappe, L. Cunha, S. Lanceros-Mendez, P. Alpuim, F. Vaz, E. Alves, C. Rousselot, J. P. Espinos, and A. R. Gonzalez-Elipe, "Influence of the chemical and electronic structure on the electrical behavior of zirconium oxynitride films", *Journal of Applied Physics,* vol. 103, pp. 104907-15, 2008.

[67] N. Jacobson, "Use of Tabulated Thermochemical Data for Pure Compounds", *Journal of Chemical Education,* vol. 78, p. 814, 2001.

[68] V. A. Lubarda, "On the effective lattice parameter of binary alloys", *Mechanics of Materials,* vol. 35, pp. 53-68, 2003.

[69] M. K. Sinha, S. K. Mukherjee, B. Pathak, R. K. Paul, and P. K. Barhai, "Effect of deposition process parameters on resistivity of metal and alloy films deposited using anodic vacuum arc technique", *Thin Solid Films,* vol. 515, pp. 1753-1757, 2006.

[70] R. K. Bhan, A. Jain, D. Kumar, and S. K. Mehta, "Conduction, dielectric and interface properties of Al_2O_3 films on GaAs deposited by the e-beam evaporation technique." *Semiconductor Science and Technology,* vol. 24, p. 095017, 2009.

[71] L. J. Van Der Pauw, "A method of measuring specific resistivity and Hall effect of discs of arbitrary shape", *Philips Research Reports,* vol. 13, pp. 1-9, 1958.

[72] L. J. Van Der Pauw, "A method of measuring the resistivity and Hall coefficient on lamellae of arbitrary shape", *Philips Technical Review,* vol. 20, pp. 220-224, 1958.

[73] D. R. Lide, *Handbook of Chemistry and Physics (72nd edition).* Boston: CRC Press, 1991.

[74] A. Kusy, "An equivalent network for resistance and temperature coefficient of resistance versus temperature and composition of thick resistive films", *Journal of Applied Physics,* vol. 62, pp. 1324-1334, 1987.

[75] G. E. Pike and C. H. Seager, "Electrical properties and conduction mechanisms of Ru-based thick-film (cermet) resistors", *Journal of Applied Physics,* vol. 48, pp. 5152-5169, 1977.

[76] P. M. Attekum and J. M. Trooster, "Bulk- and surface-plasmon-loss intensities in photoelectron, Auger, and electron-energy-loss spectra of Al metal", *Physical Review B,* vol. 18, pp. 3872-3883, 1978.

[77] D. S. McLachlan, "New models for the positive and negative temperature coefficients of resistivity for $TiO_{0.80-1.23}$ metallic oxides", *Physical Review B,* vol. 25, p. 2285, 1982.

[78] G. Reiss, J. Vancea, and H. Hoffmann, "Grain-Boundary Resistance in Polycrystalline Metals", *Physical Review Letters,* vol. 56, p. 2100, 1986.

[79] N. Martin, A. Besnard, F. Sthal, F. Vaz, and C. Nouveau, "The contribution of grain boundary barriers to the electrical conductivity of titanium oxide thin films", vol. 93, p. 064102, 2008.

[80] P. Carvalho, F. Vaz, L. Rebouta, L. Cunha, C. J. Tavares, C. Moura, E. Alves, A. Cavaleiro, P. Goudeau, E. Le Bourhis, J. P. Riviere, J. F. Pierson, and O. Banakh, "Structural, electrical, optical, and mechanical characterizations of decorative ZrO_xN_y thin films", *Journal of Applied Physics,* vol. 98, pp. 023715-8, 2005.
[81] D. Y. Smith and B. Segall, "Intraband and interband processes in the infrared spectrum of metallic aluminum", *Physical Review B,* vol. 34, p. 5191, 1986.
[82] S. K. O' Leary, S. R. Johnson, and P. K. Lim, "The relationship between the distribution of electronic states and the optical absorption spectrum of an amorphous semiconductor: An empirical analysis", *Journal of Applied Physics,* vol. 82, pp. 3334-3340, 1997.
[83] C. C. Kim, J. W. Garland, H. Abad, and P. M. Raccah, "Modeling the optical dielectric function of semiconductors: Extension of the critical-point parabolic-band approximation", *Physical Review B,* vol. 45, p. 11749, 1992.
[84] S. Dirk Poelman and Philippe Frederic, "Methods for the determination of the optical constants of thin films from single transmission measurements: a critical review", *Journal of Physics D: Applied Physics,* vol. 36, p. 1850, 2003.
[85] H. Ehrenreich, H. R. Philipp, and B. Segall, "Optical Properties of Aluminum", *Physical Review,* vol. 132, p. 1918, 1963.
[86] W. A. Harrison, "Parallel-Band Effects in Interband Optical Absorption", *Physical Review,* vol. 147, p. 467, 1966.
[87] N. W. Ashcroft and K. Sturm, "Interband Absorption and the Optical Properties of Polyvalent Metals", *Physical Review B,* vol. 3, p. 1898, 1971.
[88] M. Torrell, L. Cunha, M. R. Kabir, A. Cavaleiro, M. I. Vasilevskiy, and F. Vaz, "Nanoscale color control of TiO_2 films with embedded Au nanoparticles", *Materials Letters,* vol. 64, pp. 2624-2626, 2010.
[89] M. Torrell, P. Machado, L. Cunha, N. M. Figueiredo, J. C. Oliveira, C. Louro, and F. Vaz, "Development of new decorative coatings based on gold nanoparticles dispersed in an amorphous TiO_2 dielectric matrix", *Surface and Coatings Technology,* vol. 204, pp. 1569-1575, 2010.
[90] D. Barlev, R. Vidu, and P. Stroeve, "Innovation in concentrated solar power", *Solar Energy Materials and Solar Cells,* vol. 95, pp. 2703-2725, 2011.
[91] P. S. Nicholas, A. Mukul, and P. Peter, "Design of selective coatings for solar thermal applications using sub-wavelength metal-dielectric structures," 2009, p. 74100C.
[92] C. G. Granqvist, "Solar Energy Materials", *Advanced Materials,* vol. 15, pp. 1789-1803, 2003.
[93] A. Biswas, H. Eilers, J. F. Hidden, O. C. Aktas, and C. V. S. Kiran, "Large broadband visible to infrared plasmonic absorption from Ag nanoparticles with a fractal structure embedded in a Teflon AF matrix", *Applied Physics Letters,* vol. 88, pp. 013103-3, 2006.
[94] S. Zhao and E. Wäckelgård, "The optical properties of sputtered composite of Al-AlN", *Solar Energy Materials and Solar Cells,* vol. 90, pp. 1861-1874, 2006.

CHAPTER 9

HfSiON Films Deposited by Radio Frequency Reactive Sputtering

Li-Ping Feng* and Zheng-Tang Liu

State Key Lab of Solidification Processing, College of Materials Science and Engineering, Northwestern Polytechnical University, Xi'an, 710072, People's Republic of China

Abstract: In order to continue scaling electronic devices, an alternative gate material with high dielectric constant (high-*k*) has been proposed for capacitors, memories, and optoelectronic devices in future complementary metal-oxide-semiconductor (CMOS) generations. At present, HfSiON has gained much attention to be one of the most promising candidate materials to replace conventional SiON-based dielectrics. This chapter is focused on the synthesis and properties of HfSiON prepared by radio frequency magnetron reactive sputtering (RFMRS). This chapter is organized as follows. The scaling issue and major requirements of high-*k* gate dielectrics are described in Section 1. The fabrication process and properties of HfSiON films deposited by RFRCS are discussed in detail in Section 2. Conclusions and potential future developments are presented in Section 3.

Keywords: HfSiON films, reactive sputtering, high dielectric constant, MOS structure, sputtering pressure, deposition temperature, sputtering power, annealing.

INTRODUCTION

With the quick development of microelectronics, the size of semiconductor devices is becoming smaller and smaller. Over the past four decades, the success of the semiconductor industry relies on the reducing the dimensions of the metal-oxide-semiconductor field effect transistors (MOSFET) and the scaling the thickness of the silicon oxide (SiO_2) or nitrided silicon oxide (SiON) gate dielectric as quantified in Moore's law [1]. Table **1** shows the technology and equivalent dielectric thickness for continued CMOS scaling. Definitely, the reduction of dimensions allows the bigger integration of transistors and higher

***Address correspondence to Li-Ping Feng:** School of Materials Science, Northwestern Polytechnical University, Xi'an, Shaanxi, 710072, China; E-mail: flpmerry@yahoo.com.cn

Filipe Vaz, Nicolas Martin and Martin Fenker (Eds)

speed of electronic devices. In recent years, performance of scaled device has been compromised with fundamental material limits because of the high gate tunneling leakage currents and the loss of gate capacitance. Quantum tunneling effect can lead to the exponential increase of tunneling leakage currents with the decreasing dielectric thickness. For example, at $V_g = 1.0$ V, gate leakage currents increased from 1×10^{-12} A/cm^2 to 10 A/cm^2 when dielectric thickness decreased from 3.5 nm to 1.5 nm [2]. Additionally, the electrical device reliability is blocked by the time-dependent dielectric breakdown of the ultrathin SiON dielectric. Therefore, SiO_2 or SiON, as the present gate dielectric, will not satisfy this scaling trend in the next years. It means further scaling down of the SiON dielectric thickness is required for future CMOS technologies.

Undoubtedly, the continuously scaling down of the gate dielectric is necessary so as to increase the current drive capability of MOSFETs and the operation speed of integration circuits. The relationship between drive current and gate dielectric thickness is shown as [3]:

$$I_D = \frac{1}{2}\mu_n \frac{\varepsilon_0 W K_x}{L t_x}\left(V_G - V_T\right)^2 \quad \textbf{(1)}$$

where I_D is the drive saturation current of MOSFET, V_G the applied gate voltage and V_T the threshold voltage, ε_0 the dielectric constant of vacuum, μ_n the carrier mobility, W the tunnel width and L the tunnel length, t_x the dielectric thickness and k_x the dielectric constant of the gate capacitor. From Eq. (1), it is clear that the only solution to overcome the present difficulty and to continuously scale down the gate dielectric is to use thicker dielectric with higher dielectric constant value than SiON ($k = 3.9$ for pure SiO_2 and $k = 7.5$ for pure Si_3N_4).

The replacement of SiON layer by high-*k* dielectric can solve the scaling down problems, as can be supported theoretically by equivalent oxide thickness (EOT). Using the concept of EOT, comparisons can be made for various films and different thicknesses. EOT refers to the thickness of any dielectric scaled by the ratio of its dielectric constant to that of SiO_2 [4]:

$$EOT = t_x \left(\frac{K_{SiO_2}}{K_x} \right) \quad \textbf{(2)}$$

where t_x is the physical thickness of the alternative oxide film, k_x its dielectric constant and k_{SiO_2} the dielectric constant of SiO_2.

Table 1: Technology and equivalent dielectric thickness for continued CMOS scaling

Year	Minimum Technology (nm)	Equivalent Dielectric Thickness (nm)
1997	250	4 ~ 5
1999	180	3 ~ 4
2001	150	2 ~ 3
2003	130	2 ~ 3
2006	100	1.5 ~ 2
2009	70	< 1.5
2012	50	< 1

According to Eq. (2), one solution is to replace the conventional SiON-based dielectrics by using high-*k* material that will allow the "equivalent oxide thickness" scaling. Usually, the extraction of EOT requires to fit capacitance-voltage (*C-V*) characteristics of the MOS structure and take into account the quantum confinement effects and poly-Si depletion effects.

High-*k* gate materials can potentially reduce the leakage current flowing through the gate dielectric layer as well as improve the reliability of the electronic device. Consequently, novel high-*k* gate dielectric have been identified in the recent years. The tremendous research efforts have been focused on the investigation of high-*k* gate dielectrics for the potential MOSFET applications in the next CMOS generations.

There are many requirements for the implementation of new high-*k* gate dielectric materials in CMOS applications, which should be understood before these materials are adopted in industry. The fundamental requirements are listed as follows.

- The dielectric materials should have a bandgap greater than 5 eV and the band offsets with silicon should be sufficient.

- The dielectric constant of the materials should be higher than 10, which is less dependent on the temperature and frequency.
- High breakdown field strength and low tunnel leakage current.
- Less chemical reaction for dielectric materials in contact with silicon and gate electrode.
- Thin interfacial layer and low interface trap defect density.
- Low fixed charge and low electron mobility.
- Dielectric materials should be thermodynamic stable in direct contact with silicon and gate electrode.

In the recent years, many high-*k* dielectrics have been studied for gate oxide application such as $Si_xO_yN_z$ (k = 3.9 - 7.5), Al_2O_3 (k = 10), ZrO_2 (k = 25), Ta_2O_5 (k = 25), HfO_2 (k = 26), TiO_2 (k = 50) and *etc.* It is difficult to find an oxide dielectric that can meet all the fundamental requirements listed above. In order to match the requirements of the IC industry, researchers have combined the desirable properties from two different oxides, while eliminating the undesirable properties of each individual material. Among of them, Hf-based pseudobinary dielectric materials have caught the attention including HfAlO, HfSiO, HfON, HfYO, HfLaO and *etc.* Presently, nitrided $HfSiO_x$ (HfSiON) has gained much attention to be one of the most promising candidate materials as alternative gate dielectrics [5-8]. HfSiON has a broad compositional range, which is often considered as pseudobinary mixtures of $(HfO_2)_x(SiO_2)_{1-x}$ type materials. The performance of HfSiON high-*k* gate dielectric depends strongly on its component and the atom ratio of Si to Hf. Consequently, it is necessary to systematically investigate the characteristics of HfSiON gate dielectric so as to understand the optoelectronic processes as well as the design and analysis of optoelectronic devices. More details on the process and properties of HfSiON films prepared by the RFMRS can be found in the section 2.

DEPOSITION PROCESS AND PROPERTIES OF HfSiON FILMS

Many deposition techniques are available to form high-*k* layers such as physical vapour deposition (PVD), metallo-organo chemical vapour deposition (MOCVD), and atomic layer deposition (ALD). Each method has its own specific (dis)advantages. For the preparation of the high-*k* materials, the control of the interfacial (SiO_2) layer is very important. Less oxidation during metal oxide deposition is expected. Films prepared by CVD process may suffer easily from contaminates like hydrogen, carbon, and chlorine which originate from the precursor molecules. Opposite to CVD techniques, PVD has an advantage of not suffering from contamination during the deposition. Additionally, PVD involves the reactive oxygen from plasma and low process temperature, which allows the decent control of the interfacial layer thickness and layer composition.

RF reactive magnetron sputtering is a typical method of PVD techniques. In this chapter, the process and properties of HfSiON films deposited by RFMRS will be discussed and presented.

Experimental

P-type Si (100) substrates were cleaned and etched in a diluted hydrofluoric solution to remove native oxide on the surface, leaving an H-terminated surface. HfSiON films were deposited on the cleaned *p*-Si (100) substrates by RF magnetron reactive sputtering of HfO_2 (99.99 %) target covered with SiO (99.99 %) pieces. A mixture of Ar (99.999 %) and N_2 (99.999 %) was used as sputtering gas, which was introduced into the vacuum chamber *via* two gas flow controllers after the chamber was pumped to a base vacuum lower than 1.5×10^{-4} Pa. During the HfSiON films deposition, the applied RF power was varied from 60 to 120 W. The N_2/Ar flow ratio was changed from 0.20 to 0.65. The sputtering pressure was changed from 0.1 to 0.9 Pa. The substrate temperature was in the range from ambient temperature to 300 °C. The reflected power was kept below 0.2 W by adjusting the load and tuning coils. Before HfSiON films deposition, the targets were pre-sputtered in Ar atmosphere for 15 min to remove the surface contamination on the targets.

Film Deposition Rate

Deposition rate parameter is used to describe the film deposition speed, which is influenced by many process parameters such as RF power, sputtering pressure, N_2/Ar flow ratio, substrate temperature, and *etc.* Hence, film deposition and growth is composed by a series of complex processes. Fig. **1** shows the schematic of thin film growth mechanism.

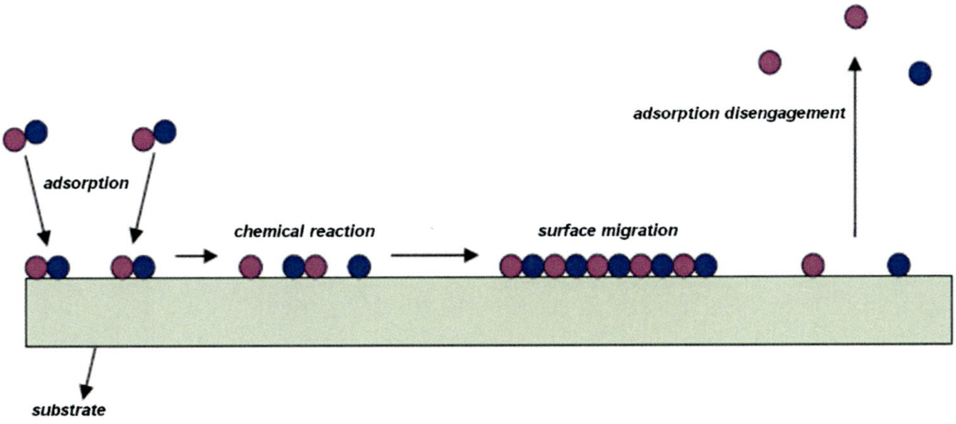

Figure 1: Schematic diagram of film growth.

The growth process usually consists of adsorption, chemical reaction, surface migration, and adsorption disengagement, *etc.* If the atoms reach the substrate they lose its momentum perpendicular to the surface of the outside, then the atoms remain on the substrate surface, known as atomic adsorption, otherwise they will leave the surface and re-enter the space. Adsorbed atoms will diffuse on the substrate surface, as was called the surface migration. When the atoms move to the steps or defects, they remain in the area and collide with other atoms to form clusters, known as nucleation. After the formation of a critical nucleus, the nucleus will grow steadily to become an island with decreasing overall energy. With the further supply of foreign atoms, the island continues to grow. Coalescence of the islands leads to the formation of network structures. The continuous supply with coating atoms will fill the open porous network structure and thus a continuous film will be formed.

In order to guide the film deposition and to understand the mechanism of reactive sputtering, it is very important to investigate the influence of the parameters on the film deposition rate. The deposition rate (R) is films thickness formed on the substrate surface per unit time, which can be expressed as:

$$R = d / t \quad \textbf{(3)}$$

where d is average film thickness and t is deposition time. General method is to measure the thickness of the deposited film under a certain period of time. Then we can get the average film deposition rate. After that, we can control the deposition time so as to control the thickness of the film.

Many processing parameters influence the HfSiON films deposition rate, in which sputtering power is one of the most obvious factors. Fig. **2** shows the relationship between the deposition rate and the RF power. It can be seen that the film deposition rate increases almost linearly with the RF power.

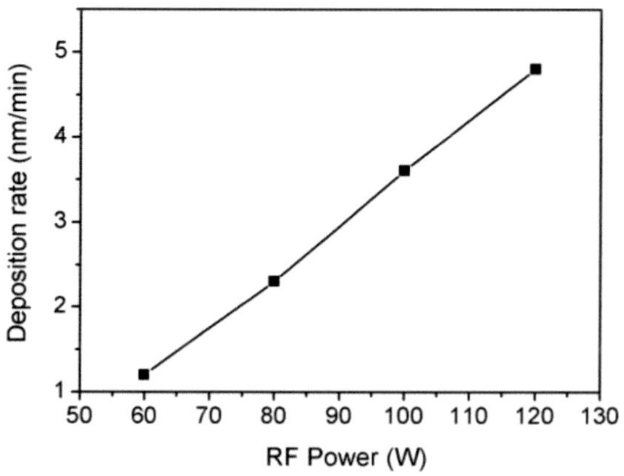

Figure 2: Deposition rate as a function of RF power.

In the sputtering process, the deposition rate R can be expressed as [9]:

$$R = C\, J\, \eta \quad \textbf{(4)}$$

where C is a constant, η is the sputtering yield, and J is the surface current density of sputtering target. When the RF power increases, the kinetic energy of the sputtered particles increases, which results in the increase of the sputtering yield η. Additionally, the RF power will also enhance the surface current density J, causing the increase of the number of ions in sputtering. According to Eq. 4, the higher the η and J, the larger is R. Therefore, the deposition rate of the films increases almost linearly with the increasing RF power.

The variation of the deposition rate with an increasing N_2/Ar flow ratio is shown in Fig. **3**.

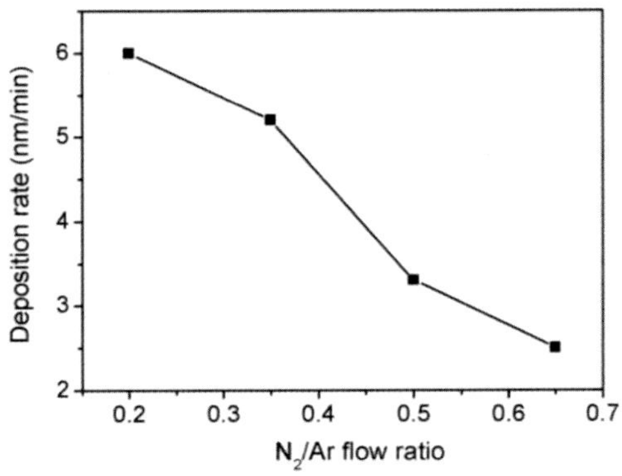

Figure 3: Deposition rate as a function of N_2/Ar flow ratio.

Argon ions have higher mass and larger momentum than nitrogen ions. Hence, argon ions can produce higher sputtering yield and greater deposition rate compared with nitrogen ions. In the case of fixed sputtering pressure, the increase of N_2 flow leads to the target poisoning and makes the reduction of the concentration of Ar gas. It reduces the number of Ar ions, which leads to the falling of the sputtering yield and the decrease of films deposition rate.

Fig. **4** shows the relationship between sputtering pressure and film deposition rate. The other process parameters are kept the same except the sputtering pressure.

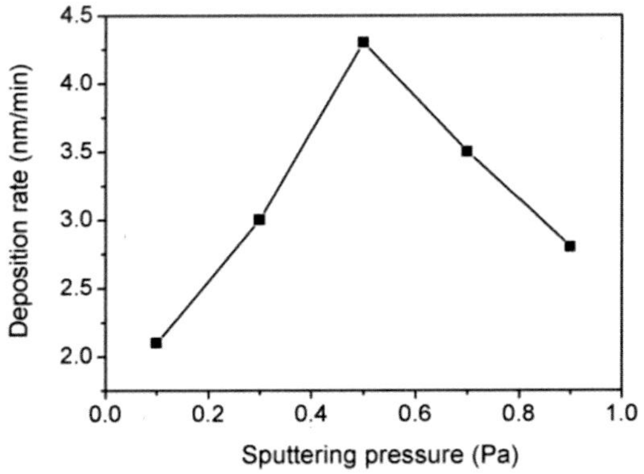

Figure 4: Deposition rate as a function of sputtering pressure.

The deposition rate firstly increases when the pressure increases from 0.1 to 0.5 Pa and then decreases when the pressure increases from 0.5 to 0.9 Pa. When the sputtering pressure is 0.5 Pa, the deposition rate of the films has a peak. In order to discuss the sputtering pressure on the deposition rate of the films, the mean free path ($\bar{\lambda}$) of gas molecules should be introduced. According to the kinetic theory of gas molecules, $\bar{\lambda}$ has a relationship with the pressure, and can be approximately expressed as follows [10]:

$$\bar{\lambda} = \frac{k_B T}{\sqrt{2}\pi d^2 p} \tag{5}$$

where k_B is the Boltzmann constant, T is the gas temperature, d is the diameter of gas molecules, p is the sputtering pressure of deposition. According to the Eq. 5, if the temperature and diameter of the gas molecular are kept the same value, the mean free path of gas molecule decreases with the increasing pressure. Based on the gas discharge theory, the current density j_o is described as follows [10]:

$$j_o = \frac{4\varepsilon_o \mu_i V_c}{d_c^3}(1+\gamma) \tag{6}$$

where γ is Townsend ionization coefficient, d_c is the thickness of the cathode-bit down area, ε_0 is the vacuum permittivity, μ_i is the ion mobility, and V_c is the cathode potential drop. With the increase of the number of the secondary electron emission, γ increases and d_c decreases. According to the Eq. 6, both the increase of γ and the decrease of d_c will lead to the enhancement of the current density j_o. As a result, the deposition rate of the films increases with the increasing current density.

The two factors produced by sputtering pressure are mutual restraint. With the increasing sputtering pressure, the deposition rate of the films increases initially and this time the number of sputtered atoms plays a decisive role. After that, with the further increase of sputtering pressure, deposition rate begins to decrease because of the poisoning effect and due to the collisions between sputtered atoms and gas molecules playing a decisive role. The combined effects of the two factors lead to a peak appearing in the Fig. **4**.

Fig. **5** shows the deposition rate of the films as a function of substrate temperature.

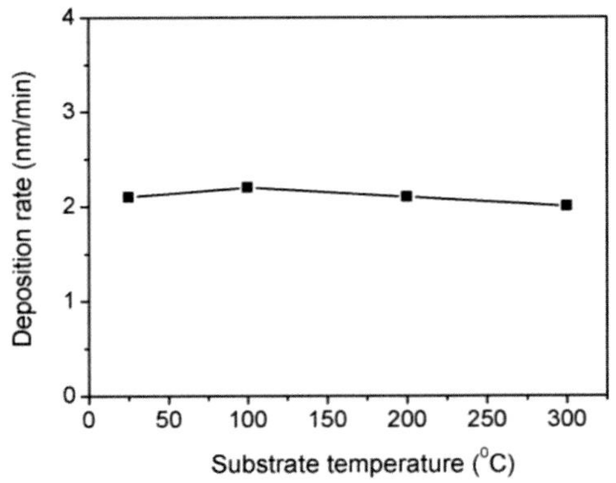

Figure 5: Deposition rate as a function of substrate temperature.

It can be seen that the deposition rate varies little with the increasing substrate temperature. The substrate temperature will be higher than the set value because part of the kinetic energy of the sputtered particles will transform to thermal energy. On one hand, the high temperature will contribute to the increase of the deposition rate. Meanwhile, the higher the substrate temperature the lower the sticking coefficient. The low sticking coefficient means that the average adsorption time is reduced and that the atoms on substrate surface are easily desorbed. Hence, on the other hand, the high temperature will also contribute to the decrease of the deposition rate. The combined two factors result in the little effect of substrate temperature on the deposition rate of the films.

Structural Characterization

Amongst the high-*k* materials, HfO_2 has emerged as a good candidate due to its relatively high dielectric constant, wide bandgap and stability on Si. However, the crystallization temperature of amorphous HfO_2 is very low [$T_g(HfO_2)$ ~500 °C (Ref. 11)]. The ideal structure, for the gate dielectric, is the amorphous state because crystalline structure will induce high leakage current and severe mass transport along the grain boundaries. Hence, it is very important to ensure the stability of dielectric films as a single amorphous phase throughout MOS thermal processing (with temperatures ranging up to ~1000 °C). In order to raise the crystallization temperature, HfO_2 is alloyed with some elements, such as Aluminum, Nitrogen, Silicon, or Lanthanum, to form HfO_2-based gate dielectrics. In this section, structural and thermal stability of HfO_2 and HfSiON films are studied and compared systematically.

The structures of HfSiON and HfO_2 films were characterized by means of grazing incidence X-ray diffraction (GIXRD) and high-resolution scanning electron microscopy (HRSEM). Fig. **6** shows the XRD patterns of HfSiON films deposited with different substrate temperatures.

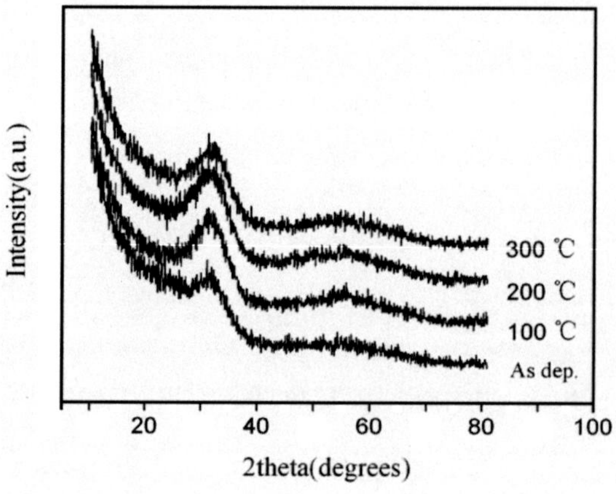

Figure 6: XRD patterns of HfSiON films deposited with different substrate temperatures.

The diffraction patterns mainly include two amorphous wave packets. One wave packet located from 28° to 35° is strong and broad, which is the diffraction peak coming from the Si substrate. The other wave packet located from 50 ° to 60 ° is derived from the films. Except the amorphous wave packets, the diffraction pattern does not appear obvious diffraction peaks, indicating that the HfSiON films deposited at ambient temperature are amorphous. With the increasing substrate temperature, peaks indicative of crystallization are not found in the XRD patterns, indicating the HfSiON films remain amorphous when the substrate temperature varies from ambient temperature to 300 °C.

The XRD patterns of HfO_2 films deposited with different substrate temperatures are shown in Fig. **7**.

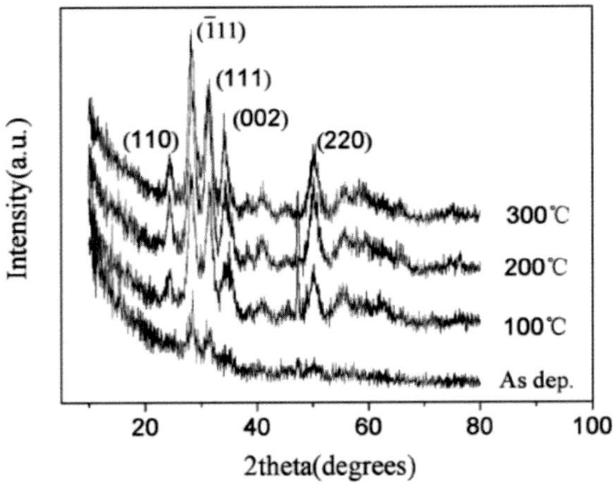

Figure 7: XRD patterns of HfO_2 films deposited with different substrate temperatures.

It can be seen that HfO_2 films deposited at ambient temperature are amorphous. However, for HfO_2 films deposited at 100 °C, peaks indicative of crystallization corresponding to the monoclinic phase HfO_2 are observed. The most intense peak around 28 ° was caused by the diffraction from ($\bar{1}11$) crystal plane of monoclinic HfO_2. As substrate temperature increases, the diffraction peaks from ($\bar{1}11$), (111), (002) crystal planes of monoclinic HfO_2 become sharp and apparent. It means that the crystallization of HfO_2 films is enhanced with the increasing substrate temperature. A comparison of HfO_2 and HfSiON films leads to the conclusion that the incorporation of N and Si in the HfO_2 films may suppress crystallization.

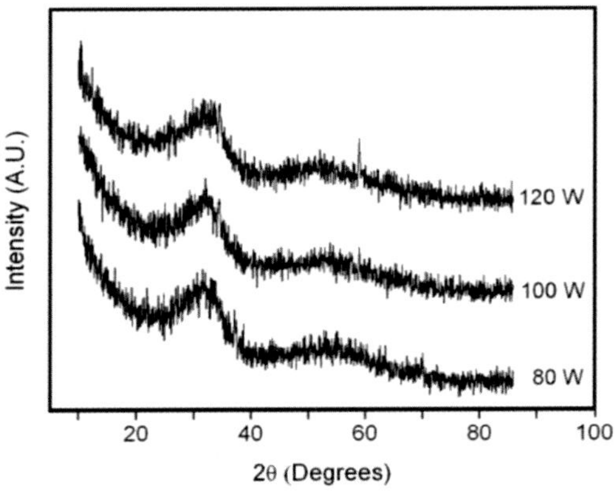

Figure 8: XRD patterns of HfSiON films deposited with different RF powers.

Fig. **8** shows the XRD patterns of HfSiON films deposited with different RF powers. It is clear that the HfSiON films are amorphous under the RF power in the range from 80 to 120 W.

Fig. **9** shows the XRD patterns of HfO_2 films deposited with different RF powers.

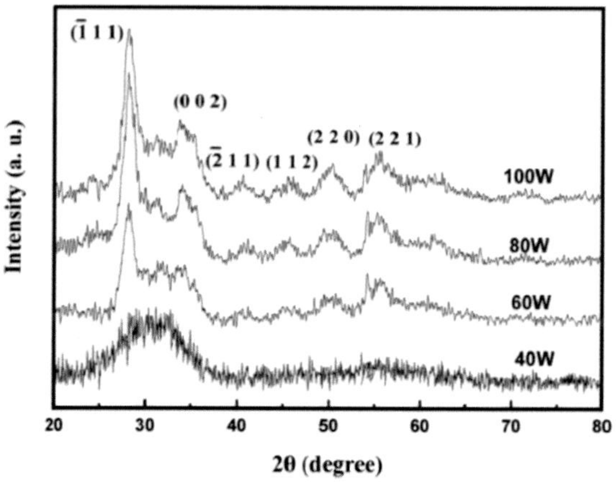

Figure 9: XRD patterns of HfO_2 films deposited with different RF powers.

It can be seen that the HfO_2 films deposited at 40 W are amorphous. However, the HfO_2 films deposited at RF power higher than 60 W exhibit obvious diffraction peaks, showing that HfO_2 films possess all a polycrystalline structure when the RF power is higher than 60 W. The kinetic energy of the sputtered particles from the target increases with the increasing RF power. The high kinetic energy can prompt efficient migration of the sputtered particles when they are transported onto the film-growing surface, which is advantageous to sputtered particles finding the proper site and forming the ordered structure. Consequently, the crystalline films can be obtained when the applied RF power is high enough. Comparing HfO_2 and HfSiON films, it is found that the HfO_2 films appear to be crystallized easily.

High-resolution SEM image of the HfSiON films deposited at RF power of 120 W is shown in Fig. **10**.

It can be seen that the surface of HfSiON films was smooth and featureless. Detectable phase separation or evidence of crystallization was not observed in the surface image of HfSiON films.

Figure 10: SEM image of HfSiON films deposited with RF power of 120 W.

High-resolution SEM image of the HfO_2 films deposited at RF power of 100 W is shown in Fig. **11**.

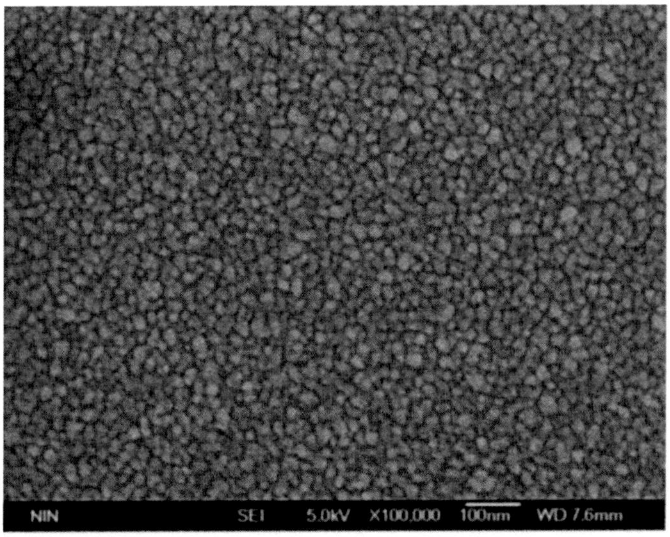

Figure 11: SEM image of HfO_2 films deposited with RF power of 120 W.

The surface of the HfO_2 films shows a granular morphology with small round nodules, indicating the evidence of film crystallization. This result is consistent with the XRD patterns.

Fig. **12** shows the XRD patterns of HfO_2 annealed at different temperatures.

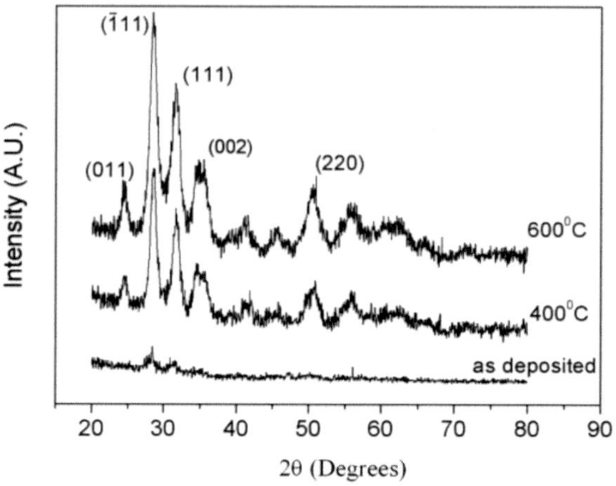

Figure 12a: XRD patterns of HfO_2 films annealed at different temperatures.

As shown in the Fig. **12**, the as-deposited HfO_2 film was amorphous. For the HfO_2 film after 400 °C annealing, peaks indicative of crystallization corresponding to the monoclinic phase of HfO_2 were observed in GI-XRD patterns. With increasing annealing temperature, the peaks became more pronounced indicating better order of HfO_2 crystal structure. Fig. **12** shows the XRD patterns of HfSiON films annealed at different temperatures. Apparently, there were no crystalline peaks in the as-deposited HfSiON films, this indicates an amorphous character. Similar to the as-deposited HfSiON films, the HfSiON films after annealing at 400 °C, 600 °C, and 800 °C were all amorphous.

Comparing Figs. **12** and **13**, shows that according to the XRD patterns crystallization in the HfO_2 films occurs after 400 °C annealing, but there is no detectable crystallization in the HfSiON films after 800 °C annealing. It was shown that HfSiON films have better thermal stability than HfO_2 films.

Compositional Analysis

The chemical states of HfO_2 and HfSiON films prepared on Si substrates were studied by XPS analysis. O1s and Hf 4f XPS spectra for the HfO_2 films are shown in Fig. **13**.

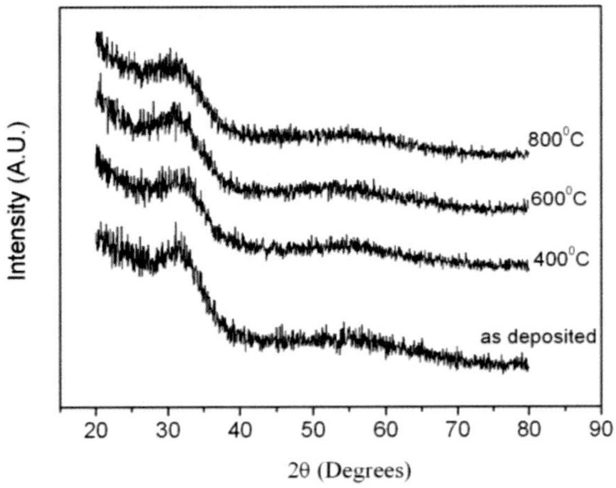

Figure 12b: XRD patterns of HfSiON films annealed at different temperatures.

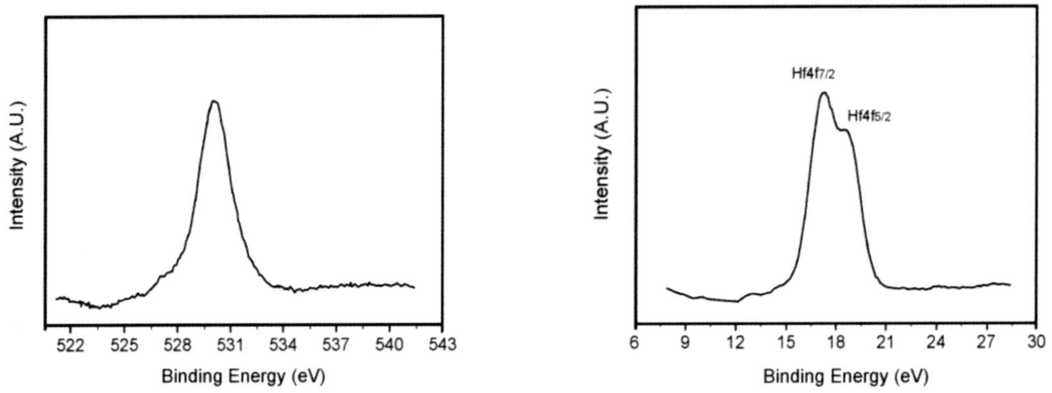

Figure 13: O1s and Hf 4f XPS spectra for HfO_2 films.

For HfO_2 film, the binding energy (BE) of O1s was 530.4 eV. Hf 4f spectra of HfO_2 were composed of two peaks at the BE of 17.2 eV and 18.7 eV, corresponding to Hf $4f_{7/2}$ and Hf $4f_{5/2}$, respectively. The BE values of O1s and Hf $4f_{7/2}$ for HfO_2 film indicate the formation of Hf-O bonds [12, 13]. According to XPS measurements, the atom ratio of Hf to O was estimated to be about 1:2 in the HfO_2 film.

O1s, Si 2p, N1s and Hf 4f XPS spectra for the HfSiON films are shown in Fig. **14**.

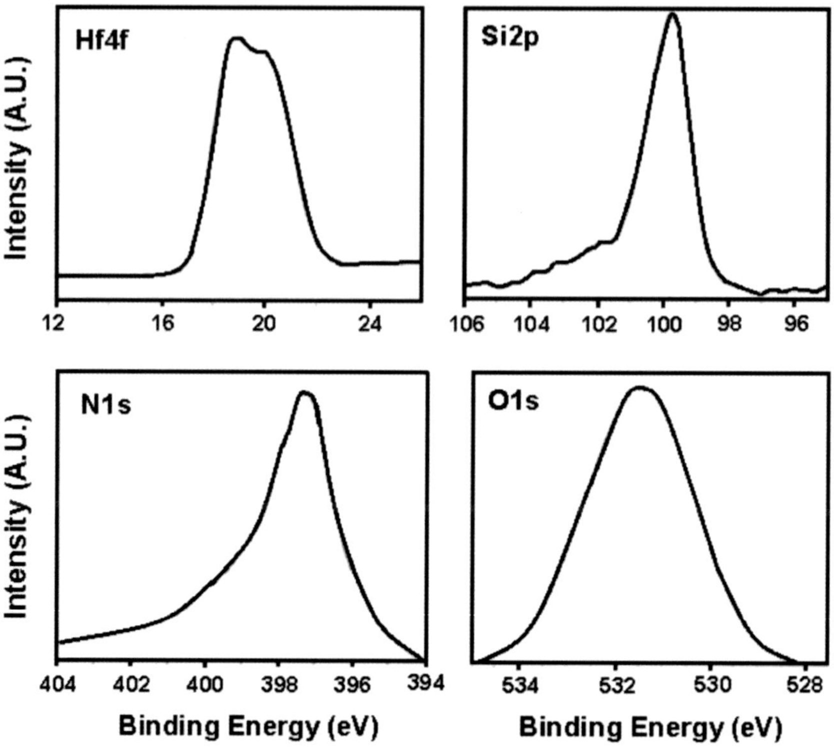

Figure 14: Hf4f, Si2p, N1s and O1s XPS spectra for HfSiON films.

For HfSiON films, the binding energies (BE) of O1s, Si 2p, N1s and Hf $4f_{7/2}$ is 531.5 eV, 99.8 eV, 397.3 eV and 18.5 eV respectively. Comparing the BE of HfO_2 and HfSiON films, shows that the BE of O1s in the HfSiON films is higher than that in the HfO_2 films. This may be related to the reaction between Si, N elements and O element and the formation of the O-Si and O-N bonds in the HfSiON films. The BE of N 1s (397.3 eV) is similar to the BE of N-Si (397.4 eV) bond in the Si_3N_4 films, indicating the formation of Si-N bond in the HfSiON films. Additionally, the peak for Hf-Si bonds was not detected in HfSiON film (BE of 14.6 eV for Hf-Si bonds [14]), indicating that a silicate may be formed.

Compositional Analysis

MOS structure is composed of the metal (M), oxide (O), and semiconductor (S). It is a common basic structure and the most important part in planar devices and integrated circuits. As we know, the stability and reliability of planar devices and integrated circuits are greatly depended on the electrical, structural and reliability of the MOS structures. Current-voltage (I-V) and capacitance-voltage (C-V) characteristics are the most important electrical properties of MOS structures. Some important parameters can be gained from electrical properties measurements, such as leakage current, flat band voltage shift, and dielectric constant, *etc.* In addition, the measurements of electrical characteristics can be used to test the preparation and processing of MOS devices Influences of processing parameters on electrical properties of MOS structures with HfSiON dielectric films are reported in this section.

Preparation of MOS Structures

Au/HfSiON/p-Si/Al MOS structures should be prepared before the electrical properties measurements. Firstly, HfSiON dielectric films were deposited on the top surface of *p*-type Si substrate by RFMRS method. In order to reduce the contact resistance, the Al bottom electrodes were prepared on the bottom surface of *p*-type Si substrate so as to form the ohmic contact of Al/Si. Then, the Au top electrodes were prepared on the surface of the HfSiON diclcctric films. The diagram of the MOS structure is shown in Fig. **15**.

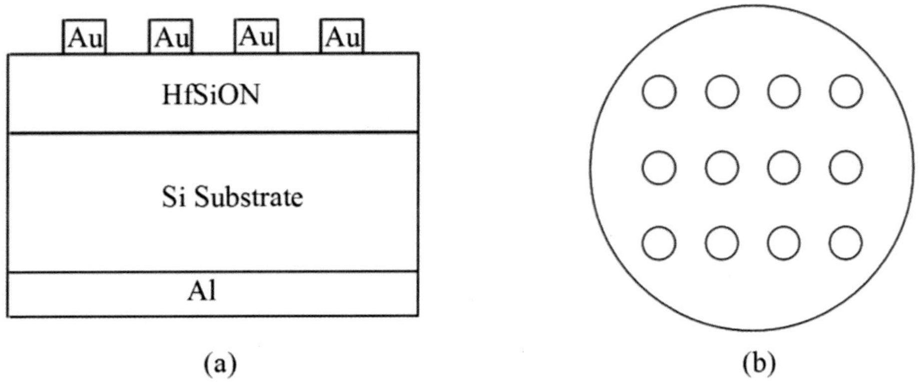

Figure 15: Schematics of MOS structure (a) side view of MOS structure, (b) top electrodes on the HfSiON film.

It can be seen that the top electrodes are distributed regularly on the surface of the HfSiON gate dielectric film. However, the bottom electrode covers the entire back surface of the Si substrate.

Preparation of MOS Structures

I-V characteristics for the MOS structures as a function of RF power are shown in Fig. **16**.

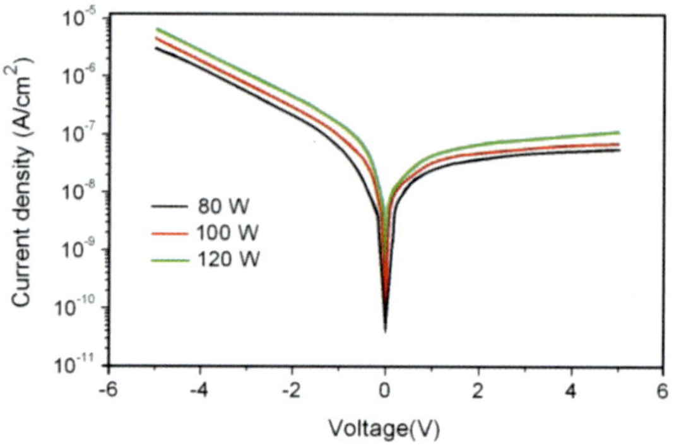

Figure 16: Leakage current density of the MOS structure as a function of RF power.

It was noticed that the leakage current was asymmetric for all the HfSiON films, indicating that the potential barrier height at the interface of Au/HfSiON films was lower than that at the interface of HfSiON/Si. Additionally, it is obvious that the leakage current density increases with the increasing RF power. Fig. **8** shows that the structure of the HfSiON films reveals no significant change as the RF power increases. Hence, the increase of the leakage current density is not owing to the crystallization of the HfSiON films. With the increasing RF power, the kinetic energy of the sputtered particles from the target increases. The bombardment of high energetic sputtered particles may lead to the increase of defects in the HfSiON films, resulting in the increase of the leakage current density.

C-V measurements for the MOS structures were performed at a frequency of 1 MHz. Fig. **17** shows a typical C-V curve for the MOS structures measured at high-frequency.

It clearly shows the accumulation region, the depletion region and the inversion region of the MOS structure. The C-V curves shift to negative gate bias voltage, showing the effective positive charges (oxygen vacancies) or that non-bonded metal atoms were formed in the HfSiON films [15]. Additionally, the curves of the forward and reverse scans are not completely overlapping, indicating that there is some amount of positive charges or interface states in the HfSiON films. Furthermore, the relative dielectric constant of the HfSiON films can be obtained from the C-V curves. The relative dielectric constant of the HfSiON films is expressed as follows:

$$\varepsilon_{ox} = \frac{d_{ox} C_{\max}}{\varepsilon_0 S} \tag{7}$$

where C_{max} is the maximal capacitance in the accumulation region, S the area of the electrode, d_{ox} the physical thickness of the films and ε_0 the permittivity of vacuum. From Fig. **17**, the maximal capacitance C_{max} in accumulation was found to be 482 pF for the HfSiON films. According to Eq. (7), the dielectric constant for the HfSiON films was estimated to be around 16.9.

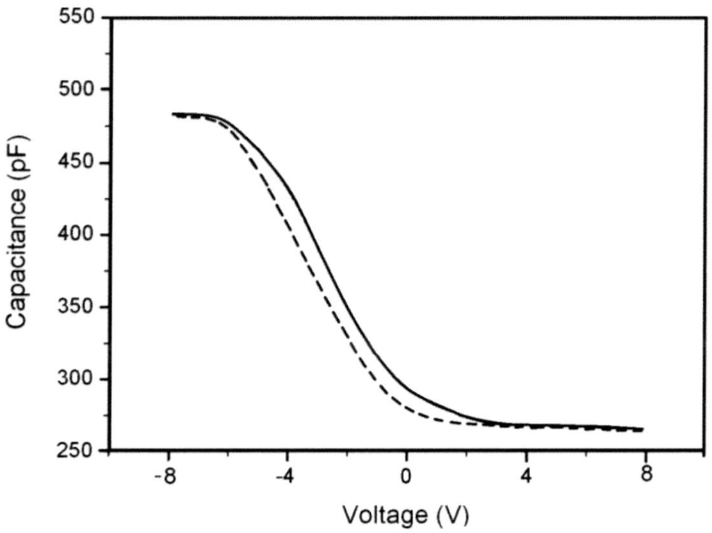

Figure 17: C-V curve of MOS capacitor at the high frequency (f=1MHz).

C-V measurements for the Au/HfSiON/p-Si/Al structures with the HfSiON films deposited at different N_2/Ar flow ratios were performed. Fig. **18** shows the dielectric constant as a function of N_2/Ar flow ratio.

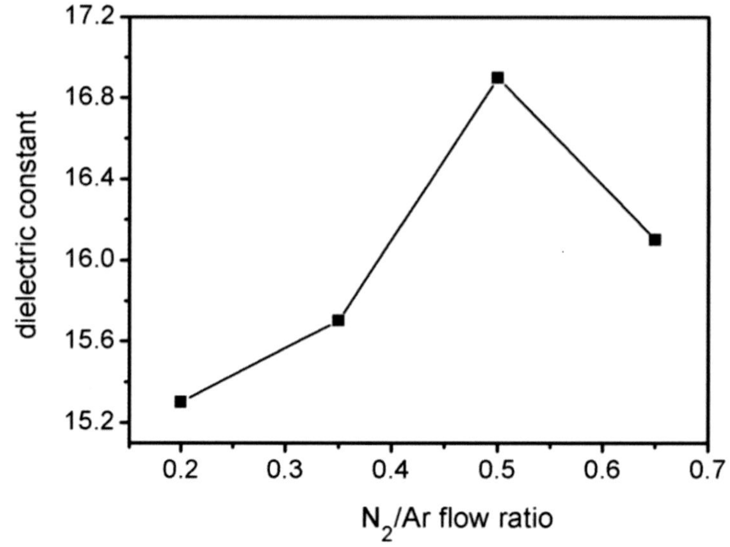

Figure 18: Dielectric constant of the MOS structure as a function of N_2/Ar flow ratio.

It is clear that the dielectric constant of the HfSiON films increases when the N_2/Ar flow ratio increases from 0.20 to 0.50. The N content in the HfSiON films has been increased with the increasing N_2/Ar flow ratio. The increase of N ions in the films may enhance the electronic polarization as well as the ionic polarization, both of which contribute to the enhancement of the dielectric constant [16]. Hence, the dielectric constant of the HfSiON films increases when the N_2/Ar flow ratio changes from 0.20 to 0.50. Nevertheless, the dielectric constant of the HfSiON films decreases when the N_2/Ar flow ratio increases from 0.50 to 0.65. It was reported that the excessive N in the HfSiON films may accumulate in the interface layer and cause the increase of the interface defects or interface state [17]. This may relate to the decrease of the dielectric constant when the N_2/Ar flow ratio increases from 0.50 to 0.65.

The value of flat band voltage (V_{fb}) shift for the films can be determined by the flat band capacitance (C_{fb}) method [18], which was expressed as follows:

$$C_{fb} = \sqrt{\frac{\varepsilon q N_A}{\Phi_t}} \qquad \textbf{(8)}$$

where ε is the dielectric constant of the Si semiconductor, q the elementary charge, N_A the acceptor impurity density and Φ_t, the thermal voltage ($\Phi_t = kT/q \approx$ 26 mV at the room temperature). Thus, through Eq. (8) and the measured C-V curves, the V_{fb} shift can be obtained. Fig. **19** shows the V_{fb} shift as a function of N_2/Ar flow ratio.

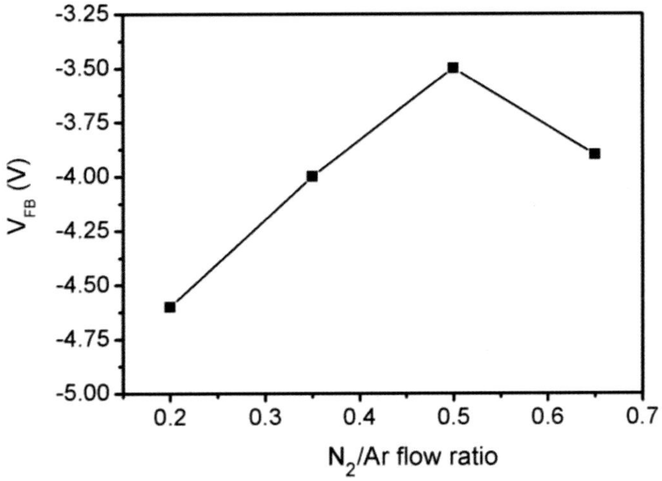

Figure 19: V_{FB} of the MOS structure as a function of N_2/Ar flow ratio.

It is obvious that the V_{fb} shift decreases initially and then increases with the increasing N_2/Ar flow ratio. The variation of the V_{fb} shift may be related to the fixed charge densities. The amount of fixed positive charge density (Q_f) could be obtained according to the calculated V_{fb} value by [19]:

$$Q_f = \frac{C_{max}(V_{ms} - V_{fb})}{qS} \tag{9}$$

where C_{max} is the maximal capacitance in accumulation, V_{ms} the work function difference between the Au (metal) and Si (semiconductor), q the electron charge and S the electrode area. Through Eq. (9), the fixed positive charge densities can be obtained. When the N_2/Ar flow ratio increases from 0.20 to 0.50, the fixed positive charge densities decrease because of the reduction of the oxygen vacancies or non-bonded metal atoms in the films. The decrease of the fixed positive charge densities may cause the reduction of the V_{fb} shift. However, the fixed positive charge densities increase when the N_2/Ar flow ratio is further increased from 0.50 to 0.65, which leads to the increase of the V_{fb} shift. These results show that the N_2/Ar flow ratio has great effect on the electrical properties of the HfSiON films.

CONCLUSION

Radio frequency magnetron reactive sputtering is a good technique to deposit oxides and nitrides thin films because of its several advantages such as low deposition temperature, good uniformity, and also the ability to deposit compact films with high mechanical integrity and stability. However, it is well known that the characteristics of the thin films depend strongly on the sputtering parameters during the magnetron sputtering process. For HfSiON thin films, its properties are affected by RF power, sputtering pressure, N_2/Ar flow ratio, substrate temperature, *etc.* Therefore, the effects of processing parameters on the properties of sputtered HfSiON thin films should be studied systematically so as to obtain high quality HfSiON thin films. Consequently, complementary investigations and analysis are required in order to improve the physical properties of this promising high-*k* gate dielectric.

ACKNOWLEDGEMENT

Declared none.

CONFLICT OF INTEREST

The author(s) confirm that this chapter content has no conflict of interest.

REFERENCES

[1] G.E. Moore, Electronics, vol. 38, pp. 114-117, 1965.
[2] S.G. Dos Santos, L.C. Hasenack, L.C. Salay, P. Mertens, J. Electrochem. Soc., vol. 142, pp. 902-907, 1995.
[3] C.C. Liao, A. Chin, C. Tsai, J. Crystal Growth, vol. 201/202, pp. 652-655, 1999.
[4] International Technology Roadmap for Semiconductors, Semiconductor Industry Association, San Jose, CA, 2003.
[5] M.R. Visokay, J.J. Chambers, A.L.P. Rotondaro, A. Shanware, L. Colombo, Appl. Phys. Lett., vol. 80, pp. 3183-3, 2002.

[6] M.A. Quevedo-Lopez, J.J. Chamber, M.R. Visokay, A. Shanware, L. Colombo, Appl. Phys. Lett., vol. 87, pp. 012902-3, 2005.

[7] S. Miwa, Appl. Surf. Sci., vol. 255, pp. 1384-1386, 2008.

[8] J. Martin-Martinez *et al.* Microelectronics Reliability, vol. 50, pp. 1263-1266, 2010.

[9] T.F. Tang, P.F. Gu, Thin film optics and technology, Machinery Industry Press, 1989.

[10] S.Z. Cheng, J.Z. Yong, General physics. Higher Education Press. 1982.

[11] D.A. Neumayer, E. Cartier, J. Appl. Phys., vol. 90, pp. 1801-8, 2001.

[12] D. Triyoso *et al.* J. Electrochem. Soc., vol. 151, pp. F220-F227, 2004.

[13] J.H. Hong, T.H. Moon, J.M. Myoung, Microelectron. Eng., vol. 75, pp. 263-268, 2004.

[14] M.H. Cho *et al.* Appl. Phys. Lett., vol. 81, pp. 472-3, 2002.

[15] M. Hiratani, K. Torii, Y. Shimamato, S. Saito, Appl. Surf. Sci., vol. 216, pp. 208-214, 2003.

[16] M. Koyama *et al.* IEDM, pp. 849-852, 2002.

[17] J.C. Lee *et al.* IEDM, pp. 441-444, 2003.

[18] S.M. Sze, Physics of Semiconductor Devices, 2nd ed., Wiley, New York, 1981.

[19] W.L. Guo, Silicon/silicon dioxide interface physics, National Defence Industry Press, Beijing, 1982.

Send Orders of Reprints at reprints@benthamscience.net
254 *Metallic Oxynitride Thin Films by Reactive Sputtering and Related Deposition Methods*, **2013, 254-264**

CHAPTER 10

Properties of Oxynitride Thin Films for Biomedical Applications

Martin Fenker*

FEM - Forschungsinstitut Edelmetalle und Metallchemie, Katharinenstraße 17, 73525 Schwäbisch Gmünd, Germany

Abstract: Several biomedical coatings on the basis of oxynitride compounds have been studied in the last 20 years. Published results will be presented on TiNbON, NbON, TiON and SiON. These coatings have been deposited by PVD methods, like cathodic arc evaporation and plasma immersion ion implantation and deposition. The aim of the studies was to improve the biomedical parts with respect to biocompatibility, corrosion resistance and mechanical strength, like hardness or wear resistance. The coatings have been characterized by elemental depth profiling, X-ray diffraction, nanohardness and wear tests. For SiON the platelet adhesion test and the enzyme-linked immunoassay method have been used. The latter two methods have proven that SiON has a better blood compatibility compared to low-temperature isotropic pyrolytic carbon (LTIC). Hence, SiON is suggested as a viable coating in blood-contacting medical products. Finally, the patent situation about oxynitride coatings for biomedical usage has been analyzed and will be briefly discussed.

Keywords: Biomedical coatings, PVD, cathodic arc evaporation, plasma immersion ion implantation and deposition, oxynitride, TiNbON, NbON, TiON, SiON, patents, diffusion barrier, chemical composition, structure, abrasive wear test, Ringer´s solution, cytotoxicity tests, platelet-adhesion test, water contact angles, enzyme-linked immunoassay method, blood compatibility.

INTRODUCTION

Parts, like implants, which have to be implemented into the body of humans or animals to replace or assist dysfunctional body functions, have to be biocompatible. However, not all structural materials fulfill this requirement and have to be coated with biocompatible coatings or surface treated. Bioinert materials are *e.g.* Ti, Ta, Nb, Al_2O_3 or ZrO_2 *etc.* The hardness of these bioinert coating materials has to be enhanced when they are used in applications where high shear forces occur, like for example for hip joint prostheses. In this case hard

***Address correspondence to Martin Fenker:** FEM - Forschungsinstitut Edelmetalle und Metallchemie, Katharinenstraße 17, 73525 Schwäbisch Gmünd, Germany; Tel: +49-7171-1006-400; Fax: +49-7171-1006-900; E-mail: fenker@fem-online.de coatings like nitrides or oxynitrides can be a solution for this problem [1]. Different oxynitride materials like TiNbON, NbON, TiON and SiON have been reported in literature and will be discussed in the following paragraphs. The insertion of oxygen leads to more biocompatible, especially osteophil characteristics and blood compatibility. The coatings should be hard with high wear resistance, should have a good passivating property and a high corrosion resistance.

BIOCOMPATIBILITY *VS.* OXYNITRIDE MATERIALS

TiNbON

Several publications on TiNbON as biomedical coating can be found in the literature [1-5]. In the publication of Probst *et al.* [1], a more detailed study of the deposition of the TiNbON coatings was presented. The main results will be briefly discussed in the following.

Probst *et al.* used the steered arc technique with a special sector target to deposit TiNbON coatings onto medical grade titanium. The aim of their study was the development and characterization of wear resistant and adhesive (oxy-)nitride refractory metal PVD coatings for biomedical applications. In implantology, titanium dominates as a biocompatible material. Another reason, asides from the biocompatibility, is its high corrosion resistance in the aggressive body environment. In hip joint prostheses, high shear strengths occur between the metallic ball and the polyethylene (PE)-socket. The resulting wear debris can be attributed to the high friction coefficient and the low hardness of titanium(oxide) of 150 $HV_{0.1}$, on average. To overcome these severe drawbacks, surface materials were developed based on hard coatings with high wear resistance, good passivating properties and high corrosion resistance. Surface modification of medical grade titanium was performed by the PVD process. For coating deposition a PVD system, type PLS 570 (Pfeiffer Vacuum, Wiesbaden, Germany) with cathodic steered arc was used. The vaporization of the metal components occurred from a bisectoral target. Half discs of pure titanium and niobium were welded together. The specimen were physically cleaned by an argon plasma (total pressure: $6.4x10^{-3}$ mbar; power: 200 W) for 30 min and heated to a starting temperature of approximately 200 °C. For the nitride and oxynitride coatings, the cleaned titanium specimens were rotated for 2 h in front of the arc cathodes in a nitrogen/oxygen atmosphere of $8.0x10^{-3}$ mbar constant total pressure. The applied substrate voltage was U_S = -300 V. The deposition rate was in the range of 1 µm/h. The ratio of the O_2/N_2-partial gas pressure was varied between 0.08 and 1.0.

Secondary neutral mass spectrometry (SNMS)-depth profile measurements presented in Fig. **1** exemplarily show the elemental concentrations of the (Ti,Nb)ON coatings as a function of the sputtering depth. The PVD sputtered coating show a depth-stable oxygen concentration in the range of 12%–15%.

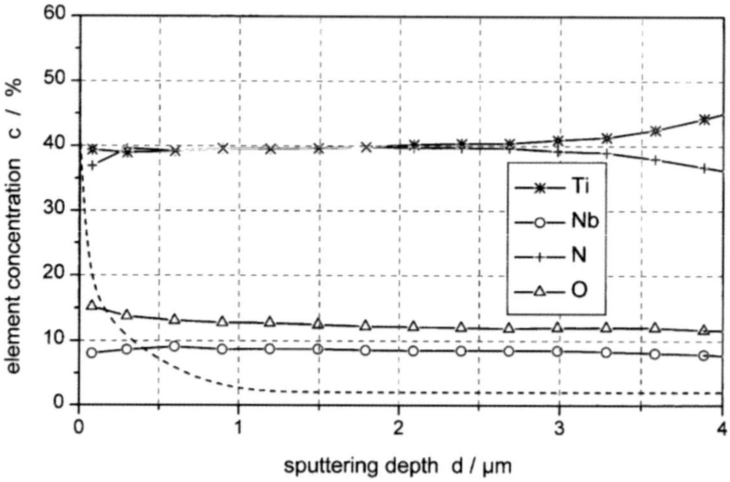

Figure 1: SNMS-depth profile of a (Ti,Nb)ON coating on titanium [1]. In contrast to the appropriate curve of the oxygen content for an oxidized (Ti,Nb)N layer (dashed line) a strong oxygen gradient at the surface is avoided with the PVD coating technique (–Δ–).

Fig. **2** presents the result of the XRD investigations of (Ti,Nb)ON films in the 2θ-region between 30° and 80°. Oxygen between 2% and 21% was introduced in the layer with all other process parameters remaining constant. The two cubic interplanar spacings (111) and (222) of the coatings are labeled with Miller indices. Likewise the spacings of the Ti substrate appear.

For the surface modification of titanium-based materials, the oxygen diffusion hardening (ODH) process is frequently applied, where a controlled diffusion of atomic oxygen into the lattice of Ti takes place. The resulting distortion of the

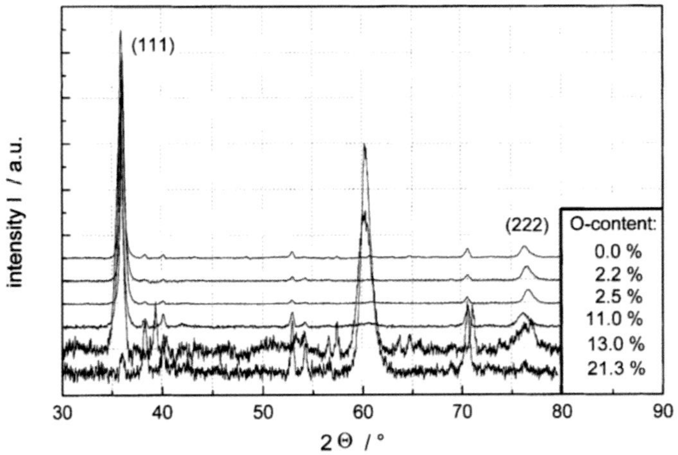

Figure 2: XRD patterns of (Ti,Nb)ON coatings on titanium [1]. The oxygen content increases from top to bottom. Beginning at 11% oxygen content, the preferred (111) growth direction of the layer lattice disappears. Macroscopically this expresses itself in an increasing layer embrittlement.

crystal lattice leads, *e.g.* for titanium alloys, to an increase of the surface hardness of 350 HV for the untreated material up to 1200 HV. However, the high oxygen gradient at the surface layer and the associated lattice distortion close to the surface are unfavorable for additional coatings. The resulting considerable compressive stresses will usually cause a spalling of the coating even during small lateral loads. Therefore, the aim of this work was the development of a coating with constant oxygen depth profile, thereby avoiding a strong oxygen gradient at the coating surface. XRD diagrams of different (Ti,Nb)ON coatings are shown in Fig. **2**. Above approximately 11.0% oxygen content, the intensity of the preferred (111) growth direction strongly decreases; until at 21.3% the (111) reflection has almost vanished. Obviously there is no successive replacement of nitrogen atoms by oxygen atoms, as in the case of the metallic components, which do not change the basic crystalline structure. The new signal appearing at $2\theta = 60.20°$ (13.0% O content) and $2\theta = 60.38°$ (21.3% O content), respectively, cannot be assigned to a cubic or any another crystalline structure. Also, the large peak width of $\beta = 1.5°$ rather indicates a lattice distortion caused by the insertion of oxygen into interstitial lattice sites. The disappearing of the preferred (111) growth direction in the presence of higher oxygen content is accompanied by the embrittlement of the material surface observed with a scanning electron microscope. Therefore, the XRD measurements can be seen as a sensitive method to estimate the beginning of the embrittlement of this type of coating material and the maximum possible oxygen that can be tolerated without functional loss of the coating.

Probst *et al.* [1] concluded that by the insertion of nitrogen into the metal lattice, the coating hardness improves, and the insertion of oxygen leads to more biocompatible, especially osteophil characteristics. With the aid of the PVD process, oxygen can be inserted into the lattice of the film without any concentration gradient, which was confirmed by SNMS-measurements. However, the oxygen atoms do not replace the nitrogen atoms in the same way the metallic components do. XRD proved to be a very sensitive method to detect the onset of embrittlement of the coating caused by excessive insertion of oxygen atoms at interstitial lattice sites.

Hunter *et al.* [5] studied the abrasive wear of TiNbON coatings in comparison to other hard coatings like TiNbN, ZrN or DLC. However, no details on the deposition process were reported. From the results presented, it seems that an arc technology was used: "coatings have a heterogeneous structure with numerous nodules and pores, which may be inherent in the vapor deposition process (perhaps due to embedment of solid particles, or chemical segregation from elevated temperatures and long processing durations)". The abrasive wear test was conducted in physical solution (Ringer´s solution) in a reciprocating pin-on-disc device. The test revealed that the ceramic-coated CoCr or Ti-6Al-4V exhibited better abrasion resistance compared to the untreated or solely surface hardened alloys. A superior performance in comparison to the ceramic-coated CoCr or Ti-6Al-4V was found for an oxidized Zr-2.5Nb alloy. The better performance of this sample was ascribed to the homogeneous microstructure of the oxide coating with very small columnar grains and no inherent pores, which leads to excellent cohesion and adhesion. However none of the surface modifications exhibited sufficiently improved performance in the study of Hunter *et al.* compared to the untreated CoCr. It should be emphasized, that other vapor deposition technologies like magnetron sputtering and a modified microstructure (*e.g.* gradient coatings) could have resulted in better wear performance of the nitride or oxynitride coating. The paper lacks of an optimization procedure for the coatings. The biocompatibility of the materials was not subject of the paper.

NbON

Recently, cytotoxicity tests have been performed on three selected Nb-O-N coatings (for more details see [6]). All coatings deposited onto glass slides were well tolerated by 3T3 mouse cells directly grown on the coatings or beside the glass slides. There were no statistical significant differences observed between the tested coatings and the control for all parameters analysed.

TiON

Wan *et al.* [7] studied graded Ti-O-N coatings for biomedical implants. The coatings have been deposited onto stainless steel by plasma immersion ion implantation and deposition (PIII&D). For more information about the PIII&D technique please have a look at reference [8]. In previous investigations the authors have noticed that Ti-O films can improve the blood compatibility of implants substantially. Furthermore, TiN films have excellent mechanical properties. Hence, to achieve higher durability as well as excellent blood compatibility, nano-structured Ti-O-N gradient films have been synthesized on stainless steel by PIII&D. Tension tests have revealed that no delamination, peeling or cracking occurred on the Ti-O-N coated SS316L sheets after relatively high plastic deformation. A high binding strength, hardness and Young′s modulus of the Ti-O-N film to the SS316L substrate was confirmed by scratch and nanoindentation testing. Microstructural investigations showed that the films consist of rutile Ti-O and cubic TiN crystals with a grain size of 40 nm. The authors hypothesize that the excellent mechanical properties of the synthesized Ti-O-N film are attributable to the nano-crystalline structure of the films, their high density, and excellent surface quality as well as a broader film/matrix interface achieved by the PIII&D process. However, no details about the values of the mechanical properties were available as only an abstract of the paper of Wan *et al.* could be obtained but not the manuscript itself. Furthermore, the authors have not studied the biomedical properties of the Ti-O-N coated SS316L samples.

SiON

Silicon nitride and oxynitride films have been deposited by Wan *et al.* [9] by filtered cathodic arc evaporation of a silicon target combined with PIII&D. As reactive gases nitrogen, oxygen or ammonia have been fed into the deposition chamber. Sample #1 has been deposited by introducing oxygen and nitrogen. For samples #2 and #3 the oxygen flow was zero and only nitrogen (sample #2) or ammonia (sample #3) has been fed into the deposition chamber. According to XPS measurements the surface chemical composition of the Si-O-N coating of sample #1 was about 56 at.% oxygen, 6 at.% nitrogen, 33 at.% silicon and about 5 at.% carbon. Surprisingly, also the silicon nitride coatings (samples #2 and #3) contain a lot of oxygen in the surface: about 50 at.% and 31 at.% for the coatings produced with N_2 and NH_3, respectively. The authors ascribe this "contamination" to the strong affinity of oxygen to silicon during deposition. As no oxygen was fed into the chamber during deposition of these coatings, it can be assumed that the oxygen originates from the residual gas of the deposition chamber.

The platelet-adhesion test was conducted on the Si-O-N samples as well as on low-temperature isotropic pyrolytic carbon (LTIC) serving as the control. LTIC is the most common clinical blood-contacting biomaterial used in artificial heart valves. The cultivation time was two hours in each test. The blood-platelet-adhesion behavior is a good indication of blood compatibility as thrombosis is predominately formed by platelet-activation reactions in a complex cascade way. The quantity and morphology of the adhered platelets on the samples have been studied by scanning electron microscopy (SEM). On LTIC, apparent aggregation of adhered platelets occurs and many of them have undergone structural change as manifested by serious pseudopodium and spreading. This phenomenon is typically accompanied by the secretion or synthesis of prothrombotic factors which activate the platelets. Much less aggregation is observed on the Si-O-N films, although it should be mentioned that all three samples exhibit some degree of pseudopodium, albeit not as severe as that on LTIC (see Table **1**). The platelets on samples #2 and #3 exhibit much less pseudopodium as demonstrated by the isolated and nearly round state, indicating less activation and thrombosis risks.

In addition to the platelet behavior, fibrinogen conformational changes measured by the enzyme-linked immunoassay method (ELISA) reveal the thrombosis tendency as the fibrinogen (called the first coagulation factor) conformational change is considered the initial triggering step to stimulate coagulation. The structural changes can be assessed as relative concentrations of the antibody marker for the transformed fibrinogen by measuring the optical density (492 nm).

Fibrinogens adsorbed on LTIC have undergone larger conformational changes than those on the Si-O-N samples. Samples #1 and #2 show obviously smaller values compared to LTIC, and sample #3 shows the least activated states.

Table 1: Platelet adherent numbers measured by SEM from platelet adhesion-tested samples (120 min) and relative concentrations of fibrinogen conformationally transformed represented by the optical density in the ELISA measurement (492 nm). In the first row, the used precursor gases are given in brackets.

Samples N°	Platelets Adherent Number ($x10^4/mm^2$)	Fibrinogen Transformed Number (Relative Concentration)
LTIC (control)	7.14 (roughly)	2.86 ± 0.1
#1 (O_2/N_2)	5.89 ± 0.03	1.99 ± 0.1
#2 (N_2)	4.15 ± 0.02	1.96 ± 0.2
#3 (NH_3)	5.53 ± 0.05	0.82 ± 0.08

The surface free energy, which is also represented by surface tension, affects the interfacial reactions of platelets and fibrinogen. Some of the reactions are less significant on hydrophilic surfaces.

Table **2** indicates that samples #2 and #3 exhibit the smallest water contact angles and relatively high surface tensions, correlating with the least degree of platelet activation results.

Table 2: Water contact angles, surface tension and its dispersive and polar components, and polar contribution ratio determined from the Si-O-N films and LTIC. In the first row, the used precursor gases are given in brackets.

Samples N°	Water Contact Angle (θ)	γ_s ($mN.m^{-1}$)	γ_s^d ($mN.m^{-1}$)	γ_s^p ($mN.m^{-1}$)	γ_s^p / γ_s
LTIC	78.4±2	44.5 ± 0.3	38.58 ± 0.5	5.83 ± 2.1	0.16
#1 (O_2/N_2)	54.1 ± 2	40.8 ± 0.2	8.37 ± 0.3	32.43 ± 0.9	0.79
#2 (N_2)	49.2 ± 3	47.0 ± 0.2	23.79 ± 0.4	23.23 ± 1.2	0.49
#3 (NH_3)	49.5 ± 2	49.9 ± 0.1	24.35 ± 0.3	25.53 ± 2.2	0.51

The high surface tensions and hydrophilicity can be attributed to the high polar contribution of the surface energy. Although sample #1 has the highest polar contribution of the surface energy, the much smaller dispersion contribution makes the surface energy lower and more hydrophobic, resulting in poorer platelets adhesion and fibrinogen conformational performance. LTIC is more hydrophobic than the Si-O-N films, showing a very low polar contribution to the surface energy, possibly resulting in stronger interactions with fibrinogen and platelets. Therefore, the relatively hydrophilic Si-O-N films might offer one suitable conditioning surface to make the surface more compatible with blood. The various hydrophilic properties of the films are related to the bonding states such as Si-N and Si-O-N, which contribute more to the higher hydrophilicity than Si-O and Si-Si, according to the obtained results.

A silicon cathodic arc combined with PIII&D was employed to synthesize Si-O-N films. The nitrogen concentrations can be adjusted using different precursors, and the films have good quality. According to the platelet behavior and fibrinogen conformational change, the films have a better blood compatibility than LTIC, making the materials viable in blood-contacting medical products such as bioMEMS, biosensors, and artificial implants.

Another paper, from Wilhelm *et al.* reports the deposition of Ta_2O_5 and SiON coatings as the two pH-sensitive layers for a pH sensor [10]. The sensor is based on the differential measurements between two electrically insulated pH-FETs. The oxynitride coating was deposited by PECVD on a PECVD-Si_3N_4 layer. However, the pH sensitivity of the oxynitride layer was too high. The authors suggests in their paper to improve the pH sensor by increasing the oxygen content in the oxynitride layer. This example shows how the behavior of oxynitride coatings can be enhanced by simply changing the metalloid concentration.

PATENTS

A US patent (US7037603 (B2), [11]) filed in 2004, discusses the usage of oxynitrides (Si-O-N, Al-O-N or Al-Si-O-N) as a hermetic protective coating against low temperature degradation of yttria-stabilized zirconia (YSZ) in biomedical implants. The invention discusses different coating materials comprising these three oxynitrides. These ceramic oxynitride coatings may be applied by combustion chemical vapor deposition, physical vapor deposition, electron beam evaporation, ion beam assisted deposition, ion implantation, or chemical vapor deposition.

For dental devices, TiON, HfON, SiON, ZrON coatings as ceramic layers are mentioned in the US patent application US2006105297 [12]. This layer which acts as a diffusion barrier against gold diffusion could also consist of the relevant metal nitrides or oxides. As a joint layer for a final porcelain layer a nanocomposite layer is claimed consisting of gold and minor constituents like nitrides, oxides or oxynitrides of In, Sn or Zn. The minor constituents of the nanocomposite bond to porcelain. The base layer is a self-supporting metallic structural layer (*e.g.* Co alloy) that provides most of the strength.

Beneath nitrides, borides and carbides, TiON and TiAlON are claimed in the US patent US7320798 [13] as a first layer on an implant followed by a final layer consisting of apatite and a binder phase. The first layer protects the metallic implants from corrosion, apatite dissolution, and interfacial reaction with apatite and the binder. The first layer comprises a dense material insoluble and inert in body fluids.

CONCLUSION

It has been shown that oxynitride coatings are studied and patents are filed for their usage on biomedical parts. However, only a few publications on this issue exist up to now in literature. Despite this fact, oxynitride coatings have been commercialized, *e.g.* TiNbON is applied on biomedical parts by job coaters. Further investigations are necessary to broaden the knowledge basis on the biocompatible properties of oxynitride coatings.

ACKNOWLEDGEMENT

Declared none.

CONFLICT OF INTEREST

The author(s) confirm that this chapter content has no conflict of interest.

REFERENCES

[1] J. Probst, U. Gbureck, and R. Thull, "Binary nitride and oxynitride PVD coatings on titanium for biomedical applications", Surf. Coat. Technol., vol. 148, pp. 226-233, July 2001.

[2] R. Thull, "Elektrochemische Prüfungen von (Ti, Nb)ON-beschichteten Dentallegierungen zur Qualitätssicherung", Biomedizinische Technik, Band 36, Heft 9, pp. 214-221, 1991.

[3] R. Thull, "Corrosion behavior of dental alloys coated with titanium niobium oxinitride [Korrosionseigenschaften mit Titan-Niob-Oxinitrid beschichteter Dentallegierungen.] ", Deutsche zahnärztliche Zeitschrift, Band 46, Heft 11, pp. 712-717, 1991.

[4] R. Thull, K.D. Handke, and E.J. Karle, "Tierexperimentelle Prüfung von Titan mit Oberflächenbeschichtungen aus (Ti, Nb)ON und (Ti, Zr)O", Biomedizinische Technik, Band 40, Heft 10, pp. 289-295, 1995.

[5] G. Hunter, V. Pawar, A. Salehi, and M. Long, "Abrasive wear of modified CoCr and Ti-6Al-4V surfaces against bone cement", Medical Device Materials - Proceedings of the Materials and Processes for Medical Devices Conference 2003, pp. 91-97, 2004.

[6] Chapter 7 of this eBook: M. Fenker, "A comprehensive study of the properties of sputtered NbO_xN_y thin films", paragraph on "Biocompatibility".

[7] J.G. Wan, N. Huang, H. Sun, P.Y. Yang, Y.X. Leng, J.Y. Chen, and J. Wang, "Ti-N-O Nano-Films Deposited by Plasma Immersion Ion Implantation and Deposition: Deformation Behavior and Mechanical Properties", IEEE International Conference on Plasma Science, p. 399, 2003.

[8] A. Anders, Ed., Handbook of Plasma Immersion Ion Implantation and Deposition, New York: Wiley, 2000.

[9] G.J. Wan, N. Huang, S.C.H. Kwok, Zh.Y. Shao, A.S. Zhao, P. Yang, and P.K. Chu, "Si-N-O films synthesized by plasma immersion ion implantation and deposition (PIII&D) for blood-contacting biomedical applications", IEEE Transactions on Plasma Science, vol. 34, no. 4, pp. 1160-1165, 2006.

[10] D. Wilhelm, H. Voigt, W. Treichel, R. Ferretti, and S. Prasad, "pH Sensor Based on Differential Measurements on One pH-FET Chip", Sensors and Actuators B, vol. 4, pp. 145-149, 1991.

[11] B.J. Lasater, "Material and method to prevent low temperature degradation of zirconia in biomedical implants", U.S. Patent US7037603 (B2), May 2, 2006.

[12] K.E. Knapp, M.J. Hedrick, and B.S. Wells, "Vapor deposited multilayer dental devices", U.S. Patent application 2006/0105297 A1, May 18, 2006.

[13] Z. Zhang, and T.D. Xiao, "Coatings, coated articles and methods of manufacture thereof", U.S. Patent US7320798 (B2), January 22, 2008.

Metallic Oxynitride Thin Films by Reactive Sputtering and Related Deposition Methods, 2013, 265-284

CHAPTER 11

Oxynitrides and Oxides Deposited by Cathodic Vacuum Arc

Jörg Vetter*

Sulzer Metaplas GmbH, Am Böttcherberg 30-38, 51427 Bergisch Gladbach, Germany

Abstract: Different methods of cathodic vacuum arc evaporation are used to deposit oxynitrides. $Me_aMe_bSi_cN_xO_yC_z$ coating systems with a+b+c+x+y+z =100 at.%. The paper summarizes the results of these coatings. Me is at least one metal of the group Cr, Al, Ti, Nb. The influence of process parameters like bias, partial pressure and evaporator current at the coating properties are discussed for different coating types. The generated oxynitride coatings show mechanical properties, oxidation behaviour, thermal stability and tribological properties which are promising for various tribological applications. The most highlighted coating system CrNO is in industrial mass production for piston rings.

Keywords: PVD, cathodic vacuum arc, oxynitride films, oxide films.

INTRODUCTION

The cathodic vacuum arc is one of the successful techniques to deposit metallic hard coatings like nitrides (CrN), carbides (TiC), borides (TiB) and its mixtures (*e.g.* CrCN, TiBNC) for about 30 years. The main applications are wear reduction, friction reduction and decoration. However the coating properties of ionic hard coatings as oxides and oxynitrides have specific properties useful for different applications. The main specific properties against the metallic hard coatings are: lower thermal and electrical conductivity, high dielectric constant, high refractive index, and high transparency in the visible and near-infrared region [1]. The coatings show also a low interaction tendency with contact partners in wear applications. Most of the oxide coatings are deposited by CVD processes, thermal evaporation and magnetron sputtering. Some coating types were also investigated

***Address correspondence to Jörg Vetter:** Sulzer Metaplas GmbH, Am Böttcherberg 30-38, 51427 Bergisch Gladbach, Germany; Tel: 02204 299 261; Fax: 02204 200 266;
E-mail: Joerg.Vetter@sulzer.com

Filipe Vaz, Nicolas Martin and Martin Fenker (Eds)

using the cathodic vacuum arc and are partially in industrial use. The present contribution reviews the metal oxynitrides (MON) deposited by the cathodic vacuum arc technique

DEPOSITION TECHNOLOGY

The coatings are deposited by evaporating the metallic components (Cr, Al ...) in a reactive gas atmosphere. The deposition of oxynitrides is realized by using reactive gas mixtures of nitrogen with oxygen. Oxides require only oxygen as the reactive gas. Three different arc evaporation methods are used:

1) Classical DC arc evaporation [2].

2) Filtered arc evaporation [3].

3) Pulsed arc evaporation [4].

The schematic arrangement including the three different methods within one and the same deposition system is shown in Fig. **1**. It has to be pointed out, that the different methods are normally used in separate coating system.

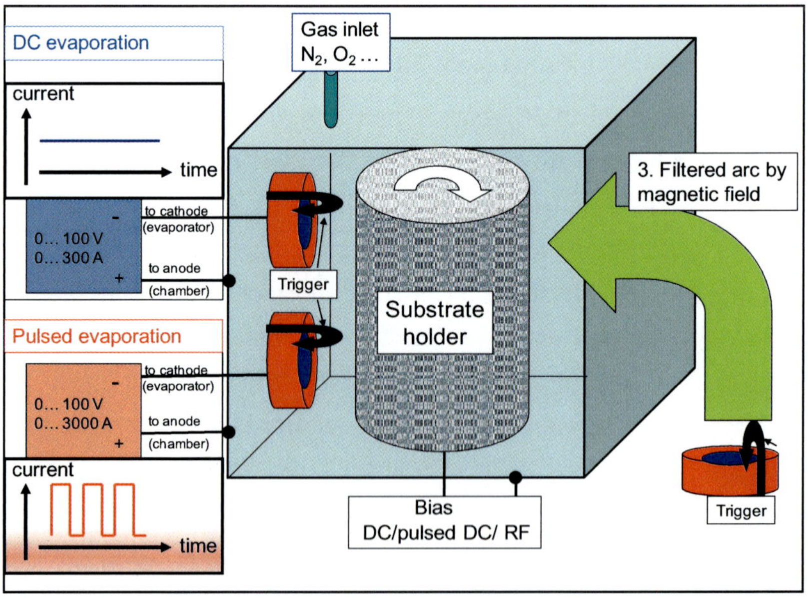

Figure 1: Schematic arrangement of types of arc evaporation for oxynitride and oxide deposition.

One difficulty during oxide deposition arises from the poisoning of the cathode material to be evaporated and another difficulty is the deposition of a layer of low electrical conductivity at the anodes. Both effects often result in process instabilities during the growth over a longer time.

The pulsed arc evaporation was developed to overcome process instabilities of the classical DC arc evaporation. One advantage of the filtered arc is the separation of droplets generated during arc running and the ion flux of the evaporated material for coating deposition [5, 6].

COATINGS

Cr-Based Coatings

$CrN_{1-x}O_x$ Coatings

Among transition metal nitrides, CrN coatings are one of the best candidate materials to fulfil multifunctional requirements for different applications due to the formation of a passivating Cr_2O_3 layer with interstitial nitrogen, which prevents oxygen diffusion through the coating [7]. Oxynitrides have been revealed technologically even more appealing, due to their superior performance as diffusion barriers in several industrial applications, such as cutting tools, rotor blades of gas turbines, *etc.* [8]. As for CrN, the corrosion resistance and diffusion barrier performance of chromium oxynitride films, have been attributed to the formation of a passivation layer of stable oxides at the surface. The $CrN_{1-x}O_x$ and Cr_2O_3 coatings show often a lower interaction tendency with the counter part and a lower friction value in wear applications.

Typical industrial applications include:

- Coatings for tools: cutting of Al-alloys and Cu-alloys, plastic moulds, die casting.
- Coatings for components: piston rings [9].

Selected properties of $CrN_{1-x}O_x$ coatings deposited by the cathodic vacuum arc evaporation will be described in the following:

Phases

Different groups investigated the phases of $CrN_{1-x}O_x$ coatings by X-ray diffraction. Results of the phase formation for different oxygen contents for coatings shows Fig. **2**:

- Deposited by arc (ARC-A) [10], (ARC-B) [11], (ARC-F) [8].
- Deposited by magnetron sputtering (MAG-D) [12], (MAG-E) [13].
- Deposited by PLD (pulsed laser deposition) (PLD-C) [14]. The cubic B1 crystal lattice of the $\mathbf{CrN_{1-x}O_x}$ coatings is stable up to an O/O+N content of at least 75 % might be also up to 90 %. Then the new Cr_2O_3-phase is generated that adopts the rhombohedral corundum structure, consisting of a hexagonal close packed array of oxide anions with 2/3 of the octahedral holes occupied by chromium.

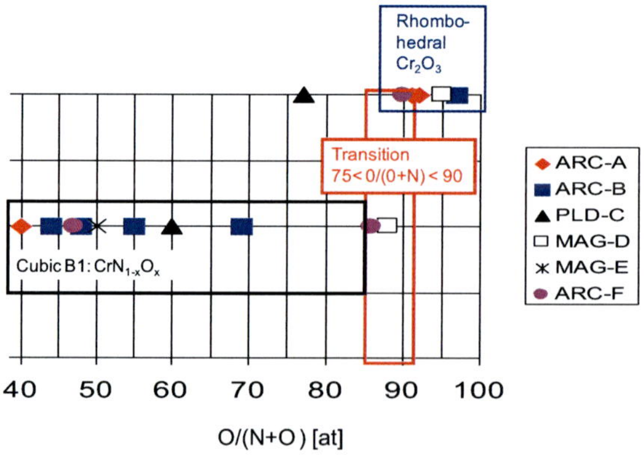

Figure 2: Phases of the $CrN_{1-x}O_x$ coatings for different O/(O+N) ratios.

Fig. **3** shows the average size D_{ave} of the coherently diffracting domains (estimated using the Scherrer formula) and the lattice parameter development [11]. The stress-free lattice parameter a_0 decreases with increasing O_{rel} similar to the calculated Vegard line using a Cr–O lattice parameter of 0.408 nm. This indicates that oxygen substitutes nitrogen within the cubic CrN structure. D_{ave} decreases with increasing O_{rel} up to 70%, and then increases when the Cr_2O_3-phase is formed.

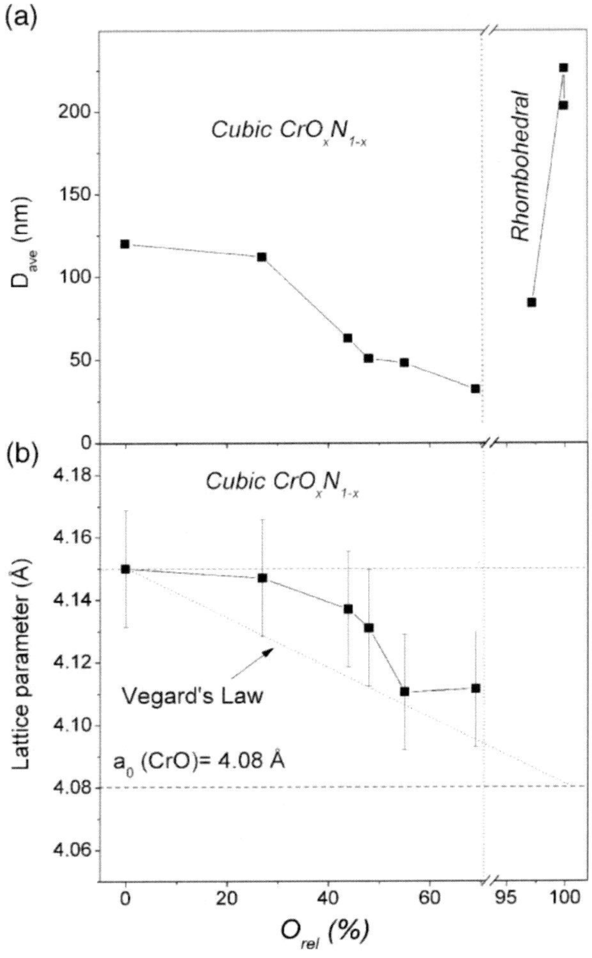

Figure 3: (a) D_{ave}, (b) Stress-free lattice parameter of the $CrN_{1-x}O_x$ coatings as a function of O_{rel} = O/(O+N) in %.

Hardness and Young's Modulus

Fig. **4** shows the dependence of the relative hardness H on O/(O+N) for the PLD coatings [14] and coatings deposited by the cathodic vacuum arc evaporation [11]. As can be seen, both deposition methods exhibit a similar trend: the hardness increases with the O content to a maximum. The hardness increase was measured to be in the region of the factor 1.5 to 1.9. A maximum value was measured for the rhombohedral $Cr_2(O_{0.97}N_{0.03})_3$ samples of 28 GPa for the cathodic vacuum arc deposited coatings. Then the hardness decreases if pure Cr_2O_3 is formed. A similar sudden decrease is observed for PLD-deposited coatings.

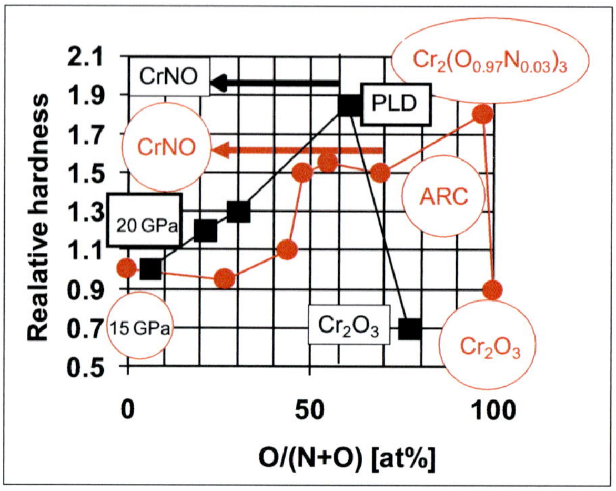

Figure 4: Relative hardness values as a function of O/(O+N) for coatings deposited by PLD and cathodic vacuum arc.

The absolute hardness values and the corresponding values for the Young's modulus are shown in Fig. **5** [11]. It can be seen that the Young´s modulus measured by the indentation method shows the same tendency like the hardness.

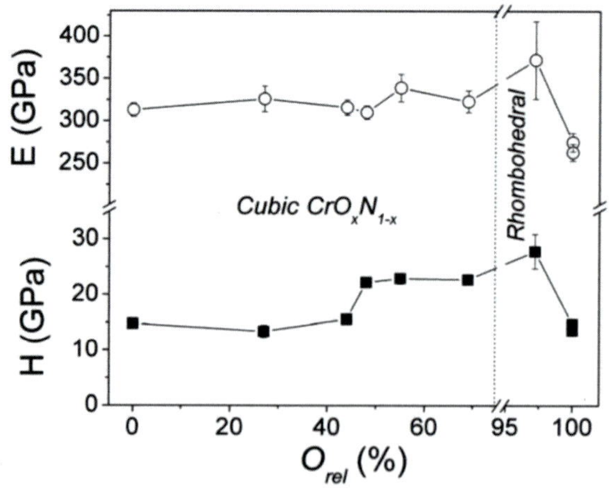

Figure 5: Hardness (H) and Young's modulus (E) as a function of O_{rel} = O/O+N for the $CrN_{1-x}O_x$ coatings.

Temperature Stability

The temperature stability was investigated by annealing experiments up to 1000°C in air for 40 min and in vacuum [11]. The results for the grain size evolution (measured by *in situ* X-ray diffraction) are shown in Fig. **6**. The annealing of CrN coatings resulted in an increase of the grain size even at 400°C both for air and vacuum. The onset temperature for a change of the grain size increases with the oxygen content: O_{rel} = 44 % up to 800°C in air and 1000°C in vacuum, O_{rel} = 69 % up to 900°C (or higher) in air and 800°C in vacuum. It was also shown that coatings with O_{rel} = 69% show the presence of the cubic B1 structure over the whole investigated temperature range.

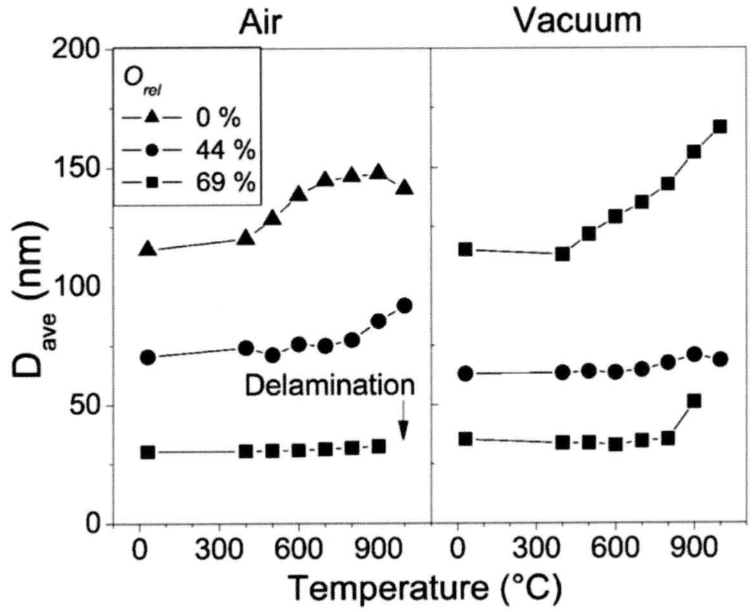

Figure 6: Average grain size of $CrN_{1-x}O_x$ as a function of annealing temperatures in air and vacuum.

An increased stability against grain growth with increasing O-content was also reported for RF sputtered $CrN_{1-x}O_x$ with O_{rel} = 10% and with O_{rel} = 50% [12].

Oxidation

The structure and phases for annealed $CrN_{1-x}O_x$ coatings under air are shown in Table **1** [11]. A significant oxidation of CrN was detected by X-ray diffraction indicating the Cr_2O_3 phase at 800°C. There is no evidence of the oxidation for the **$CrN_{1-x}O_x$** coatings with the cubic structure with higher O_{rel}. The coatings exhibit a progressively increasing oxidation resistance, which become excellent for O_{rel} = 50%.

Table 1: $CrN_{1-x}O_x$ coatings: composition and temperature for oxidation (Cr_2O_3 formation) in air

T (°C) for Oxidation	Composition	O/(N + O) (%)	Structure
800	CrN	0	B1
900	$CrO_{0.27}N_{0.73}$	27	B1
1000	$CrO_{0.44}N_{0.56}$	44	B1
> 1000	$CrO_{0.48}N_{0.52}$	48	B1
> 1000	$CrO_{0.55}N_{0.45}$	55	B1
> 1000	$CrO_{0.69}N_{0.31}$	69	B1

Tribological Investigations

A double layer – 2µm CrN with 1µm Cr_2O_3 on top – was deposited by cathodic vacuum arc evaporation [15]. The Cr_2O_3 coating was grown in a crystalline phase state. The measurement of the friction value of that coating was done by pin-on-disc under the following conditions: 6.35 mm diameter WC ball, load 10 N, speed 0.3 m/s, RH 65%. The friction value as a function of the sliding distance shows Fig. **7**. In addition, a previous study of a double layer CrN/Cr_2O_3 coating deposited by unbalanced magnetron sputtering (UBM) is also shown in the Fig. **7** [16].

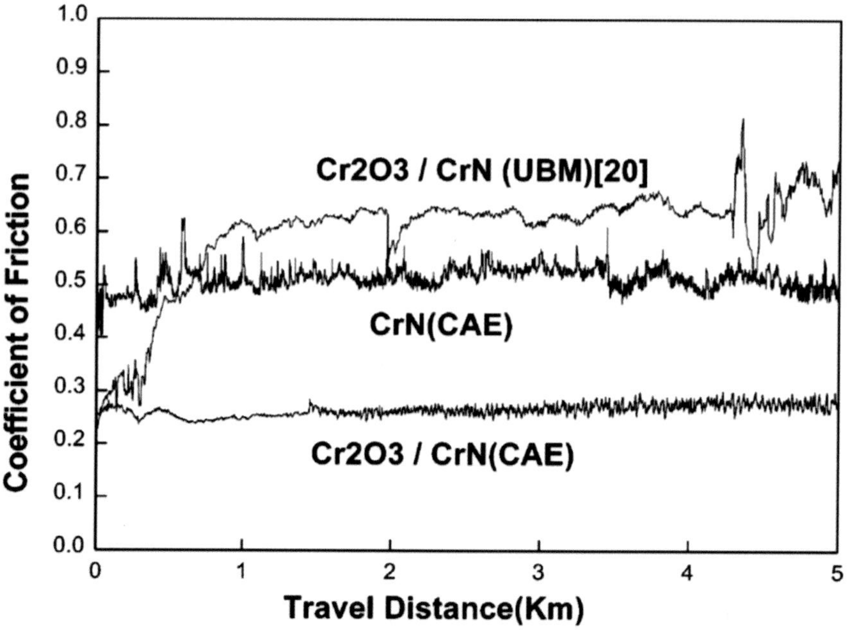

Figure 7: Comparison of friction values of double layers CrN/Cr_2O_3 deposited by UBM-sputtering and cathodic vacuum arc evaporation (CAE).

It should be mentioned, that the sputtered Cr_2O_3 coating was in an amorphous state. This sputtered coating reduced the friction value against CrN only at the initial stage of sliding. The oxide coating was rapidly worn out. The coating deposited by the vacuum arc evaporation showed a stable friction value during the sliding distance of 5 km. The friction value was about 0.25.

The tribological performance for wear reduction of $CrNO_x$ coatings are proved in mass production of piston rings. The applied coating thickness is in the range between 15μm and 50μm. The morphology of the cross section of a 30 μm thick coating is shown by light optical microscope and SEM in Fig. **8**.

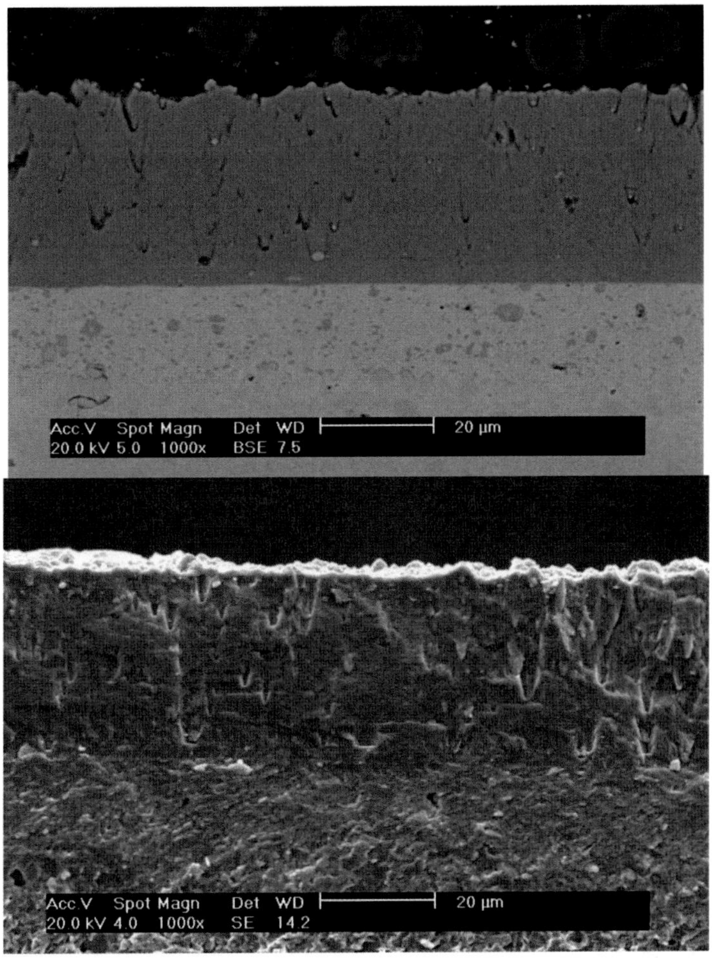

Figure 8: Cross section of $CrNO_x$ coatings applied for piston rings; upper image: light optical microscope, lower image: SEM.

Alloyed Cr-Based Coatings ($Cr_{1-a}Si_a$) ($N_{1-x}O_x$) Coatings

CrSiN Coatings

First, some specific properties of the Cr-based coatings alloyed with Si are shown to demonstrate the influence of the Si addition. Second, selected properties of the oxygen-containing coatings are discussed.

Several studies have been performed to obtain Cr based nanocomposites of the type nc-CrN/a-Si_3N_4 by PVD [17-21]. Composition, structure and morphology of $Cr_{1-x}Si_xN$ coatings (relative Si atomic concentrations Si/(Cr+Si) within 0 and 15 %) deposited by cathodic vacuum arc evaporation are characterized by [22]:

1) All $Cr_{1-x}Si_xN$ coatings consist of nanocrystalline CrN grains.

2) The crystallite sizes decreases with increasing Si content. The sharp decrease even at low Si content is shown in Fig. **9**.

3) The morphology of all $Cr_{1-x}Si_xN$ coatings is columnar.

4) Higher Si contents result in a densification of the morphology.

5) The hardness and Young's modulus of the coatings increase with increasing Si concentration up to a maximum value of 26 GPa and 430 GPa, respectively, for the $Cr_{0.94}$ $Si_{0.06}N$ coatings, followed by a progressive decrease. Fig. **10** shows the hardness *vs.* the Si content. A significant hardness maximum is visible at 6 at.% for the coatings deposited with the high power of the evaporators. The dependence of the hardness from the Si-content doesn't show a decrease at higher Si contents for the coatings deposited by low power of the arc evaporators.

6) The best oxidation resistance was obtained for the hardest samples with 6 at.% Si for which the formation of Cr_2O_3 has not been observed even at 1000°C.

7) Recrystallization becomes significant above a temperature of approximately 800°C both in vacuum and in air.

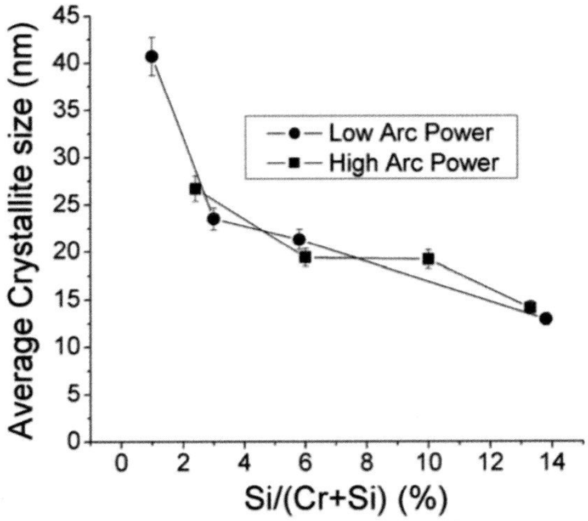

Figure 9: Average crystallite size of the $Cr_{1-x}Si_xN$ coatings as a function of the Si/(Cr + Si) ratio for different powers of the evaporators.

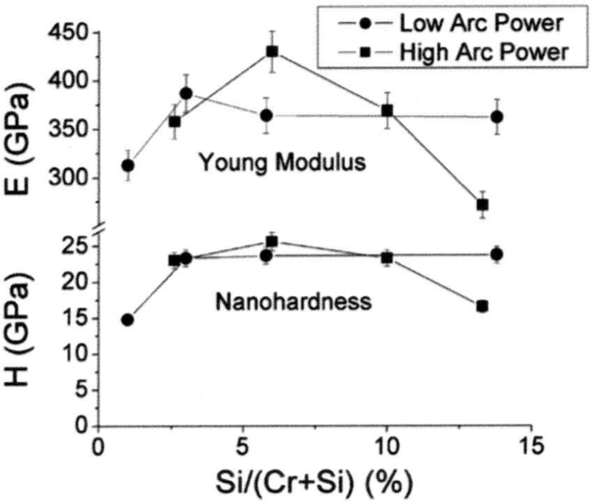

Figure 10: Hardness (H) and Young's modulus (E) *vs.* Si/(Cr+Si) of $Cr_{1-x}Si_xN$ coatings as a function of different powers of the evaporators.

CrSiNO Coatings

Castaldi *et al.* investigated the influence of oxygen addition to the CrSiN system for a fixed Si-Content of 8 at.% [23]. The main results are:

1) The cubic phase was stable up to $O/(O+N) \leq 77$ %. The X-ray diffraction result shows Fig. **11**.

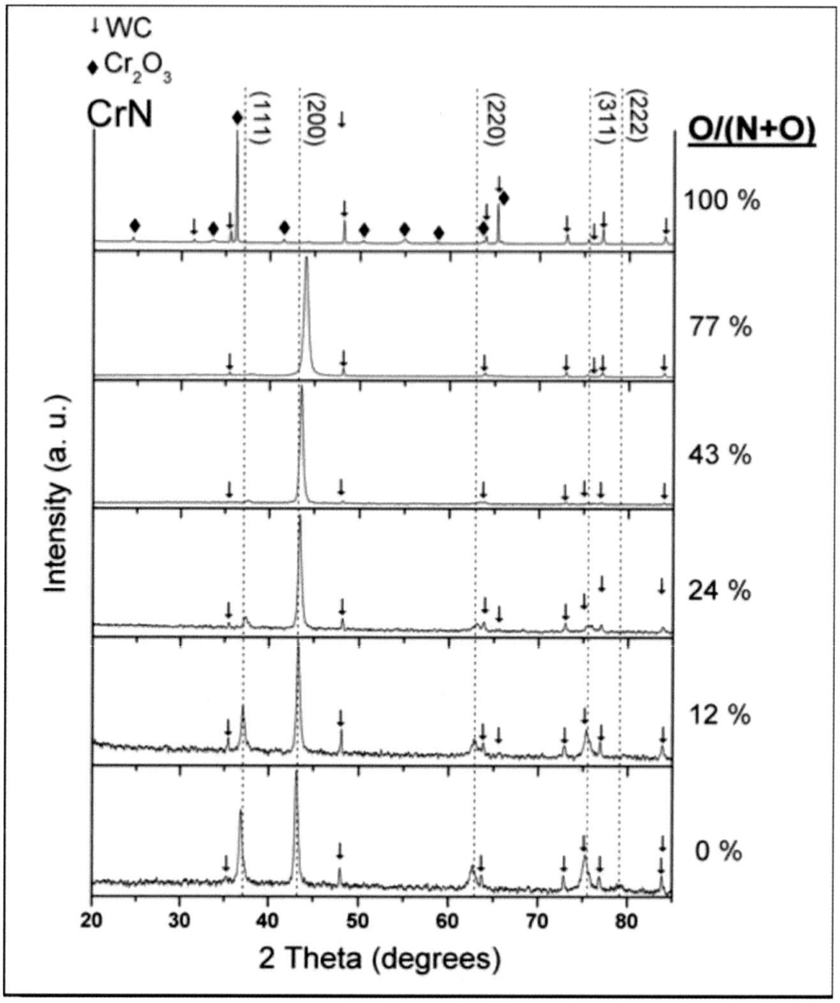

Figure 11: X-ray diffraction diagrams for $Cr_{0.92}Si_{0.08}NO$ in dependence on the O/(O+N) ratio.

2) The lattice parameter of the cubic solid solution decreases with increasing oxygen content from 0.416 nm for the $Cr_{0.92}Si_{0.08}$ N coating to 0.41 nm for the $Cr_{0.92}Si_{0.08}O_{0.77}N_{0.33}$ coating. The measured values and the calculation according to the Vegard's law are shown in Fig. **12**.

3) Smooth and dense morphologies were observed at high O/(O+N) values.

4) The cubic coatings exhibit excellent oxidation resistance, phase and nanocrystalline stability.

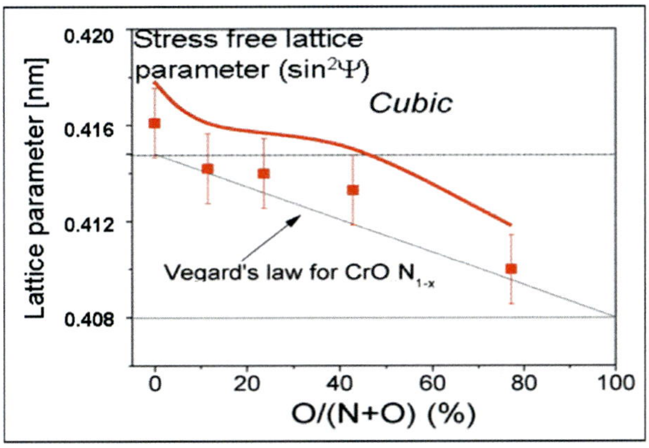

Figure 12: Lattice parameter of $Cr_{0.92}Si_{0.08}NO$ in dependence on the O/(O+N) ratio.

5) The $Cr_{0.92}Si_{0.08}N_{0.57}O_{0.43}$ coating showed the highest hardness with H = 27.2 GPa (E = 327 GPa, H^3/E^2 = 0.19). The variation of hardness and Young's modulus is demonstrated in Fig. **13**.

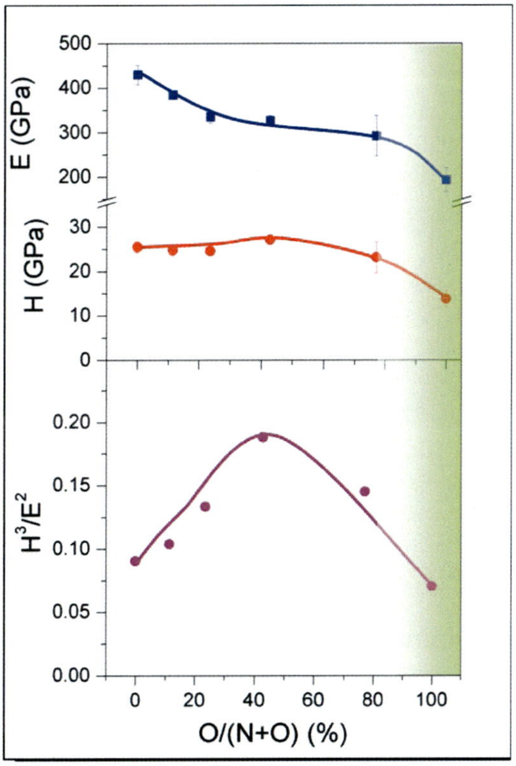

Figure 13: Hardness and Young's modulus for $Cr_{0.92}Si_{0.08}NO$ in dependence on the O/(O+N) ratio.

6) The $Cr_{0.92}Si_{0.08}N_{0.57}O_{0.43}$ coating combines the improved mechanical properties with a high oxidation resistance as well as phase and nanocrystalline stability for temperatures T > 1000 °C both in air and in vacuum.

Cr-C-O System as a Nanocomposite

Cr_2O_3 and Cr–C–O/a-C coatings were synthesized by cathodic arc evaporation using CO_2 as the reactive gas [24]. XPS and TEM analysis revealed the formation of a nanocomposite Cr–C–O/a-C coating coexisting of crystalline Cr_2O_3, Cr_7C_3 and amorphous carbon. The coating morphology is dense and compact. The hardness and Young's modulus for different CO_2 partial pressures is shown in Fig. **14**. The highest hardness of 23 GPa was measured for the highest deposition pressure.

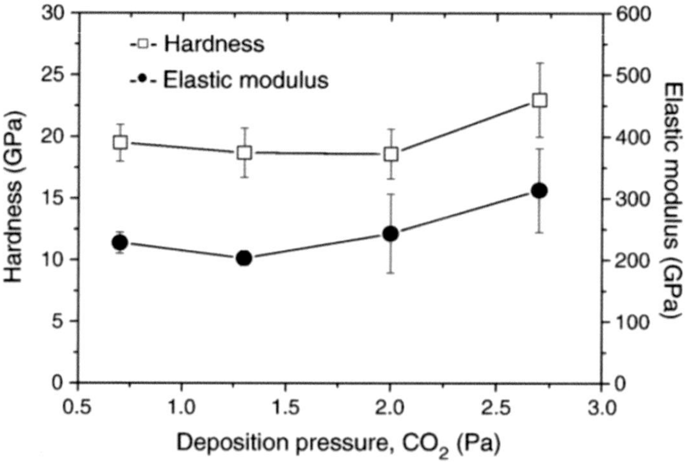

Figure 14: Hardness and Young's modulus for Cr-C-O coatings in dependence on the deposition pressure.

Al-Based Coatings

The Al_xO_y coatings can contain different structural modifications depending on the process conditions. Three main modifications are observed: amorphous Al_2O_3, γ-Al_2O_3 and α-Al_2O_3. Aspects to generate the important α-Al_2O_3 including doped types will be discussed in the following. Furthermore selected results for AlTiNO will be given.

Alumina Al_2O_3

Filtered arc deposition of alumina was carried out to investigate the phase generation of alumina depending on process parameters (ion flux, temperature, *etc.*). It was shown that α-alumina can be grown at substrate temperatures as low as 460°C on vanadium substrates [25]. Another group verified the formation of α-alumina on Inconel substrates at substrate temperatures of about 600°C utilizing substrate bias [3]. The role of the ion energy on the structure and composition of alumina was detailed investigated by Rosen *et al.* for the deposition on stainless steel substrates [26]. The phase formation to α-alumina started at substrate temperatures of about 600°C. Lower temperatures and non optimized ion fluxes result in mixtures of amorphous Al_xO_y, γ- Al_2O_3 and α-Al_2O_3.

Alumina of $(Al_{1-x}Me_x)_2O_3$ Type

Bergmann *et al.* reported about the possibility to deposit α-Al_2O_3 type coatings by DC arc evaporation at low temperature if the Al is doped by Cr or other selected elements [27]. A pulsed arc process for the deposition of metal oxides in a pure oxygen environment was developed [4]. Corundum-type solid solutions of $(Al_{1-x}Cr_x)_2O_3$ were deposited for different Cr-contents at 550°C deposition temperature. The oxide layers were synthesized as single layers and in combination with standard PVD layers.

Ti–Al–O–N Coatings

The structure, mechanical properties, and machining performance of arc evaporated Ti–Al–O–N coatings on cutting inserts have been investigated for a target of $Al_{0.66}Ti_{0.34}$ composition. The O/(O+N) gas flow ratio varied between 0 to 24% [28]. It was found out:

1) That the cubic (Ti,Al)N coatings can incorporate up to 35 at.% of O in exchange for N.

2) The cubic crystal structure of (Ti,Al)(O,N) was stabilized by a pseudomorphic growth onto a TiAlN buffer (nucleation) layer.

3) Renucleation and competitive growth makes the cubic layer evolve into a nanocomposite consisting of crystallites of TiN and amorphous

Al_2O_3, Ti(O,N), and AlN at steady-state growth. Thus, TiAlON coatings are characterized by a dual layer structure. The dual structure demonstrates Fig. **15**. It shall be mentioned that the multilayer structure of the buffer layer is caused by rotation effects.

4) The addition of oxygen decreases the hardness from 26 GPa to 20 GPa for 35 at.% oxygen. At oxygen concentrations larger than 13 at.%, however, the cutting performance is dramatically reduced as a result of increased crater wear.

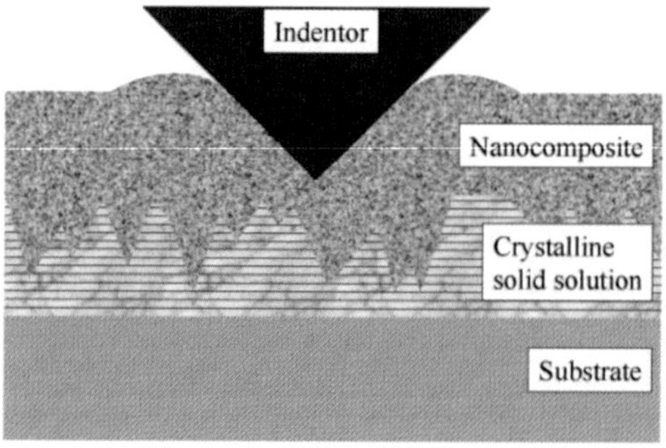

Figure 15: Schematic drawing of the dual layer structure of Ti-Al-O-N coatings. Large crystalline grains, with a multilayered structure grow initially near the substrate ($Ti_{66}Al_{34}N$ buffer layer) and an amorphous/fine-grained (nanocomposite) structure is evolved closer to the top of the coatings.

Al-Ti–Cr-N–O and Al-Ti-N-C-O Coatings

Vetter *et al.* reported about the deposition of oxygen containing coatings deposited by the vacuum arc for tool applications: AlTiCrNO [29], AlTiNCO [30]. The authors pointed out that the addition of small oxygen content has a positive influence due to a reduction of intrinsic compressive stress.

Ti Based Coatings

TiNO

TiNO coatings were deposited feeding a fixed O_2/N_2 flow ratio of 2:1 into the deposition chamber [31]. The hardness values and the Young's modulus in dependence on the applied pulsed bias voltage are shown in Fig. **16**. The hardness varied between 9 GPa for 0 V bias and 20 GPa for -500 V. It was reported, that the lowest friction value (against hardened steel) of 0.15 was measured for a bias voltage of -300 V.

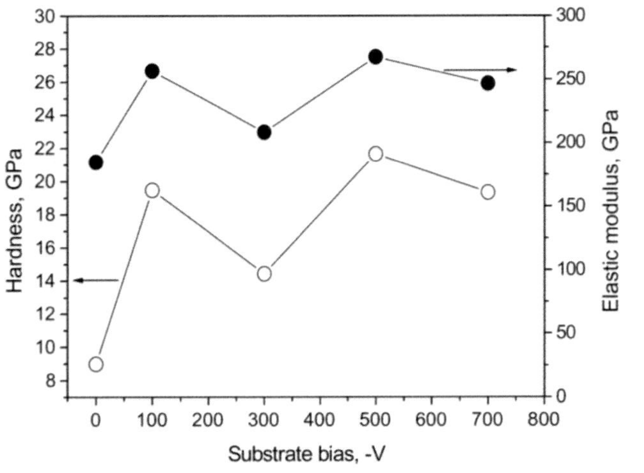

Figure 16: Hardness and Young's modulus for TiNO coatings in dependence on the substrate bias.

Different compositions of TiNO coatings are industrially applied for decorative coatings based on interference colours.

TiSiNO

$TiSi_xN_yO_z$ coatings with the cubic B1-type structure were deposited using $Ti_{0.7}Si_{0.28}O_{0.02}$ targets [32]. An analysis of the influence of the deposition parameters (working gas pressure, substrate temperature and bias voltage) on the chemical and phase compositions of Ti-Si-N(O) films shows that, the upper concentration boundary of formation of the monophasic B1-type structure of $TiSi_xN_yO_z$ is determined by the limiting value of (N+O)/Ti = 1.1. Increasing (N+O)/Ti results in the formation of an additional amorphous phase based on silicon oxides and nitrides.

TiNbNO

The development and characterization of wear resistant and adhesive oxynitride refractory metal PVD coatings is also directed to biomedical applications, *e.g.*

(Ti,Nb)ON [33]. More information can be found in chapter 10 "Properties of oxynitride thin films for biomedical applications".

CONCLUSIONS

1) Three different methods of cathodic vacuum arc evaporation are used to deposit oxynitrides: classical arc evaporation, filtered arc evaporation and pulsed arc evaporation.

2) Different types of $Me_aMe_bSi_cN_xO_yC_z$ coating systems with a+b+c+x+y+z =100 at.% were successfully generated. Me is at least one metal of the group Cr, Al, Ti, Nb and X is Si. However other metals like *e.g.* Zr or V can be used too.

3) The investigated coating types show properties promising for various tribological applications.

CrNO coatings are used in mass production for piston rings.

ACKNOWLEDGEMENT

Declared none.

CONFLICT OF INTEREST

The author(s) confirm that this chapter content has no conflict of interest.

REFERENCES

[1] P. Baumeister, and O. Arnon, "Use of hafnium dioxide in multilayer dielectric reflectors for the near UV", Appl. Opt., vol. 16, pp. 439-444, 1977.

[2] C. Gautier, and J. Machet, "Study and elaboration of ternary chromium based compounds (Cr, O, N) deposited by vacuum arc evaporation", Surf. Coat. Technol., vol. 94-95, pp. 422-427, 1997.

[3] R. Brill, F. Koch, J. Mazurelle, D. Levchuk, M. Balden, Y. Yamada-Takamura, H. Maier, and H. Bolt. "Crystal structure characterisation of filtered arc deposited alumina coatings: Temperature and bias voltage", Surf. Coat. Technol., vol. 174–175, pp. 606-610, 2003.

[4] J. Ramm, M. Ante, T. Bachmann, B. Widrig, H. Brändle, and M. Döbeli, "Pulse enhanced electron emission (P3e™) arc evaporation and the synthesis of wear resistant Al-Cr-O coatings in corundum structure", Surf. Coat. Technol., vol. 202, pp. 876-883, 2007.

[5] R.L. Boxman, D.M. Sanders, and P.J. Martin, eds., Handbook of Vacuum Arc Science and Technology - Fundamentals and Applications. New Jersey: Noyes Publications, 1995.

[6] A. Anders, Cathodic Arcs: From Fractal Spots to Energetic Condensation. New York: Springer, 2008.

[7] F. Esaka, K. Furuya, H. Shimada, M. Imamura, N. Matsubayashi, H. Sato, A. Nishijima, A. Kawana, H. Ichimura, and T. Kikuchi, "Comparison of surface oxidation of titanium nitride and chromium nitride films studied by X-ray absorption and photoelectron spectroscopy", J. Vac. Sci. Technol. A, vol. 15, pp. 2521-2528, 1997.

[8] Q.M. Wang, Y.N. Wu, M.H. Guo, P.L. Ke, J. Gong, C. Sun, and L.S. Wen, "Ion-plated Al–O–N and Cr–O–N films on Ni-base superalloys as diffusion barriers", Surf. Coat. Technol., vol. 197, pp. 68-76, 2005.

[9] S. Miyazaki, N. Yamashita, S. Tanaka, H. Fukutome, and H. Tamagaki, "Bearing element and process for its production", German Patent 4,419,713, December 8, 1993.

[10] D.H. Huang, C.H. Hsu, Y.C. Lin, C.L. Chang, K.W. Wong, and W.Y. Ho, "Thermal stability behaviours of Cr(N,O)/CrN double-layered coatings by TGA/DTA analysis", Surf. Coat. Technol., vol. 201, pp. 6681-6685, 2007.

[11] L. Castaldi, D. Kurapov, A. Reiter, V. Shklover, P. Schwaller, and J. Patscheider, "Effect of the oxygen content on the structure, morphology and oxidation resistance of Cr–O–N coatings", Surf. Coat. Technol., vol. 203, pp. 545-549, May 2008.

[12] St. Collard, H. Kupfer, G. Hecht, W. Hoyera, and H. Moussaoui, "The reactive magnetron deposition of CrN_xO_y films: first results of property investigations", Surf. Coat. Technol., vol. 112, pp. 181-184, 1999.

[13] P. Wilhartitz, S. Dreer, and P. Ramminger, "Can oxygen stabilize chromium nitride? Characterization of high temperature cycled chromium oxynitride", Thin Solid Films, vol. 447-448, pp. 289-295, 2004.

[14] T. Suzuki, J. Inoue, H. Saito, M. Hirai, H. Suematsu, W. Jiang, and K. Yatsui, "PLD Influence of oxygen content on structure and hardness of Cr–N–O thin films prepared by pulsed laser deposition", Thin Solid Films, vol. 515, pp. 2161-2166, June 2006.

[15] W.Y. Ho, D.H. Huang, L.T. Huang, H. Hsu, and D.Y. Wang, "Study of characteristics of Cr_2O_3/CrN duplex coatings for aluminum die casting applications", Surf. Coat. Technol., vol. 177-178, pp. 172-177, 2004.

[16] D.Y. Wang, and M.J. Chiu, "Characterization of Cr_2O_3/CrN duplex coatings for injection moulding applications", Surf. Coat. Technol., vol. 137, pp. 164-169, 2001.

[17] E. Martinez, R. Sanjinès, A. Karimi, F. Estève, and F. Lévy, "Mechanical properties of nanocomposite and multilayered Cr-Si-N sputtered thin films", Surf. Coat. Technol., vol. 180–181, pp. 570-574, 2004.

[18] E. Martinez, R. Sanjinès, O. Banakh, and F. Lévy, "Electrical, optical and mechanical properties of sputtered CrN_y and $Cr_{1-x}Si_xN_{1.02}$ thin films", Thin Solid Films, vol. 447–448, pp. 332-336, 2004.

[19] J.H. Park, W.S. Chung, Y.R. Cho, and K.H. Kim, "Synthesis and mechanical properties of Cr-Si-N coatings deposited by a hybrid system of arc ion plating and sputtering techniques", Surf. Coat. Technol., vol. 188–189, pp. 425-430, 2004.

[20] A. Thobor-Keck, F. Lapostolle, A.S. Dehlinger, D. Pilloud, J.F. Pierson, and C. Coddet, "Influence of silicon addition on the oxidation resistance of CrN coatings", Surf. Coat. Technol., vol. 200, pp. 264-268, 2005.

[21] J.W. Kim, K.H. Kim, D.B. Lee, and J.J. Moore, "Study on high-temperature oxidation behaviours of Cr-Si-N films", Surf. Coat. Technol., vol. 200, pp. 6702-6705, 2006.

[22] L. Castaldi, D. Kurapov, A. Reiter, V. Shklover, P. Schwaller, and J. Patscheider, "High temperature phase changes and oxidation behaviour of Cr–Si–N coatings", Surf. Coat. Technol., vol. 202, pp. 781-785, 2007.

[23] L. Castaldi, D. Kurapov, A. Reiter, V. Shklover, P. Schwaller, and J. Patscheider, "High oxidation resistance and phase stability of Cr-Si-O-N coatings", presentation at ICMCTF 2010 conference, San Diego, CA, USA.

[24] Y.Y. Chang, S.J. Yang, and D.Y. Wang, "Structural and mechanical properties of Cr–C–O thin films synthesized by a cathodic-arc deposition process", Surf. Coat. Technol., vol. 202, pp. 941-945, 2007.

[25] Y. Yamada-Takamura, F. Koch, H. Maier, and H. Bolt, "Characterization of α-phase aluminum oxide films deposited by filtered vacuum arc", Surf. Coat. Technol., vol. 142-144, pp. 260- 264, 2001.

[26] J. Rosén, S. Mraz, U. Kreissig, D. Music, and J.M. Schneider, "Effect of ion energy on structure and composition of cathodic arc deposited alumina thin films", Plasma Chem. Plasma Process., vol. 25, pp. 303-317, 2005.

[27] H. Schulz, and E. Bergmann, "Hard layer, work piece coated with such a layer and process for coating with the layer", European Patent 513,662, November 11, 1992.

[28] J. Sjölén, L. Karlsson, S. Braun, R. Murdey, A. Hörling, and L. Hultman, "Structure and mechanical properties of arc evaporated Ti–Al–O–N thin films", Surf. Coat. Technol., vol. 201, pp. 6392-6403, 2007.

[29] J. Vetter, T. Krug, and V. von der Heide, "AlTiCrNO coatings for dry cutting deposited by reactive cathodic vacuum arc evaporation", Surf. Coat. Technol., vol. 174-175, pp. 615-619, 2003.

[30] J. Vetter, T. Ishikawa, and N. Shima, "Nanocomposite AlTiNCO coatings deposited by reactive cathodic arc evaporation", Plasma Processes and Polymers, vol. 4, pp. S668-S672, 2007.

[31] M. Zhang, G. Lin, C. Dong, and K.H. Kim, "Mechanical and optical properties of composite TiO_xN_y films prepared by pulsed bias arc ion plating", Current Applied Physics, vol. 9, pp. S174-S178, 2009.

[32] M.V. Kuznetsov, E.V. Shalaeva, S.V. Borisov, B.V. Mitrofanov, A.L. Ivanovsky, "Metastable $TiSi_xN_yO_z$ films of B1-type structure prepared by the arc process", Thin Solid Films, vol. 279, pp. 75-81, 1996.

[33] J. Probst, U. Gbureck, and R. Thull, "Binary nitride and oxynitride PVD coatings on titanium for biomedical applications", Surf. Coat. Technol., vol. 148, pp. 226-233, 2001.

CHAPTER 12

Application of Oxynitrides for Microelectronic Devices and Gas Barriers

Yung-Hsien Wu and Jia-Hong Huang*

Department of Engineering and System Science, National Tsing Hua University, Hsinchu, Taiwan, R.O.C.

Abstract: As the microelectronic devices enter into nano-scale, dielectrics used in integrated circuits become versatile. Extensive study of these newly introduced materials is necessary to optimize the circuit performance. Oxynitride provide dielectrics with additional latitude to tailor their properties by adjusting the incorporated nitrogen content. Currently oxynitrides have found wide applications in the gate dielectric for CMOS devices, the charge-trapping layer and tunnel/inter-poly dielectric for non-volatile memories and the storage dielectric for DRAM cell capacitors. With the progress of microelectronic industry, new devices will certainly be developed and oxynitride will undoubtedly play a crucial role in exploring the next-generation devices. Gas barrier thin film is another promising area for the applications of oxynitrides. Currently AlO_x and SiO_x are the most widely used materials for gas barriers. However, AlO_xN_y has been demonstrated to have better gas barrier performance than its counterpart oxide. In addition, recent research results showed that transition metal oxynitrides ($TMeO_xN_y$) displayed comparable gas barrier performance as other conventional oxides, and the adjustable optical and electrical properties are especially attractive. To enhance the gas barrier performance, there are two major approaches: a) increasing the packing density of the barrier film and b) multilayer structures. For applications requiring extremely low gas permeation, such as flexible OLED, multilayer structures comprised of alternating polymer and inorganic layers are much better than the single-layer films. However, finding barriers with satisfying performance at a cost that is compatible with large-scale, low-cost manufacturing still remains a challenging issue.

Keywords: Oxynitride, microelectronic device, CMOS technology, gate dielectric, non-volatile memory, charge-trapping layer, tunnel dielectric, inter-poly dielectric, Ge MOS, gas barrier, multilayer structure, interlayer, packing density, OLED, PECVD, magnetron sputtering, oxygen transmission rates (OTR), water vapor transmission rate (WVTR), transition metal oxynitrides, gas permeation, polymer substrate.

***Address correspondence to Jia-Hong Huang:** Department of Engineering and System Science, National Tsing Hua University, 101, Kuang Fu Road, Section 2, Hsinchu, Taiwan, R.O.C.; Tel: (886)35715131 ext 34274; Fax: (886)35720724; E-mail: jhhuang@mx.nthu.edu.tw

Filipe Vaz, Nicolas Martin and Martin Fenker (Eds)

PART A: APPLICATION OF OXYNITRIDE FOR MICROELECTRONICS DEVICES

INTRODUCTION TO SCALING

Motivation of Scaling

The invention of the first transistor by J. Bardeen and W. Brattain in late 1947 ushered in the era of microelectronics. In the following decades the industry of microelectronics developed at an astonishing pace, which not only makes electronic devices affordable and ubiquitous but also has an unprecedented impact on human life. One of the major enablers that renders the rapid progress of microelectronics possible is the continuous scaling down (shrink) of transistors. The motivation to scale down transistors lies in the fact that it results in a larger driving current and therefore a higher operation speed for logic circuits since the charging and discharging time for parasitic capacitances can be reduced. Most importantly, further scaling of transistors allows a higher packing density, which provides more functions in a chip while reducing production costs.

Transistors are the core components of electronic devices and are mainly formed on silicon (Si). Two types of Si-based transistors, the metal oxide semiconductor field effect transistor (MOSFET) and the bipolar junction transistor (BJT), prevail in the microelectronic industry. For logic circuits, MOSFETs are the dominant transistors since they can form complementary MOSFET (CMOS) technology which includes both n-channel and p-channel MOSFETs and provides lower power consumption as compared with BJT technology. Over the past decades, thanks to the great effort made by engineers and researchers, the technology node characterized by feature size of MOSFETs advances from 10 μm in 1970 to 45 nm in 2009. As the feature size shrinks, the number of transistors in a chip increases at an incredible rate from very larger scale integration (VLSI) technology with $3x10^4$~$1x10^6$ transistors per chip in 1980 to ultra-large scale integration (ULSI) technology with more than 10^6 transistors per chip in 1990. In 2006, the industry formally entered into giga scale integration (GSI) era in which more than 10^9 transistors are encompassed in a chip. Apparently, in the past 30 years, the scaling trend of MOSFETs abides by Moore's Law shown in Fig. **1** which states that the number of transistors in a chip doubles every two years [1] and was first introduced in 1965 by Intel co-founder, Gordon Moore.

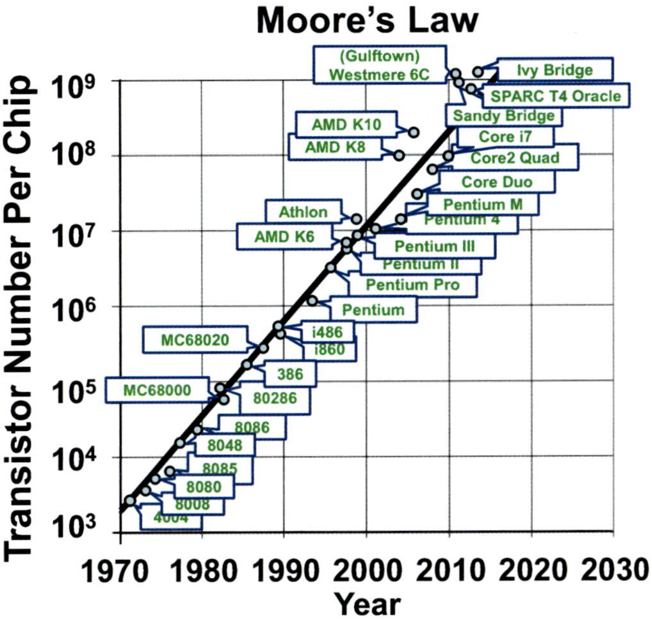

Figure 1: Evolution of number of transistors in Intel's microprocessor chip. This evolution follows Moore's law which predicts that the complexity of integrated circuit chips is doubling every two years. This trend has lasted in the microelectronic industry for decades and similar trends exists for other companies and for both memory and logic circuitry. Copied from Wikipedia website at http://ja.wikipedia.org/.

Issues of CMOS Scaling

During the development of new technology to shrink transistors to meet Moore's Law, a roadmap that provides a guideline for the continuing innovation or revolution in process, material, testing, assembly and system design is a prerequisite. The International Technology Roadmap for Semiconductors (ITRS) jointly developed by experts from the US, Europe, Japan, Korea and Taiwan provides such a guideline to determine the technology trajectory for the next 15-years. In addition, ITRS also addresses the technological challenges the industry may encounter and advises possible solutions. A typical form of ITRS [2] can be found in Fig. **2** which predicts that the feature size and gate length of a MOSFET will be 25 nm and 10 nm respectively in 2015 (definition of gate length of a MOSFET can be found in the next section).

Year of Production	*2011*	*2012*	*2013*	*2014*	*2015*	*2016*	*2017*	*2018*
MPU/ASIC Metal 1 (M1) ½ Pitch (nm) (contacted)	38	32	27	24	21	18.9	16.9	15.0
L_g: Physical Lgate for HP Logic (nm) [1]	24	22	20	18	17	15.3	14.0	12.8
V_{dd}: Power Supply Voltage (V) [2]								
Bulk/FD SOI/MG	0.90	0.87	0.85	0.82	0.80	0.77	0.75	0.73
EOT: Equivalent Oxide Thickness (nm) [3]								
Extended Planar Bulk	0.88	0.84	0.79	0.73	0.67	0.61	0.55	
FD SOI		0.88	0.84	0.8	0.76	0.72	0.68	0.63
MG		0.92	0.88	0.84	0.8	0.76	0.72	0.68
Channel Doping (10^{18}/cm^3) [4]								
Extended Planar Bulk, FD SOI/MG	4.5	5	6	7	7.7	8.4	9	0.1
Junction Depth or Body Thickness (nm) [5]								
Extended Planar Bulk (junction)	10	8.8	8	7.2	6.4	5.7	5	
FD SOI (body)		7.4	6.6	5.8	5.2	4.6	4.1	3.7
MG (body)		15	13.6	12.3	11.3	10.3	9.4	8.5
T_{BOX}: Buried Oxide Thickness for UTB FD (nm) [6]								
UTB FD		18	17	16	15	14	13	12
EOT_{elec}: Electrical Equivalent Oxide Thickness (nm) [7]								
Extended Planar Bulk	1.2	1.16	1.1	1.04	0.98	0.92	0.86	
FD SOI		1.28	1.24	1.2	1.16	1.12	1.08	1.03
MG		1.32	1.28	1.24	1.2	1.16	1.12	1.08

Figure 2: Selected scaling trend for CMOS devices from 2012 ITRS. Copied from [2].

In scaling MOSFETs, not only the dimensions are scaled, applying voltage and dielectric thickness should be scaled down as well. It is not difficult to imagine that continuous scaling is a formidable task since the so-called short channel effect always plays a critical role in determining device performance. Short channel effect refers to all adverse effects which cause threshold voltage reduction (leakage current increase) in scaled devices or lead to phenomena that are not observed in long channel devices. This effect may incur channel length modulation, drain-induced barrier lowering, punch-through, subthreshold leakage current, field dependent charge carrier mobility, gate oxide breakdown, velocity saturation and parasitic BJT operation.

Owing to many innovative processes proposed by researchers, the microelectronic industry successfully pushes the scaling trend according to Moore's Law in the past 30 years. However, severer challenges lie ahead to keep up Moore's Law as the feature size of transistors advances to nanometer regime. Besides the process challenge such as photolithography technology to define an ever smaller pattern, two of the major challenges exist even though tinier MOSFETs and chips with higher transistor density can be fabricated. These challenges include power dissipation and device performance. For power dissipation concerns, active power along with standby power should be considered. Active power refers to power consumption in the active state and is defined as:

$$P_{active} = n\, f\, C_{load}\, V_{cc}^2 \quad (1)$$

where n is the number of transistors in a chip, f is the operating frequency, C_{load} is the switching capacitance of the gate and wire load, and V_{CC} is the supplying voltage. As process technology advances and a chip with versatile functions becomes a prerequisite, the number of transistors and the operating frequency increase at an extraordinarily high rate and therefore results in soaring power density of a chip even though capacitance and supplying voltage reduces per generation. What is worse is the increasingly higher standby power stemming from leakage current. According to calculation, the power density of a processor chip processed with 100 nm technology node approaches 100 W/cm^2 and will be comparable with that of a nuclear reactor in the near future. For the concern of device performance, it arises from that fact that as MOSFETs scale into nanometer regime, conventional scaling cannot well control electrostatic channel, suppress off-currents and achieve higher driving capability. Therefore the behavior of the scaled devices will deviate from expectation and require in-depth research to obtain the desirable performance.

Beyond CMOS Scaling

Although challenges are expected in fulfilling transistors with lower power consumption and high performance in nanometer regime, there exist some opportunities behind it that we can seize to explore in the nano-electronics era. Recently new device structures and materials have been proposed to provide opportunities to achieve desirable transistors. The promising device structures include silicon-on-insulator (SOI), multi-gate MOSFET, nanowire devices, strain engineering and 3D integration. On the other hand, germanium (Ge) or III-V compound material, high-permittivity or low-permittivity dielectric and carbon nanotubes are the potential material for next generation devices.

Evolution of Dielectrics for CMOS Devices

Dielectrics have a wide variety of applications in ULSI technologies; they can be employed as the gate dielectrics for MOSFETs, the charge-trapping layers/tunnel dielectrics for non-volatile memories, storage dielectrics for volatile memories and inter-metal dielectrics for metal-insulator-metal (MIM) capacitors. In addition, dielectrics can also be employed in interconnect dielectrics, device isolation and masking material. As the process technology advances, it also has a great impact on the evolution of dielectrics. In the case of gate dielectrics, not only thinner, transition to the one with higher permittivity and low leakage current is an inevitable trend. As for charge-trapping layer in non-volatile memories, dielectrics obviously evolve to possess higher permittivity, deeper trapping level and better reliability. With regard to volatile memories and MIM capacitors, changing to dielectrics that can accomplish a higher capacitance density with minimum leakage is a must. Aiming at such evolution and specific requirement of dielectrics, oxynitride films play an indispensable role and deeply influence the development of state-of-the-art ULSI technologies. In the following sections the applications of oxynitride films to mainstream ULSI devices will be explored and the outlook for microelectronics will be discussed.

APPLICATION OF OXYNITRIDE FOR LOGIC DEVICES

Migration from SiO_2 to SiON

In the development of MOSFET technology, highly-doped polycrystalline Si has served as the gate electrode material for decades. Basically doped polycrystalline Si is not a good conductor as its precedent gate electrode, aluminum. Nevertheless polycrystalline Si has some favorable properties such that it is still employed in many integrated circuits.

1. Unlike aluminum for which the maximum allowed process temperature is only 660 °C to avoid melting, polycrystalline Si can withstand higher process temperature that is required for high-performance MOSFETs.

2. For aluminum-gated MOSFETs, the work-function between gate electrode and channel material is hardly altered. It would consequently render the threshold voltage of a MOSFET not easy to be modulated and limit the driving current according to equation (1). On the other hand, the work-function of polycrystalline Si can be controlled by adjusting the dopant type and concentration.

For polycrystalline Si-gated MOSFETs, the process can be classified into single-work-function and dual-work-function technology as shown in Fig. **3** [3].

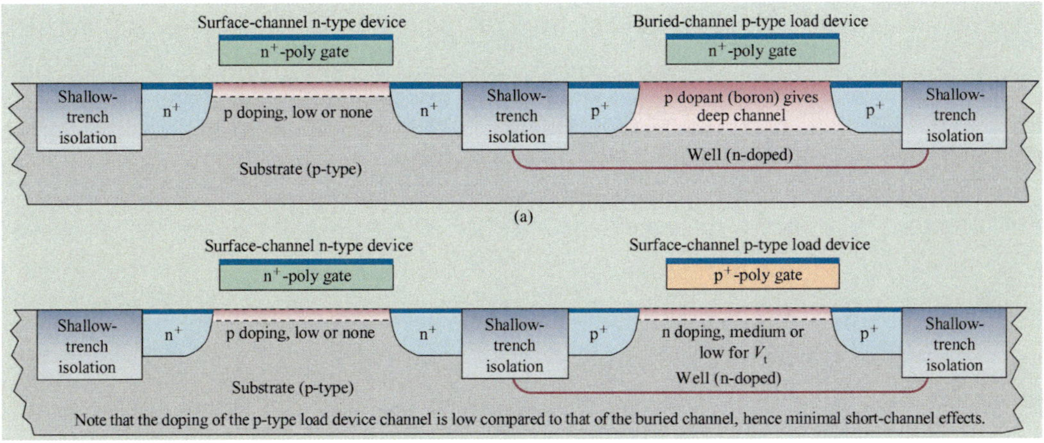

Figure 3: CMOS devices with (a) single-work-function and (b) dual-work-function technology. In single-work-function technology, n+ polycrystalline silicon gates are used for both n-channel and p-channel MOSFET. To obtain a lower threshold voltage, trivalent dopant such as boron is implanted into the surface region of the channel, resulting in a buried channel device with associated short channel degradation effects. For dual-work-function technology, n+ and p+ polycrystalline silicon gate are respectively employed for n-channel and p-channel MOSFET. Both types of devices are with surface channel [3]. Reproduced by permission of IBM Technical Journals.

For single-work-function technology, n^+-polycrystalline Si gate is applied to both n-channel and p-channel MOSFETs. In this technology scheme, p-type dopant (boron) should be implanted into the surface region (n-type substrate) of the p-channel MOSFETs to achieve a lower threshold voltage. This counterdoping implantation makes the p-channel MOSFET a buried-channel device. Although a buried-channel device enjoys a higher carrier mobility since the interface scattering effect can be avoided, its short-channel effect is worse than that of a surface-channel device. Currently this single-work-function technology is used for circuits in which cost rather than performance is of great consideration. On the other hand, for dual-work-function technology, n^+- and p^+ polycrystalline Si gate are respectively used for n-channel and p-channel MOSFETs. In this technology scheme, MOSFETs are of normal surface channel without the aforementioned issues observed in single-work-function technology and can be scaled to even smaller feature size. Although higher performance can be obtained by dual-work-function technology, the most serious problem in process integration is boron penetration from the p^+ polycrystalline Si gate to the SiO_2 gate dielectric for p-channel MOSFETs. This penetration effect comes from the fact boron is an extremely small atom, and has a very high diffusion coefficient in both silicon and silicon dioxide at high process temperature. Such boron penetration not only deteriorates the gate oxide reliability but also degrades the subthreshold swing as well as transconductance.

The most straightforward approach to alleviate the adverse effect of boron penetration is the employment of silicon oxynitride (SiON) to replace conventional SiO_2. SiON is formed by the introduction of nitrogen into SiO_2 and how the nitrogen is incorporated into oxide critically affects the device operation. Small amount of nitrogen incorporation is desirable to control channel hot-electron degradation effects in MOSFET characteristics perhaps because nitrogen lessens the diffusion of hydrogen in the oxide [4]. However, such a low nitrogen concentration is not sufficient to block boron penetration. High nitrogen concentration is effective in reducing boron penetration presumably because silicon nitride's higher density acts as a diffusion barrier [5]. However, large amount of nitrogen incorporation also causes threshold voltage shift, mobility and transconductance degradation which is due partially to the resultant positive charges in the SiO_2. A desirable nitrogen distribution profile would be one with an extremely high nitrogen concentration at the polycrystalline Si gate/SiO_2 interface to effectively retard boron penetration and a very small concentration at the Si substrate/SiO_2 interface to maintain the required oxide reliability.

Approaches to form SiON Gate Dielectric

Methods of introducing nitrogen into SiO_2 include thermal nitridation and plasma nitridation by nitrogen containing gas such as nitric oxide (NO), nitrous oxide (N_2O), ammonia (NH_3) and nitrogen (N_2). Nitridation refers to annealing of a pre-grown SiO_2 film in the aforementioned ambient. Note that thermal nitridation by N_2 is rarely used since it is relatively inert as the process temperature is below 1200 °C. For the SiON film formed by NO or N_2O thermal nitridation, nitrogen is distributed almost near the Si substrate/SiO_2 interface. The main difference between these two gases is that higher nitrogen concentration is obtained by NO nitridation. Besides nitridation of a pre-grown SiO_2 film, a SiON film can also be thermally grown by exposing Si substrate in a NO or N_2O ambient [6]. For the NO-grown SiON film, nitrogen is distributed almost uniformly in the film except for the very near-surface region. On the other hand, nitrogen pileup near the interface with Si substrate is observed for the N_2O-grown SiON film and again N_2O results in less nitrogen incorporation than NO under equivalent conditions. Obviously, neither nitridation nor direct growth by NO or N_2O can the desirable nitrogen profile be achieved. However, through an additional oxidation of the NO-grown SiON film, a preferred nitrogen profile is obtained. Nitrogen distributions for different NO or N_2O processes are summarized in Fig. **4** [6].

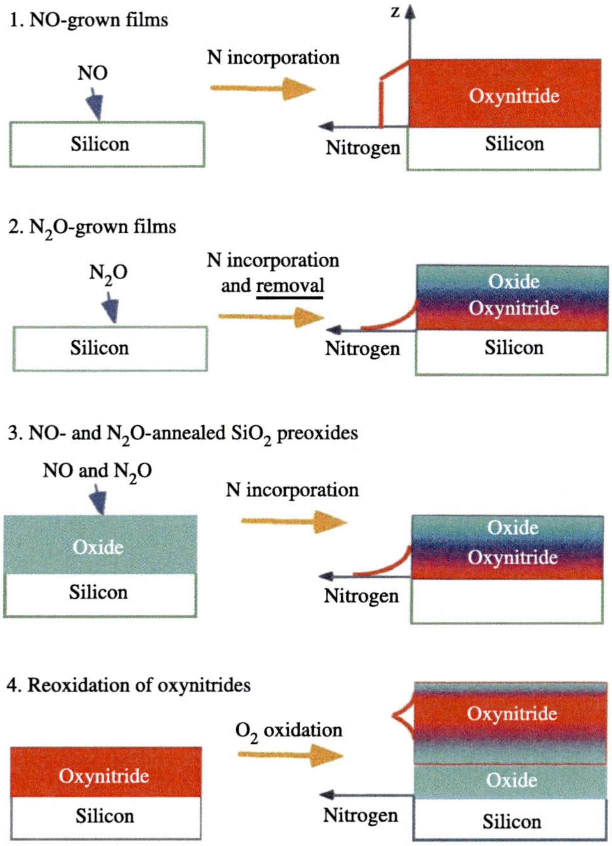

Figure 4: Nitrogen distribution for SiON films formed by different nitrogen incorporation schemes [6]. Reproduced by permission of IBM Technical Journals.

Nitridation in NH_3 was one of the first methods employed to incorporate relatively high (10~15 at. %) concentrations of nitrogen into SiO_2 films. During the initial stages of NH_3 nitridation, nitrogen is mainly distributed at the surface and the interface with Si substrate [6]. Longer time and higher temperature also lead to the incorporation of nitrogen into the bulk SiO_2. Although high nitrogen concentration caused by NH_3 nitridation is helpful to resist boron penetration, this nitridation scheme is rarely adopted by the industry because it also incorporates a large amount of hydrogen which then acts as electron traps and has detrimental effects on device reliability. Recently researchers have found that through a subtle change in process sequence a desirable SiON film can be obtained by using NH_3. In this promising process, the SiON is formed by NH_3 nitridation of Si substrate to grow an ultra-thin Si_3N_4 film followed by an oxidation step [7]. This oxidation not only pushes nitrogen atoms in the ultra-thin Si_3N_4 toward the oxide surface as shown in Fig. **5** but also eliminates the annoying hydrogen effect.

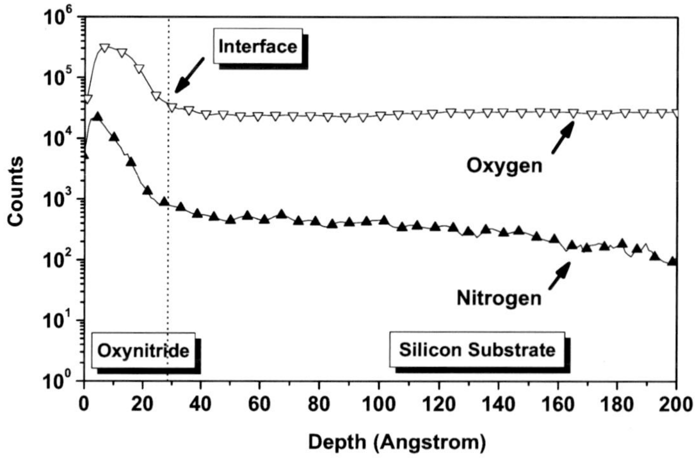

Figure 5: For the SiON film formed by a thermal NH_3 nitridation of Si substrate and a subsequent oxidation, nitrogen profile with a peak located at the SiON surface is observed [7]. Reproduced by permission of ECS-The Electrochemical Society.

In addition to thermal nitridation, remote plasma nitridation by hydrogen-free N_2 can also incorporate high concentration nitrogen at the surface of an ultra-thin SiO_2 film [8] as shown in Fig. **6** without suffering from the undesirable hydrogen effect and this process has been introduced in some mass production fabs.

In addition to suppress boron penetration and enhance gate oxide reliability, SiON also possesses higher permittivity than conventional SiO_2 and it would be beneficial to engender a higher driving current due to a larger gate oxide capacitance. Permittivity of SiON is between 3.9 (permittivity of SiO_2) and 7.8 (permittivity of Si_3N_4) and assumed to increase linearly with increasing nitrogen content [9]. It is worth mentioning that the increase of permittivity is concomitant with a lower bandgap and therefore a higher gate leakage current due to a smaller barrier height for electron/hole tunneling. Besides permittivity, refractive index (measured at 635 nm wavelength) also increases with the amount of incorporated nitrogen from 1.46 (refractive index of SiO_2) to 2.0 (refractive index of Si_3N_4). The correlation between nitrogen content in SiON and permittivity, conduction band offset and refractive index is summarized in Fig. **7** [9].

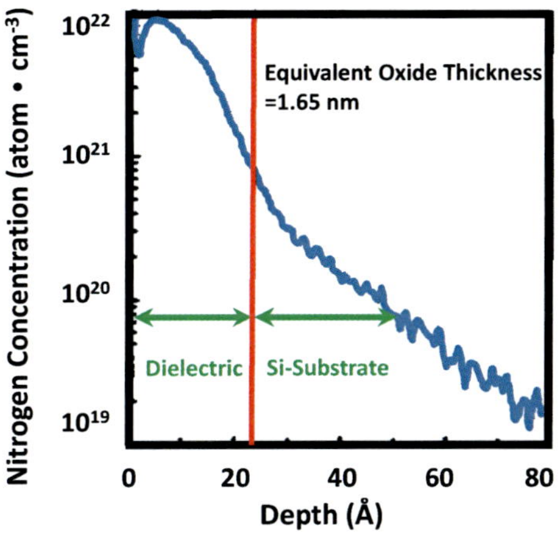

Figure 6: Nitrogen profile for the SiON film formed by a remote plasma nitridation. A high nitrogen concentration of almost 10^{22} atoms/cm^3 is effectively piled up at the top surface of the dielectric, data from reference [8].

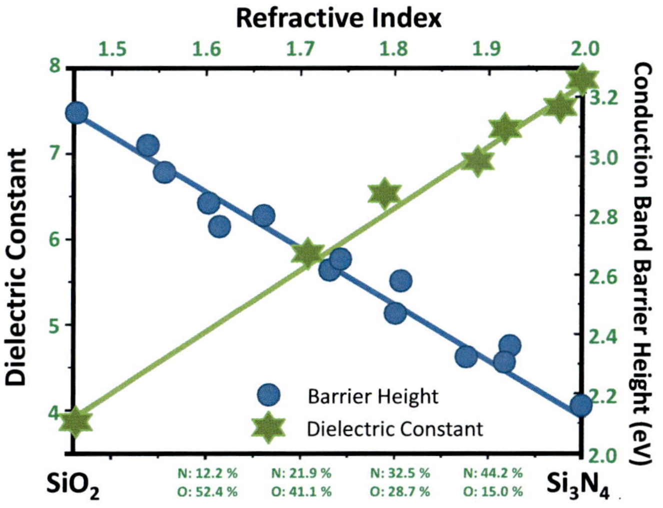

Figure 7: Impact of nitrogen content in a SiON film on its permittivity (dielectric constant), refractive index and conduction band barrier height, data from reference [9].

Transition from SiON to High-k Dielectric

Over the last four decades, the performance of modern CMOS devices has been improved primarily driven by physical scaling. One of the most important scaling factors to achieve a higher drive current is to obtain a smaller equivalent oxide thickness (EOT). EOT of a dielectric is given by:

$$EOT = t_{ox} \frac{k_{SiO_2}}{k_{dielectric}} \tag{2}$$

where t_{ox} is the physical thickness of a dielectric, k_{SiO2} is the relative permittivity (dielectric constant) of silicon dioxide, which is 3.9, and $k_{dielectric}$ is the relative permittivity of the dielectric.

Motivation to Replace Silicon Dioxide or Silicon Oxynitride

SiO_2 or SiON-based gate dielectrics, although employed for CMOS devices for decades, have encountered their scaling limit as EOT approaches 1 nm because of the issues of high leakage current, increased power consumption, poor process controllability and worse device reliability. In principle, both the physical and technological limits of SiO_2 or SiON can be solved by replacing it with a high-permittivity (high-k or high dielectric constant) gate dielectric which possesses permittivity higher than that of SiO_2 or SiON-based dielectrics. These high-k gate dielectrics can accomplish a higher gate capacitance, a larger driving current as well as a lower leakage current. By using high-k dielectrics, MOSFETs with EOT below 1 nm have demonstrated gate leakage current several orders lower than those gated by SiO_2 or SiON. This reduction in the leakage current reduces power consumption and inspires the continuous scaling.

Requirement of High-k Gate Dielectric

Development of CMOS-compatible high-k gate dielectric has recently received much attention in the industry and a less than exhaustive list of the requirements of a potential high-k dielectric for CMOS devices is given [10, 11]:

1. High permittivity ($k > 10$) value to replace SiO_2.

2. Larger than 1 eV conduction band and valence band offsets to silicon.

3. Good electrical interface with silicon, typically with interface trap density less than 10^{11} cm^{-2} eV^{-1}.

4. Few bulk electrically active defects.

5. Comparable or better reliability performance than SiO_2.

6. Good thermodynamic stability when in direct contact with silicon channel.

7. Low oxygen diffusion coefficients to suppress uncontrolled interfacial layer (re)growth.

8. Kinetically stable and compatible with CMOS process.

9. Remain amorphous during process since crystallization would degrade leakage current.

Choice of High-k Materials

Aluminum oxide (Al_2O_3) is a very stable and robust material, and its application to CMOS devices dates back to 1999 [12]. Even as Al_2O_3 has many favorable properties such as high band gap, thermodynamic stability on Si up to high temperature and is amorphous under the conditions of interest, it cannot meet the industry's needs since it has a permittivity of only about 10.

Titanium oxide (TiO_2, $k = 80{\sim}100$, depending on the crystal structure and method of deposition) and tantalum pentoxide (Ta_2O_5, $k \sim 25$) have been extensively used in discrete capacitors and memory capacitors; however, they are not suitable to serve as the gate dielectric for CMOS devices because of the low conduction band offset and the possible reaction with the silicon substrate. For yttrium oxide (Y_2O_3), although with moderately high permittivity value of 15 and bandgap of 6 eV, its integration with CMOS device is constrained by the low crystallization temperature (< 400 °C).

Among the potential high-k gate dielectrics, hafnium oxide (HfO_2) and zirconium oxide (ZrO_2) have recently attracted a lot of attention because of having much promise in overall materials properties [13]. For HfO_2, its merits include a relatively high k value (~20) as compared to silicon nitride (Si_3N_4) and Al_2O_3, the high free energy of reaction with Si (47.6 kcal/mole) as compared to TiO_2 and Ta_2O_5, a larger bandgap (5.7 eV) than most high-k competitors. One major concern to integrate HfO_2 with CMOS devices is the crystallization to monoclinic phase during subsequent high temperature processes since the grain boundaries in the crystalline phase would degrade gate leakage current. To circumvent this issue, incorporating SiO_2 to form $HfSiO_x$ is a common approach to raise its crystallization temperature. However, $HfSiO_x$ reveals a lower permittivity than pure HfO_2 and this effect can be compensated by introducing nitrogen to form hafnium silicon oxynitride (HfSiON). To simplify process, direct incorporation of nitrogen into HfO_2 to form hafnium oxynitride (HfON) is a promising alternative gate dielectric. HfON thin films are usually formed by reactive sputtering with a hafnium target in a gas mixture of argon, oxygen, and nitrogen. Nitrogen leads to a decrease in average atomic coordination, which is helpful to enhance the resistance to crystallization and ability to withstand high temperature processing [14]. The permittivity of HfON thin films increase nonlinearly with an increase in nitrogen concentration as shown in Fig. **8** [15].

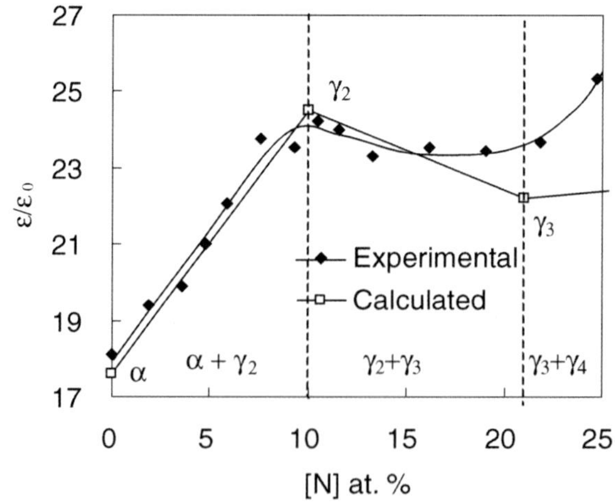

Figure 8: Correlation between nitrogen concentration and permittivity for a HfON film [15]. Reproduced by permission of The Japan Society of Applied Physics.

The inflection point of the permittivity at nitrogen concentration of approximately 9 at. % coincides with the disappearance of the monoclinic phase. The optimal nitrogen concentration for a HfON film as the gate dielectric ranges from 7.6 to

16.7 at. %. For the impact of nitrogen incorporation on the band structure, it has been found that as compared with HfO_2, conduction band offset keeps unchanged while valence band offset reduces by 1.1 eV [14]. Fortunately, this valence band offset is still large enough to serve as a good hole barrier. Since ZrO_2 is considered to have analogue structures of HfO_2, the effect of nitrogen incorporation on the properties of zirconium oxynitride (ZrON) is expected to be similar to that of HfON. Besides reactive sputtering used for ZrON deposition, a ZrON film can also be formed by NH_3 thermal nitridation of a ZrO_2 film [16] or by oxidation of a ZrN film [17, 18]. In fact, HfON or ZrON also help suppress boron penetration, however, this advantage becomes less favorable since metal gate instead of p+ polycrystalline Si is dominantly used in advanced ULSI technology as the electrode material to reduce resistance.

Searching for the Next Generation of High-k Gate Dielectric

Owing to the great effort in developing Hf-based dielectric, Intel and IBM alliance (AMD, Sony and Toshiba) respectively announced the successful introduction of Hf-based high-k gate dielectric and metal gate in their 45-nm process technology in January 2007. Even though the great advance in Hf-based dielectric has been made, the pursuit of high-k dielectric with higher performance for next-generation technology never stops. Among the possible high-k candidates, lanthanum oxide (La_2O_3) demonstrates the greatest potential to replace HfO_2-based one as the next-generation high-k dielectric for future nano-CMOS devices [19-21] as evidenced by listing in ITRS. Lanthanum belongs to the rare earth elements which were believed to be scarce in abundance as minerals. However, the term "rare earth" is now deprecated by the International Union of Pure and Applied Chemistry (IUPAC), as these elements are in fact relatively abundant in the Earth's crust. The advantages of La_2O_3 over the very popular HfO_2-based high-k material are summarized below:

1. Higher permittivity (~27) which facilitates further scaling EOT below 1 nm.

2. Larger conduction band offset of 2.3 eV while that for HfO_2 is 1.3~1.5 eV. This characteristic makes the leakage current of La_2O_3 much lower than HfO_2.

3. Increased free energy of reaction with Si (98.5 kcal/mole) as compared to 47.6 kcal/mole for HfO_2. This property implies that La_2O_3 has better thermal stability with Si and interfacial layer may not be formed during subsequent process which is beneficial to achieve EOT lower than 1 nm.

4. A very high mobility can be achieved for La_2O_3 directly deposited on Si substrate

5. Interface trap density as low as less than 1×10^{11} cm^{-2} eV^{-1} is obtained.

6. No Fermi level pinning is observed when in contact with metal electrode.

Regardless that La_2O_3 is generally perceived as the next-generation gate dielectric, its thermal stability is still a key concern which includes crystallization and the difficulty in controlling composition caused by diffusion of Si into the La_2O_3 film. It is well known that La_2O_3 has a high permittivity and Al_2O_3 has good thermal stability with Si. $LaAlO_3$, the compound of La_2O_3 and Al_2O_3, has been proven to combine their desirable chemical and electrical properties while eliminate the deficiencies of each individual material. $LaAlO_3$ is a novel high-k gate dielectric with the unique property of maintaining a high permittivity value of 25.1 which is close to that for La_2O_3, even after combining with Al_2O_3 (permittivity of 10). This is in contrast to a reduced permittivity value of 10~15 for HfAlO(N) [22] and HfSiON [23], where HfO_2 (permittivity of 22) is respectively combined with Al_2O_3 and Si_3N_4. Recently, it has been found that lanthanum aluminum oxynitride (LaAlON) formed by the introduction of nitrogen into $LaAlO_3$ shows a higher permittivity of 33~37 without reducing the bandgap [24]. This promising property of LaAlON opens a new frontier to high performance CMOS technology because it results in a higher driving current without degrading leakage current.

Oxynitride for Ge-Based MOS Devices

As mentioned in the first section, other than scaling down the dimension of MOSFETs and adoption of a high-permittivity gate dielectric, local introduction of strain on MOSFETs to boost carrier mobility for high-speed devices has been widely reported. However, the mobility enhancement by strain is subject to limitation. Recently replacing conventional Si with a high channel mobility material has been regarded as another promising avenue to further improve the device performance. Because bulk hole and electron mobility of germanium (Ge) are 4.2 and 2.6 times higher than that of Si respectively, it is one of the most appealing candidates to accommodate the ever-stringent performance requirement for future technology nodes [25]. In fact, the first bipolar junction transistor was made on germanium wafers, because of the limited availability, high cost and the most importantly, the lack of appropriate stable oxide to passivate the surface, the microelectronic industry quickly switched to silicon as the prime substrate material. Unlike SiO_2 which has a good interface quality with Si, GeO_2 is not only water soluble but also desorbs as gas-phase GeO at a temperature of ~400°C when it contacts directly germanium, and therefore GeO_2 is not a proper passivation layer for device operation. With the rapid progress of process technology, high-k gate dielectric provides the solution to circumvent the passivation issue and may once again push germanium device to the forefront. The integration of high-k gate dielectrics with germanium has been extensively explored by many research groups, the major challenge of a high-k gate dielectric on germanium lies in how to obtain a good interface quality with low electrical defect density between the high-k gate dielectric and germanium. Direct depositing high-k gate dielectrics on germanium cannot achieve a desirable interfacial quality since germanium would unavoidably out-diffuse into the high-k dielectric in the subsequent thermal processes. Therefore an intentional interfacial layer between a high-k dielectric and germanium becomes a prerequisite to prevent out-diffusion and maintain good interfacial properties. A germanium oxynitride (GeON) film is widely perceived as an appropriate interfacial layer to be integrated with a high-k dielectric, it can be formed by thermal oxidation of germanium followed by a NH_3 annealing [26-28] or by direct nitridation of the native GeO_2 already existing on the germanium surface [29, 30]. Recently, a thermally grown SiON film has been proposed to effectively passivate the germanium surface and holds great potential to serve as the interfacial layer for a high-k dielectric. Growth of the thermal SiON film on germanium is not that straightforward, it can be formed by oxidation of a SiGe layer followed by a forming gas annealing which results in a stack structure of SiO_2/Ge layer/Si substrate. Through an additional NH_3/N_2O treatment, the top SiO_2 would transform into a SiON film [31, 32]. Apparently, the most intriguing point of this process lies not only in the high quality thermal SiON film but also the formation of a virtual germanium substrate on a silicon wafer which is more compatible with existent ULSI technology. Even though GeON or SiON films reveal potential as the intentional interfacial layer, the relatively low permittivity (< 10 for GeON and < 5 for SiON) of these oxynitride films would limit further EOT scaling. A tremendous research effort is required to develop an interfacial layer with higher permittivity while well passivating germanium surface.

APPLICATION OF OXYNITRIDE FOR NON-VOLATILE MEMORY

Introduction to Non-Volatile Memory

Memory devices can be divided into two main categories: volatile and non-volatile memory. Volatile memory loses its data when the power is turned off and therefore constant power is required to keep its normal function. Dynamic random access memory (DRAM) used in computer system is one example. On the other hand, non-volatile memory retains stored data even without any power. The commonly used optical discs, magnetic computer storage medium (*e.g.* hard disk and floppy disk), and charge storage device (*e.g.* flash memory for digital camera, cell phone, and flash drive, *etc.*) fall into this category. In addition, some emerging non-volatile memory technologies such as magnetoresistive random access memory (MRAM), resistive random access memory (RRAM), ferroelectric random access memory (FeRAM) and phase change memory (PCM) have been proposed with the aim to realize the ultimate memory that integrates the advantages of high operation speed, small cell size and low production cost of volatile memory. Since charge storage-based flash memory is currently the mainstream non-volatile memory in microelectronic industry, this section will focus on the application of oxynitride to this memory type.

A conventional flash memory cell is an n-channel MOS transistor with two gates made of polycrystalline Si. The upper gate is called control gate which acts in the same manner as a usual gate of a MOSFET while the lower gate is called floating gate which is electrically isolated by an underlying tunnel dielectric and a inter-poly dielectric (IPD) beneath the control gate. The structure of a conventional flash memory cell is shown in Fig. **9**.

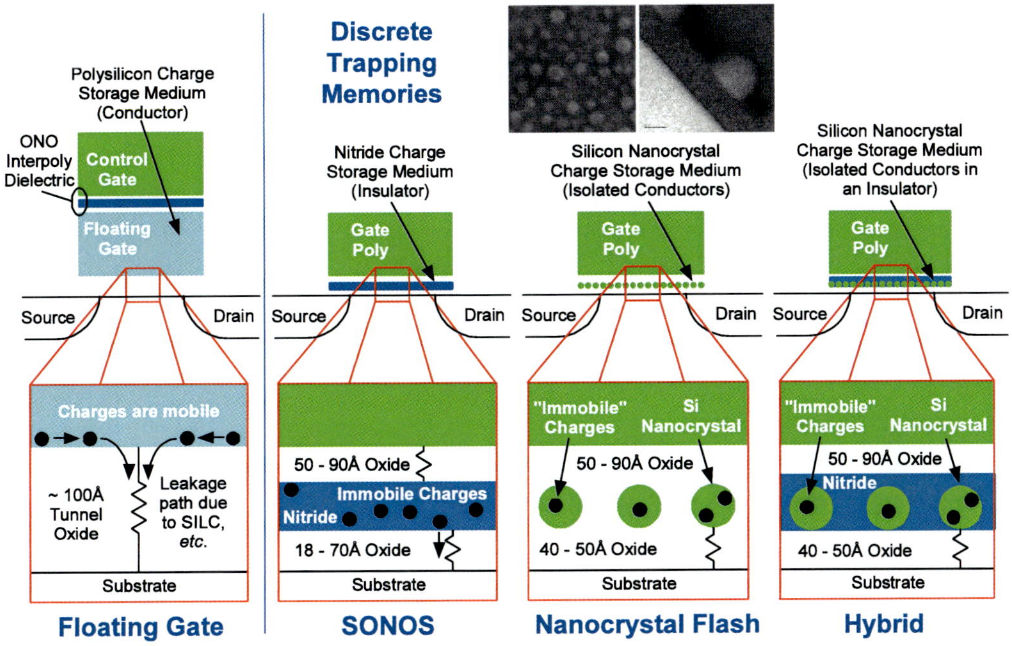

Figure 9: Various types of non-volatile memories. (a) Conventional flash memory with polycrystalline Si as the floating gate for charge storage. (b) SONOS flash memory with Si_3N_4 as the charge storage medium. (c) Nanocrystal memory with Si nanocrystals as the charge storage medium. (d) Hybrid memory with Si nanocrystals embedded in Si_3N_4 film for charge storage. Copied from digital DNA laboratory at Motorola.

The material of IPD can be SiO_2 or a $SiO_2/Si_3N_4/SiO_2$ (ONO) stack and the tunnel dielectric is usually SiO_2. When an appropriate voltage is applied on the control gate, source and drain, the electrons in the channel can be injected into the floating gate by Fowler-Nordheim (FN) tunneling or by channel hot electron (CHE) injection. The former refers to quantum process of particles penetrating a potential barrier in a wave-like behavior even though they do not have sufficiently high energy to surmount the barrier. The latter indicates that a fraction of channel electrons gain enough kinetic energy from the lateral electric field near the drain region so that they can overcome the 3.2 eV barrier height set by the tunnel SiO_2 and reach the floating gate. No matter what electron injection mechanism is adopted, once electrons are injected into the floating gate the threshold voltage of the MOSFET would be altered and therefore one can determine logic "1" or logic "0" is stored in the device by checking the threshold voltage value.

Although a huge commercial success has been obtained, conventional floating gate devices have faced their development limitation and the most prominent limitation lies in the tunnel SiO_2 scaling. Scaling tunnel SiO_2 will help enhance operation speed, however, the charge leakage issue becomes pronounced for the thinner tunnel SiO_2 since even one weak spot in the tunnel SiO_2 is sufficient to create a fatal leakage path and would cause leak away of all the stored charges because charges are stored in the continuous energy level (conduction band) of the floating gate [33]. To alleviate this issue, discrete charge storage medium that are more robust to leakages through oxide defects have been proposed to replace the conventional floating gate. Among the memory structures with discrete charge storage medium, charge-trapping memory and nanocrystal memory are the promising ones and the structure of these memory types are shown in Fig. **10**. In the charge-trapping memory, charges are stored in the discrete traps of Si_3N_4 and memory of this type is referred to SONOS (Silicon Oxide Nitride Oxide Semiconductor) memory. For the nanocrystal memory, charges are stored in the mutually isolated semiconductor or metal nanocrystals ranging from 5 to 10 nm in diameter and these nanocrystals can be formed by direct growth, deposition and ion implantation. A new memory with even superior performance has been proposed recently by combining the individual advantages of each memory type to form a hybrid memory which is also shown in Fig. **9**.

Oxynitride as Tunnel Dielectric

Typically, SiO_2 is used as the tunnel dielectric for SONOS memory devices. As devices continue to scale, a Si_3N_4 film used as the charge-trapping layer with more uniform thickness on the tunnel SiO_2 is required to achieve high performance. However, the surface morphology of Si_3N_4 degrades as it is deposited on SiO_2 since the deposition starts with an isolated Si_3N_4 nucleus on SiO_2 film. To circumvent this problem, it has been reported that if SiON is substituted for SiO_2 the surface roughness can be significantly improved [34]. In addition, adoption of a SiON film as the tunnel dielectric can also greatly improve the erase speed as compared with a conventional SiO_2 film since the introduction of nitrogen induces lowering of valence band offset between the tunnel dielectric and Si substrate, which increases the hole current and consequently results in a higher erase speed [35]. However, SiON also deteriorates retention characteristics since it causes larger electron and hole currents during data retention. Nevertheless, the nitrogen content in a SiON film can be fine-tuned to obtain the optimum erase and retention performance.

Using SiON as the tunnel dielectric is also helpful to boost the performance of nanocrystal memory. For Si nanocrystal-based memory, nanocrystals are grown on the tunnel dielectric to serve as charge storage medium and the denser the nanocrystals are the larger is the memory window that can be achieved. It has been found that replacing SiO_2 tunnel dielectric with a SiON film would result in an increased Si nanocrystal density by a factor of 3.2 which presents a better charge storage capability without degrading reliability [36]. The increased nanocrystal density can be attributed to the lower activation energy for the Si nanocrystal nucleation growth on the nitrogen-containing surface of SiON. For other types of nanocrystals, the activation energy for nucleation growth may be modulated by SiON and deserves further investigation since it is not only useful in memory devices but also in some nanocrystal-based optoelectronic devices [37-39].

For an ideal tunnel dielectric for memory, high tunneling probability of electrons at the electric fields of ~10 MV/cm for fast program and erase operations along with low transparency at the fields of 1-3 MV/cm for good retention are required. For a tunnel dielectric formed by SiO_2, because of the fixed barrier height regardless of the applied electric field, it is difficult to meet the above requirement. To achieve a tunnel dielectric with more desirable properties, band engineering by multilayer tunnel dielectric is proposed. Due to the moderate barrier height of the high-k dielectric with respect to that of low-k dielectric such as SiO_2, tunnel dielectric composed of a two-layer dielectric stack with a low-k/high-k dielectric combination (asymmetric structure) or a three-layer stack with a low-k/high-k/low-k dielectric combination (symmetric structure) has been proved to enhance the operation speed since as gate bias is applied the barrier thinning effect would occur and facilitate tunneling of electrons [40]. Based on this principle, oxynitride of high-k dielectric can also be used in the multilayer dielectric structure to achieve high-speed memory devices.

Oxynitride as Charge-Trapping Layer

Si_3N_4 used as the charge-trapping layer for SONOS-type memory has successfully exhibited better performance than conventional floating gate memory; however, some issues including erase saturation [41] and vertical migration of stored charges [42] are encountered in further scaling. More seriously, the retention characteristic is not satisfied as well due to the shallow trap energy which results from the relatively low conduction band offset of ~1.1 eV between Si_3N_4 and tunnel SiO_2. To overcome these difficulties, high-k dielectric(s) employed as the charge-trapping layer has demonstrated superior memory performance to Si_3N_4 because of the following reasons: (1) The high-k dielectric could provide deeper trap energy due to the larger conduction band offset which helps improve retention by preventing leakage from tunneling. (2) The high-k dielectric allows a higher electric field (E) over the tunnel SiO_2 due to electric flux density (D) continuity. The higher electric field across the tunnel SiO_2 would lead to a smaller operation voltage and higher operation speed [43]. (3) The high-k dielectric results in a modified FN tunneling due to the smaller conduction band offset with a Si substrate. This modified FN tunneling makes electrons easier to be trapped and consequently a higher operation speed can be achieved [44]. Hafnium oxynitride (HfON) has been widely discussed in the literature and is regarded as an appropriate charge-trapping layer for memory application [45, 46]. HfON used for memory application is quite different from that used in gate dielectric for CMOS logic technology. The former is optimized to trap a large amount of charges in deep levels by introducing high nitrogen concentration into HfO_2, whereas only a small nitrogen concentration is required to improve thermal stability for the latter. HfON can be formed by reactive sputtering from an Hf target in an ambient of O_2/N_2 mixture and the permittivity is found to be ~22 with nitrogen concentration of 20 at.%. As the nitrogen concentration becomes higher, the trap density will be enhanced accordingly while the trap energy will be also increased due to the lowered bandgap. Therefore it has been shown that a HfON film with nitrogen concentration of 20 at.% demonstrates a larger memory window along with better retention characteristics in comparison with that with 10 at.% nitrogen.

In addition to HfON, tetragonal zirconium oxynitride (ZrON) has been recently considered as a prominent dielectric for the charge-trapping layer. Crystalline ZrO_2 has drawn intensive attention because of the higher permittivity as compared to its amorphous phase. Among the crystalline phases of ZrO_2, tetragonal phase enjoys the highest permittivity which is theoretically predicted to be 46.6 [47] which is dependent on the process condition and finds its applications to high-k gate dielectrics in CMOS logic technology [48, 49] and high-density metal-insulator-metal (MIM) capacitors [50-52]. Applying a tetragonal ZrO_2 film to the charge-trapping layer of a memory device would realize a low voltage operation because of its relatively high permittivity [53], however, the retention issue caused by leakage current from grain boundaries may impose limits for further use. It has been reported that through a NH_3 nitridation of the tetragonal ZrO_2 to form ZrON, the leaky paths from grain boundaries can be effectively passivated and therefore a satisfactory retention performance is obtained [53]. This research result enlightens the investigation of nitridation for crystalline high-k dielectric rather than only for amorphous one.

Oxynitride as Inter-Poly Dielectric

Unlike the tunnel dielectric through which electrons can be injected into charge-trapping layer or Si substrate to perform program and erase operation, IPD is not expected to support charge transfer during program and erase operation. The basic requirement of IPD is to provide good capacitive coupling to the charge-trapping layer from the control gate and minimize any electrons injection from the control gate during erase operation while suppress stored charge leakage from charge-trapping layer during retention mode. Conventional SiO_2 no longer demonstrates competence to be an appropriate IPD as device and supplying voltage continue to scale because it cannot provide enough capacitive coupling. A straightforward means is to deploy high-k dielectrics to improve the capacitive coupling. From an electrostatics principle, a high-k IPD will cause a simultaneous reduction of electric field across itself and an increase of electric field across the tunnel SiO_2, leading to more efficient program and erase operation. However, a dielectric with higher permittivity usually accompanies a smaller bandgap which aggravates the retention performance. Furthermore, a high-k oxynitride film is a promising IPD material for memory devices. Besides thermal stability improvement, introduction of nitrogen into a high-k dielectric to form a high-k oxynitride also provides an effective approach to achieve the optimum performance between operation speed and retention because permittivity and bandgap of a high-k dielectric can be controlled by the amount of incorporated nitrogen. An example of high-k oxynitride for IPD is hafnium lanthanum oxynitride (HfLaON) [54, 55] which is formed by sputtered HfLaO with additional plasma nitridation and demonstrates good memory performance in terms of high operation speed along with satisfactory retention characteristics.

APPLICATION OF OXYNITRIDE FOR VOLATILE MEMORY

Introduction to Volatile Memory

Volatile memory refers to devices that could retain stored information only when power is supplied. It mainly includes DRAM and static random access memory (SRAM). SRAM is basically composed of p-channel and n-channel MOS transistors without any intentional capacitors. The aforementioned application of oxynitride to gate dielectric of logic circuits can be applied to SRAM circuits and therefore the discussion on volatile memory in this section will concentrate on DRAM. DRAM is historically a process technology driver and has the largest market share among semiconductor memories even though the DRAM industry experienced a large extent of restructuring in 2008. The core data storage unit of DRAM is called a cell which consists of an access transistor and a storage capacitor (1T-1C). DRAM stores each bit of data in a separate storage capacitor. Since a practical capacitor has some non-ideal properties which make charge leakage unavoidable, a periodical refresh of a DRAM cell to restore the data is required and therefore it is a dynamic memory as opposed SRAM or other static memory. As the functions of electronic products become more versatile and complex, DRAM capacity dramatically increases while the process technology aggressively scales. For the access transistor of a DRAM cell, it follows the evolution of that employed in the aforementioned logic circuit. For the storage capacitor in each technology node, maintaining storage capacitance more than 30 fF/cell and leakage currents less than 1 fA/cell is required to respectively ensure enough charge storage and adequate retention time. To achieve a high capacitance, DRAM chip makers currently adopt three-dimensional cell structures such as trench cell and stacked cell in order to increase the surface area. For a trench cell, storage capacitor is buried in the silicon substrate and is beneath the access transistor. On the contrary, a capacitor is formed on top of the access transistor for a stacked cell. The major advantages for trench cell technology are the inherently better topology and better performance of access transistors. The former comes from the fact that capacitors are buried in the substrate while the latter is due to the process consequence in which capacitors are formed prior to transistors and therefore thermal budget is not a concern [56]. Nevertheless one of the main obstacles trench cell technology faces in further scaling is the difficulty to form the extremely deep trench and it consequently makes stacked cell technology the mainstream one in the DRAM industry.

Oxynitride as Storage Dielectric for DRAM Trench Capacitor

Currently Si_3N_4/SiO_2 (silicon nitride/oxide or simply NO) dielectric stack is the main storage dielectric for trench capacitors. The conventional NO storage dielectric is formed by thin nitride film deposition in a low-pressure chemical vapor deposition (LPCVD) furnace and a subsequent wet oxidation in a furnace operated at one atmosphere in H_2O ambient. This NO dielectric stack has long been perceived to approach the scaling limit because of its relatively low permittivity and excessive tunneling current when a thinner EOT is desired. Fueled by this possible scaling limit, high-k dielectrics such as Al_2O_3 [57, 58], AlSiO [59] and HfSiO [60] have been proposed. Unfortunately, the high production cost and some process integration issues have caused delays to the introduction of high-k dielectric in trench DRAM technology. Thus how to better exert the properties of SiON to extend the scaling limit has recently stimulated intensive research. Nitridation of the SiO_2 film of a NO dielectric stack by NH_3 in a LPCVD furnace to convert it to a SiON film is a straightforward approach to enhance the permittivity and therefore the cell capacitance. The advantage of adopting NH_3 in a nitridation process lies in the fact that a large amount of nitrogen atoms can be incorporated in the SiO_2 film as compared with other nitrogen containing gas such as N_2O and NO. Unfortunately, NH_3 nitridation not only introduces nitrogen but also electron traps which aggravates the leakage current due to the trap-assisted tunneling. Degradation of leakage current would consequently cause poor retention characteristics of a DRAM cell and makes pure NH_3 nitridation infeasible for practical products. Recently it has been investigated that an additional N_2O re-oxidation in a LPCVD furnace after the NH_3 nitridation process could greatly improve the leakage current issue without sacrificing the cell capacitance since N_2O has a great capability to remove electron traps induced by NH_3 [61]. With this N_2O re-oxidation the SiON film becomes more appropriate to be the storage dielectric in terms of suppressed leakage current as well as better reliability performance.

Although NO storage dielectric with extra NH_3 and N_2O proves its prominent properties and potential to extend the employment of NO-based dielectric, this SiON film achieves such promising electrical characteristics at the price of adding extra cost because the additional nitridation and re-oxidation are conducted under low pressure conditions, much different from the growth condition of the SiO_2 layer which is in an atmosphere. This implies that an additional furnace process and consequent expenses are inevitable. To circumvent this cost issue, a new approach to form a SiON film on the thin Si_3N_4 film has been proposed [62]. In the newly developed approach, the SiON film was formed by *in situ* N_2O wet oxidation of the thin Si_3N_4 film. This *in situ* N_2O wet oxidation process includes a N_2O gas injection in the steam ambient and a post-oxidation N_2O treatment at high temperature at one atmosphere. The development of this process arose from the elimination of electron traps induced by hydrogen-containing gas such as NH_3 and the higher nitridation degree than a simple post-oxidation N_2O treatment since nitrogen is expected to be incorporated in the bulk rather than only on the SiO_2 interface during the *in situ* N_2O wet oxidation process. Compared with the conventional NO dielectric, it demonstrates a higher cell capacitance while preserving comparable leakage current. Moreover, this newly developed dielectric is eligible for a storage dielectric since it reveals good reliability performance with its failure rate less than 438 ppm after 10 years of operation. In contrast to previous work that meliorates the NO-based dielectric by NH_3 nitridation and N_2O re-oxidation, the appealing point of this process lies in the simplicity and economy in the SiON formation because only one furnace process is required and such an economical way to promote performance will become much more crucial in the keenly competitive DRAM market.

Aforementioned approaches have been proved to be highly suitable to be integrated into the current process with prominent cell performance enhancement, these techniques focus on the investigation of the SiO_2 film of a NO storage dielectric. To further extend the use of a NO storage dielectric to trench DRAMs, a quality improvement of the Si_3N_4 film is indispensable. In contrast to the case of SiO_2 films, it is a formidable task to increase the permittivity of the Si_3N_4 film with the current process tool; hence, the tunneling leakage current suppression without compromising its permittivity is the pivot of its quality improvement. Recently a Si_3N_4 film properly treated by N_2O to form a SiON film is proposed to greatly restrain the tunneling leakage current while preserving the cell capacitance, which implies that further scaling of the thickness of the Si_3N_4 film is possible for 90 nm technology and beyond [63]. The main difference between this process and conventional NO storage dielectric lies in an additional N_2O treatment in a LPCVD furnace at high temperature after the Si_3N_4 film deposition. During the N_2O treatment oxygen atoms can be introduced into the Si_3N_4 film to form the high-nitrogen-concentration SiON. In addition, reduced excess Si atoms and hydrogen-related species in the Si_3N_4 film can also be achieved through the N_2O treatment and therefore much lower leakage current than conventional NO storage dielectric was observed. In terms of cell capacitance, with identical physical dielectric thickness, the capacitance value was slightly smaller than conventional NO storage dielectric by 1 % since the incorporated oxygen lowers the permittivity of the Si_3N_4 film. However, because of the greatly reduced leakage current for the N_2O-treated dielectric, this capacitance degradation can be compensated by thinning down the physical thickness. In brief, the prominent electrical characteristics demonstrate the competence to further scale down the nitride thickness, which paves a new avenue to extend the scaling limit of the NO-based storage dielectric. Most importantly, without using a new tool investment, this technique can be fully integrated into the current furnace process in order to achieve the augmented cell performance, which is essential for trench DRAM manufacturers to maintain their cost advantage before a high-k dielectric is established.

The aforementioned discussions on storage dielectrics for trench capacitors focus on the development of SiON to either enhance permittivity of SiO_2 or suppress leakage current of Si_3N_4. For trench capacitors, the development of a high-k dielectric is not as flourishing as stacked capacitors because of the following reasons. (1) Its geometric structure makes an intrinsically high capacitor area. (2) Trench capacitors are formed prior to the access transistors and therefore a high-k storage dielectric should withstand high annealing temperature required for transistor process. This specific requirement sets a tough criterion for selection of high-k material. As mentioned previously, high-k materials such as Al_2O_3, AlSiO and HfSiO have been explored as the storage dielectric for DRAM trench capacitors. In addition to these materials, hafnium silicon oxynitride (HfSiON) shows the great potential to be integrated in trench capacitors because of its high thermal stability up to 1000 °C. The permittivity of HfSiON is about 13 which is relatively lower than the commonly used HfO_2 due to the incorporation of silicon and nitrogen. HfSiON for trench capacitor can be formed by HfSiOx deposition followed by a NH_3 annealing with 15 at. % nitrogen incorporated [64]. HfSiOx is formed by atomic layer deposition (ALD) with the precursors of tetrakis(ethylmethylamino) hafnium (TEMAHf) and tetrakis (ethylmethylamino) silicon (TEMASi). Because of the higher permittivity and therefore a thinner EOT, Al_2O_3/HfSiON/Si_3N_4 stack reveals a capacitance enhancement of 50 % as compared with conventional NO dielectric at the same leakage current level.

Oxynitrides as Storage Dielectric for DRAM Stacked Capacitors

With the onset of mega-bit DRAM era, DRAM stacked capacitors employed a conventional NO stack film as the storage dielectric. Since 130 nm technology node, high-k dielectrics such as Al_2O_3 [65, 66], Ta_2O_5 [67-69], ZrO_2 [70, 71], and HfO_2 [72, 73] were introduced to achieve the sufficiently high cell capacitance as the cell size continuously shrinks. Research on storage dielectric for DRAM stacked capacitors has been concentrated on the selection of appropriate dielectric and the process to suppress interfacial layer formation between the dielectric and top/bottom electrode. Incorporation of nitrogen into a high-k dielectric to form an oxynitride film for DRAM stacked capacitors has been rarely discussed. One possible application of nitridation of high-k dielectric can be found in Ta_2O_5 in which rapid thermal N_2O nitridation after Ta_2O_5 deposition was performed at high temperature [74]. During the nitridation process, reactive atomic oxygen species (O*) by the dissociation of N_2O gas can be generated and help reduce leakage current by repairing the oxygen vacancies and removing organic contamination in Ta_2O_5 film. Note that this nitridation could greatly improve leakage current without sacrificing cell capacitance and therefore may be applied to other high-k dielectrics.

It is worth mentioning that besides the development of storage dielectric to accommodate the ever-stringent scaling requirement, the capacitor structure should be evolved from SIS, MIS to MIM where "S" denotes the heavily doped polycrystalline silicon electrode, "I" indicates the insulating dielectric and "M" represents the metal electrode. Metals are used to replace doped polycrystalline silicon as the top and bottom electrode since the poly-silicon depletion effect can be alleviated. For those still employing doped poly-silicon as the bottom electrode, special care should be taken on the polycrystalline silicon and high-k interface since a SiO_2-like layer would inevitably be present due to the thermodynamic instability of high-k materials in direct contact with Si. This lower-k interfacial layer will degrade the overall effective permittivity, and deteriorate the leakage current and reliability performance. For a Ta_2O_5 dielectric, one common approach to solve this issue is to nitride the polycrystalline silicon electrode to form a thin SiON or Si_3N_4 film before Ta_2O_5 deposition. The formation of the SiON or Si_3N_4 interface layer not only enhances the stacked capacitance [69, 75] but also reduces the strain at the interface and minimizes charge traps. The SiON or Si_3N_4 interface layer can be accomplished by a thermal nitridation or plasma nitridation of the polycrystalline silicon electrode. For other high-k dielectrics such as ZrO_2, HfO_2 and TiO_2, this intentional nitridation of polycrystalline silicon electrode may also be applied to keep the desirable capacitance.

CONCLUSION

As the microelectronic devices enter nanometer regime, dielectrics used in integrated circuits become versatile and it necessitates in-depth investigation of these newly introduced material to optimize the circuit performance. Oxynitride gives dielectric additional latitude to modulate its properties by incorporating different amount of nitrogen. Currently oxynitrides have found wide applications in the gate dielectric for CMOS devices, the charge-trapping layer and tunnel/inter-poly dielectric for non-volatile memories and the storage dielectric for DRAM cell capacitors. As the microelectronics industry continues to progress, new types of devices will certainly emerge and oxynitrides will undoubtedly play a more critical role in developing the next-generation devices.

PART B: APPLICATION OF OXYNITRIDE FOR GAS BARRIERS

INTRODUCTION

In the last two decades, inorganic barrier coatings on organic polymer substrates have attracted research interests on the application of gas barrier layer in food packaging as an alternative to the traditional aluminium foil. The advantages of these barrier coatings over their metallic counterparts are microwave compatibility and recyclability. The inorganic gas barrier has also been applied in pharmaceutical and medical packing. Recently, due to the demand in portable and flexible electronic devices, such as liquid crystal displays (LCD), thin film batteries, organic thin film transistors (OTFT), solar cells and flexible organic light-emitting-diodes (OLED) [76, 77], the gas barrier performance has caught attention from both industries and academia. In order to fulfil the requirements in flexible devices, polymers with features being thin, light, impact resistant and flexible is the most promising candidate for the substrate material. The gas barrier performance of the polymer substrates, however, is not satisfactory for the devices and hence adding an effective gas barrier layer on the polymer substrate is essential to impede the penetration of oxidizing species [78]. Fig. **10** shows the required range of water vapor transmission rates (WVTR) for various applications in electronic devices [79].

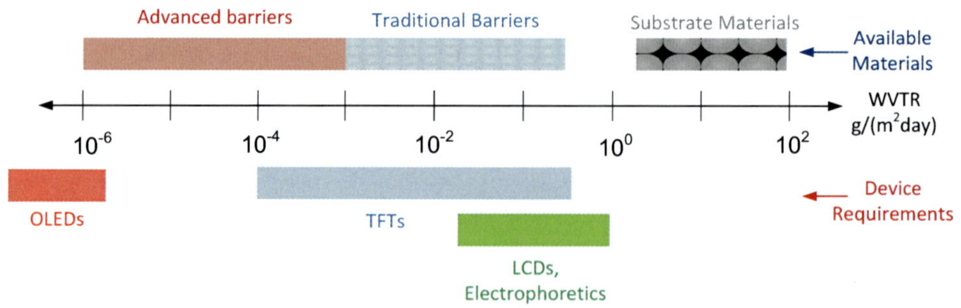

Figure 10: The requirements of WVTR of barrier coatings for different applications, data from reference (by permission of Elsevier) [79].

It can be seen that the required WVTR cover a wide range from 10^{-6} to 1 g/(m^2 day); nevertheless, the polymer substrate cannot meet the WVTR requirement without using barriers.

In this part, the characteristics of polymer substrates and gas barrier materials, deposition methods, the mechanisms of gas permeation, and the effects of film/PET interface, film thickness, and interlayer on the related properties of thin films will be reviewed. Finally, the application of gas barriers on flexible OLED will be briefly described.

ORGANIC POLYMER SUBSTRATE AND INORGANIC GAS BARRIER MATERIALS

Owing to their distinctive features being thin, light, flexible and capable for roll-to-roll process, polymer substrates are the most promising candidate materials for the application on flexible electronic devices. However, some intrinsic issues of polymers, such as high thermal expansion coefficient, poor gas barrier performance and low melting point, could limit their applications. Several polymer substrates including polyethylene terephthalate (PET), polycarbonates (PC), polyethylene naphthalate (PEN), polyethersulfone (PES), cyclic olefin resin (JSR ARTON) and polyimide (PI) have been utilized in the recent research and their physical properties are tabulated in Table **1** [80].

PET is one of the most important commercial polyesters. In comparison to other commercial polymers, PET possesses a few advantages such as higher optical transmittance, better toughness, lower thermal expansion coefficient and lower price. Furthermore, more than 5 billion pounds of PET have been produced per year for the industrial applications, mostly for the blow-molded bottle for beverages and other food products due to its better gas barrier properties [81]. As a result, PET substrate has been utilized as the major substrate in many previous studies on flexible electronic devices [82-101].

PES is also a good candidate for flexible plastic substrates. The high glass transition temperature of PES (T_g 223°C) gives good thermal stability during high temperature processes and its optical transmittance is also satisfactory for display applications, which is above 88% over the visible wavelength. Therefore, PES has been reported as the substrate material in the recent studies [102-104].

Table 1: The physical properties of different polymer substrates, data from reference [80].

Physical Properties		PET	PC	PEN	PES	PI	ARTON
Thickness (μm)		100	150	75	100		188
Density (g/cm^3)		1.4	1.2	1.36	1.37	1.43	1.08
Transmittance (%)		90.4	92.4	88	90.2	<30	92
Water Absorption (%)		0.3	0.2	0.3	1.4		0.4
Refraction Index		1.4	1.59	1.74	1.65		1.51
Retardation(nm)		68	41	TBD	13		6~7
Haze (%)		0.6	0.8	1.0	1.4		1.5
T_g (°C)		78	150	121	223	410	171
Working Temperature (°C)		105	125	180	180	240	180
CTE (ppm/°C)		15	70	13	54	20	62
Chemical Resistance	Methanol						
	Ethanol						
	Acetone		x				
	MEK		x				
	Cellosolve		x	x			

The most widely used materials for thin film gas barriers on polymer substrates are Al, AlO_x and SiO_x. Silicon nitride and oxynitride [82, 105-108], diamond-like carbon (DLC) [109-112], and carbon-based [113, 114] thin films have also been considered to be gas barriers for polymer substrates used in portable devices by enhancing their resistance to abrasive wear, gas permeation, and corrosion. Recently, transition-metal oxynitrides ($TMeN_xO_y$) emerging as a new class of coating material that may merit consideration for the application in gas barriers. The feature characteristics of $TMeN_xO_y$ coatings are the adjustable optical properties ranging from opaque color to transparent film and the electrical properties ranging from conductor to dielectrics [115-117]. The corrosion resistance of the $TMeN_xO_y$ coatings is comparable or better than the metal nitride counterparts [118]. Some recent studies have been carried out to evaluate the gas barrier performance of TiN_xO_y [87, 119] and ZrN_xO_y [120] deposited on polymer substrates.

DEPOSITION METHODS

Inorganic thin films have been deposited on polymer substrates by several methods such as electron beam evaporation [100], ion beam assisted deposition

(IBAD) [121, 122], magnetron sputtering [86-88, 95-97, 123-131], sol-gel [132], roll-to-roll processes [133], and plasma enhanced chemical vapor deposition (PECVD) [82-84, 89, 94, 101, 134-136].

For silicon oxide and oxynitrides barrier films, PECVD is the most commonly used deposition method. The popularity of PECVD is mainly due to the fact that this technique can be scaled up to industrial-scale deposition of high-quality barrier coatings with good uniformity and good adherence to the substrate. In addition, the application of ion bombardment to the surface can not only increase the film density and improve the adhesion, but also lower the deposition temperature [137]. However, the precursors used in PECVD such as SiH_4, NH_3 and $SiCl_4$ are flammable or toxic, which increases the cost in hazard control.

Among the methods mentioned above, magnetron sputtering has become a major alternative technique other than PECVD for the deposition of silicon oxides and oxynitrides. Magnetron sputtering is a deposition method that may deposit thin films at room temperature, which is required for depositing gas barrier thin films on polymer substrates, and the process can be scaled up to industrial scale. Furthermore, magnetron sputtering has advantages such as easy operation, less environmental pollution, target versatility, moderate set up costs and dense deposited films, which are more competitive than PECVD in producing gas barrier coatings. The commercialization of unbalanced magnetron sputtering (UBMS) further increased the deposition rate of the processing [138-140]. Therefore the usage of magnetron sputtering in producing SiN_xO_y gas barrier coatings is expected to increase and will become prevailing in the future.

For the deposition of AlO_xN_y thin films, the most common deposition methods are IBAD [121, 122] and reactive sputtering [95, 96, 123-126, 130]. Both techniques are physical vapor deposition (PVD) and have the advantages mentioned earlier. Since magnetron sputtering is a commercially available technique, considering the scaled-up capability and cost of the facilities, magnetron sputtering will be more competitive than IBAD. Moreover, once the deposition parameters are optimized, the process in producing gas barrier thin films can be controlled and thereby the products may maintain stable quality. This was the reason that the newly developed gas barrier thin films such as titanium oxynitride and zirconium oxynitride were chosen to be deposited by magnetron sputtering [87, 119, 120].

MECHANISMS OF GAS PERMEATION THROUGH GAS BARRIERS

To improve the gas barrier performance, it is crucial to know the mechanisms of gas permeation through the barrier film. For the widely used SiO_x and AlO_x barrier coatings, previous studies indicated that the film microstructure and composition may significantly affect the water vapor permeation [94, 95, 141]. In addition, the inclusion of nitrogen in silicon nitride and oxynitride barrier films can offer a substantial effect on the improvement of barrier performance for oxygen and water vapor permeation [142, 143]. Furthermore, the characteristics of the permeant gas will influence the mechanism or path of gas permeation. The detailed factors are illustrated in Fig. **11**.

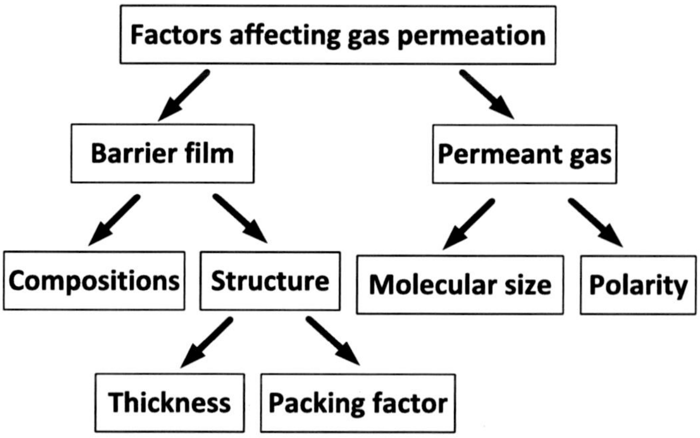

Figure 11: The factors affecting the gas permeation of a barrier film.

Three major routes for gas permeation through the oxide layer, as depicted in Fig. **12**, were proposed by Roberts *et al.* [144].

These routes included the unhindered transport through 'macro-defects' (>1 nm) in the oxide layer, the hindered transport through 'nano-defects' (<1 nm) and hindered transport through the amorphous lattice of the oxide (interstice < 0.3 nm). Considering the characteristics of the gases, however, the preferred permeation routes of water vapor (polar molecules) and oxygen (non-polar molecules) are essentially different.

It has been well established that the permeation of oxygen through the inorganic films is purely governed by the density of nano- and micron-scale defects [145].

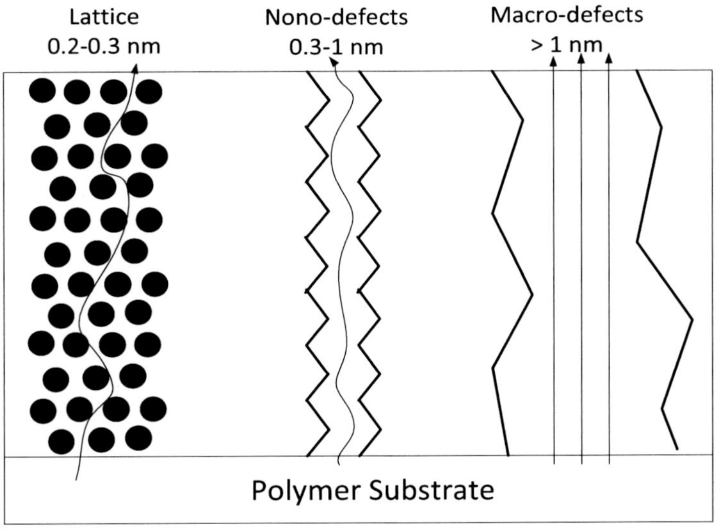

Figure 12: The proposed gas transport routes through the oxide layer (by permission of Elsevier) [144].

For the water vapor permeation, however, several studies [86, 130, 135, 141] have indicated that water molecules interact with the inorganic films and consequently the permeation mechanism of water vapor through inorganic films is more complex. Erlat *et al.* [130] studied the water vapor transport through PET/AlO_xN_y and found that most of the water vapor permeation was not defect-controlled. They proposed that the nitrogen-containing species in the AlN_xO_y film may be the main site where the chemical interaction occurs. The interaction is a dissociative process and can be represented as [146]:

$$OH_{(a)} + H_{(a)} \rightarrow H_{2(g)} + O_{(a)} \quad \textbf{(3)}$$

$$2H_{2(g)} + O_{2(g)} \rightarrow 2H_2O_{(g)} \quad \textbf{(4)}$$

where a and g represent "adsorbed" and "gas" phases, respectively. In the above reactions, a portion of oxygen species are locally trapped by the AlO_xN_y barrier film while the H_2 permeated essentially unhindered. This interaction, however, merely belongs to one of which influence the mechanism of water vapor permeation. Generally, the water vapor transmission through a barrier film/polymer structure involves many different processes, such as adsorption on the surface, solubilization in the polymer matrix, diffusion through the polymer and oxide layers and desorption at the opposite surface [92]. However, the effect of individual process on the mechanism of water vapor permeation has not been well understood yet and further research is necessary. Erlat *et al.* [130] suggested that the key requirements for producing better barrier materials are to increase the packing density such that the size and number density of the gas permeation pathways can be reduced.

FILM/PET INTERFACE

For the most widely used barrier films, AlO_x and SiO_x, previous studies indicated that the microstructure and compositions of thin films have significant effects on the gas barrier performance [94, 95, 141]. Other properties such as electrical resistivity, corrosion resistance and mechanical properties are also influenced. Deng *et al.* [141] pointed out that the microstructure of the deposited AlO_x thin film is closely related to the surface of the underneath polymer substrate. However, the growth mechanism of oxide thin films on polymer substrates has not been well understood yet. Cueff *et al.* [147] noted that the carbonyl group of PET is the primary reaction site for the formation of chemical bonds between the AlO_x thin film and PET substrate. In addition, the H_2O molecules adsorb on the surface or exist in the interior of PET substrates may supply O atoms to the coating during deposition [148]. Consequently, a "transition zone" may form on the film/polymer interface, and Cueff *et al.* [147] believed that the properties of thin films can be governed by this zone.

For the SiO_x and SiN_x films on polymer substrates, the film/substrate interface may be different for different deposition methods. The interface of evaporated SiO_x films on PET was found to consist of Si-C and Si-O-C strong covalent bonds. N_2 plasma treatment of the substrate prior to deposition is commonly used on less reactive polymers, such as polyethylene, which may facilitate uniform coverage and the formation of Si-O-C and Si-N-C bonding at the interface [149]. The film/substrate interface for SiO_x and SiN_x films on PET is found to be different when deposited by PECVD and PVD methods [83, 150]. The plasma-related deposition processes usually produce an "interphase" region of 40-100 nm for both SiO_x and SiN_x, where the composition continuously varies between the PET and the deposited film. This region possesses a crosslinked organosilicon nature, which may improve the mechanical stability and release the stress across the interface. For the PVD films this interphase region is no more than several nanometers, which is much smaller than that in PECVD films.

The crystallinity of thc coatings on PET substrates is also worth noting. Lin [119] found that the TiN_xO_y/Ti thin film on PET substrates revealed an amorphous structure. On the other hand, fine crystallites formed on PET substrates were observed by Lin *et al.* [87] using high-resolution transmission electron microscope (HRTEM) for the film deposited by RF magnetron sputtering. This was attributed to the increase of substrate temperature during deposition, which supplied sufficient driving force for the formation of fine crystallites. The crystallization of ZrN_xO_y on PET was also observed by Chu using X-ray diffraction [120]. Compared with TiN_xO_y, ZrN_xO_y film deposited on PET was easier to crystallize at similar deposition conditions.

FILM THICKNESS AND PACKING FACTOR

The characteristics of thin films, such as mechanical, optical, electrical and corrosion-resistance properties, may be affected by the film thickness, and hence altering the gas barrier performance. In general, bulk ceramic materials are impermeable to O_2, H_2O and N_2 [151]. On the other hand, the gas permeation behavior of a thin film may be quite different from its bulk counterpart. Felts *et al.* [152] observed that the transparent SiO_2 coating deposited by PECVD appeared a residual gas permeation phenomenon, which was attributed to the defects (*e.g.* pinholes, or micro-cracks) in the coating by da Silva Sobrinho *et al.* [83]. It has been pointed out by Huang *et al.* [153, 154] that penetration of a corrosion medium through a coating is mainly determined by the synergistic effect of packing factor and film thickness. The packing factor is defined as the ratio of film density to the bulk density of the same material, which represents the relative denseness of the film compared to its bulk counterpart. A film with high packing factor may reduce the penetration of the corrosive medium onto the substrate material; in addition, when the coating becomes thicker, the likelihood of having the through-thickness pinholes is lowered. Packing factor is closely related to the lattice defects in a thin film. Generally, an increase of packing factor is equivalent to a decrease of lattice defect density in the thin film. Therefore, it is reasonable that a thin film with higher packing factor possesses better corrosion resistance [153, 154], lower electrical resistance and better gas barrier performance [119]. Normally, the packing factor of a thin film increases with increasing film thickness, which is due to the increase of energy delivery into the deposited film with increasing deposition time such that more energy is available for the rearrangement of adatoms. As a result, the increase of film thickness accompanied with increasing packing factor is beneficial to the gas barrier performance.

The effect of film thickness on the gas permeation of barrier thin films has been reported and a critical thickness was proposed. Wuu *et al.* [90] suggested that the decrease of gas permeation rate with the increasing gas barrier thickness was due to the covering of substrate roughness and the filling of pinholes. Moreover, da Silva Sobrinho *et al.* [83] observed that as the film thickness exceeded a "critical thickness", the gas permeation rate continued to decrease but the trend became less distinct. Further increasing the film thickness to enhance gas barrier performance would result in a rougher surface which is not suitable for the application of flexible OLED displays [90].

The internal stress is a major concern when considering the mechanical properties of the gas barrier thin films. High compressive stress in the coatings may enhance the film density, barrier performance, and adhesion [155]. However, high internal stress tends to cause various forms of failures [156, 157]. It was found that the compressive stress increased with film thickness and then decreased; the decrease of stress was attributed to the stress relaxation with film growth [92], and the stress relaxation may be resulted from nanoscopic or microscopic cracks [90, 91]. These cracks can be favorable permeation channels for gas molecules and hence weaken the gas impermeability of the barrier thin films [155, 158]. The internal stress can be controlled by adjusting the RF power in the PECVD process, or using interlayer or multilayer structure [92].

The measurement of residual stress may be somewhat difficult for the gas barrier thin films on flexible substrates. The common residual stress measurement technique is by measuring the change of specimen curvature using optical means, which is a convenient and low cost method. However, since thermal stress from the change of deposition temperature can lead to the distortion of the substrate material, the change of specimen curvature may not truly reflect the stress in the gas barrier film. For the gas barrier films with crystalline structures, X-ray diffraction (XRD) may be a supplemental method for determining the residual stress of the films [159].

INTERLAYER AND MULTILAYER

Within the applicable film thickness, it is problematic to deposit a single-layer film on polymer substrates with satisfying barrier properties. Using a multilayer structure can improve the gas impermeability of the deposited films. A few factors contribute to the gas barrier performance. First, an interlayer can smooth the substrate surface to reduce surface flaws. This effect is often observed on thin films deposited by reactive sputtering and by electron-beam evaporation [78, 160, 161]. Secondly, the decoupling of defects in neighboring films within the multilayer structure will improve their gas impermeability [92, 162]. Thirdly, the existence of an interlayer will increase the diffusion barrier length of water vapor and/or gas molecules [163], which seems to be supported by recent measurements [164]. Therefore, it has been commonly agreed that the multilayer diffusion barrier will have lower gas permeability.

Several multilayer barrier structures have been proposed to enhance the barrier properties [91-93, 119, 165]. Yun *et al.* [165] suggested a double-layer structure consisting of SiO_x/Al_2O_3 on PET substrates, respectively deposited by PECVD and sputtering technique, which showed a very low WVTR of 10^{-3} g/(m^2 day). The enhancement of the gas barrier performance is mainly from the improvement of the film structure. They observed that granules and pinholes in SiO_x barriers can be effectively eliminated by introducing an Al_2O_3 interlayer on the PET surface prior to the SiO_x PECVD process. Moreover, plasma-induced reconstruction of PET surfaces was prevented by applying a reactive sputtering process to grow the Al_2O_3 interlayer. Wuu *et al.* [92] developed a multilayer composed of parylene/SiO_x/SiN_x...parylene/SiO_x/SiN_x(PON.PON) deposited on PC substrates. Under optimum thickness combinations, the WVTR and oxygen transmission rates (OTR) of SiO_x(50 nm)/SiN_x(50 nm) barrier coatings on PC at 80°C decreased to values near 0.01 g/(m^2 day) and 0.1 cm^3/(m^2 day), respectively. To further reduce the WVTR and OTR values, parylene layers are used as a smoothing, defect-decoupling, and protective medium in the multilayer.

Since the oxides and oxynitrides of Si and Al are dielectric materials, for the gas barrier applications requiring electric conduction, transition metal oxynitrides ($TMeN_xO_y$) may be better candidates. Lin *et al.* [87] deposited TiN_xO_y films on PET substrates by RF magnetron sputtering. They reported that the WVTR and OTR of the TiN_xO_y films reached values as low as 0.98 g/(m^2 day) and 0.60 cm^3/(m^2 day), respectively, which were about 6 and 47 times lower than those of the uncoated PET substrate. These transmission rates are comparable to those of DLC, carbon-based and Al_2O_3 barrier films.

It has been reported that TiN coatings with Ti interlayer possess stronger adhesive strength than those without Ti interlayer [153, 166, 167]. This may be ascribed to the better contact at the interfaces of substrate/Ti and Ti/TiN. In addition, the interdiffusion zone can be broadened due to the existence of Ti interlayer. The broader the interdiffusion zone, the lower is the bonding mismatch. Therefore, the adhesion between the films and substrates can be improved. Our earlier studies [153, 154, 167, 168] also found that the multilayer coatings have lower pinhole density than single-layer coatings, and can perform better corrosion resistance. This was attributed to the interruption of the through-thickness pinholes by the Ti interlayer and therefore the corrosive media could not reach the substrate by going through these interrupted defects. Moreover, the interlayer may also relieve the residual stress in the upper layer [169] and hence prevents the formation of micro- or nano- cracks which may increase the gas permeability. Therefore the metal nitride film/pure metal interlayer/polymer substrate multi-layer structure may provide an effective method to improve both gas barrier performance and electrical conduction.

Lin [119] deposited TiN_xO_y barrier films with Ti interlayer on PET substrate and found the WVTR can be lowered down to 0.232 g/(m^2 day). The sheet resistance of the barrier films was 53.6 Ω/□, which is much lower than the requirement of 100 Ω/□ for flexible electrodes used in the touch panels, inorganic solar cells and collectors of thin film batteries. The adhesion of the films evaluated by cross-hatch tests exhibited the best adhesive performance of grade 5. He concluded that the deposition of a Ti interlayer is useful in decreasing the WVTR and in reducing the sheet resistance of the TiN_xO_y/Ti thin films, which is attributed to the fact that the interlayer can interrupt the through-thickness defects and improve the packing factor of the upper TiN_xO_y layer. Similar results were reported on ZrN_xO_y gas barrier films with Zr interlayer [120]. The lowest WVTR of the barrier films was 0.295 g/(m^2 day) with the sheet resistance of 38 Ω/□.

APPLICATIONS TO OLEDS

This section can only briefly cover those issues and progress that are relevant for the applications to OLEDs. More detailed information can be found in two review articles by Lewis [79, 170]. OLED device assembled on flexible polymer substrates is considered as the next generation display technology. Fig. **13** illustrates three schematic diagrams for OLED encapsulation structures [170].

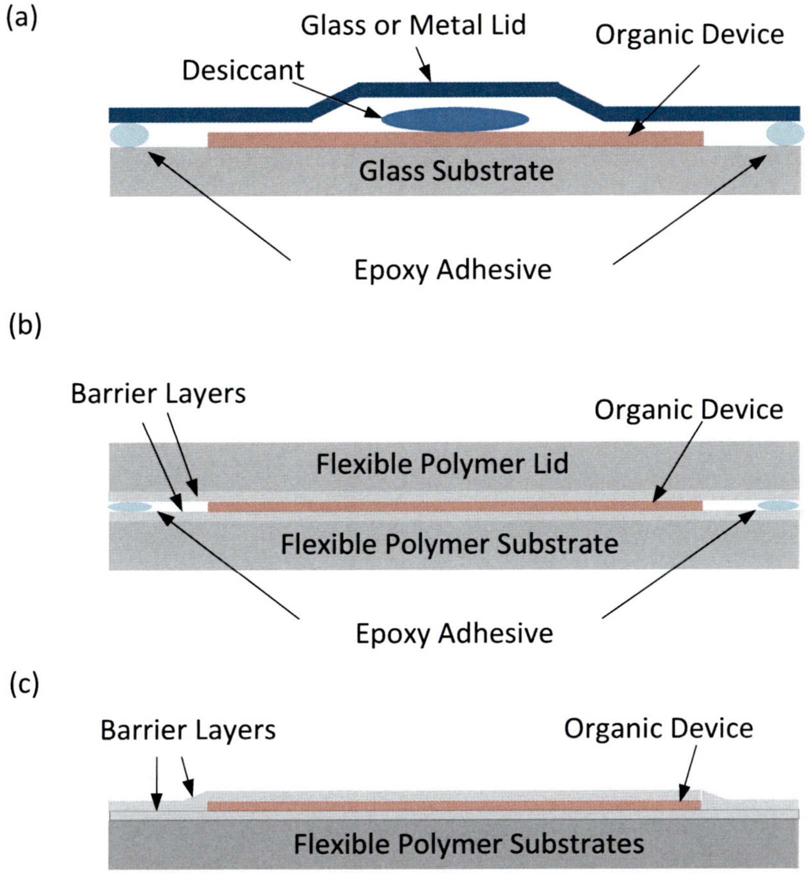

Figure 13: The schematic diagrams of encapsulation for OLEDs: (a) traditional rigid lid, (b) coated flexible lid, and (c) monolithic thin film [170].

Among the three structures, the glass-based OLED structure, as shown in Fig. **13** (a), is inappropriate for flexible display applications due to the rigidity of the lid and glass substrate, while the other two polymer-based OLED structures, as shown in Fig. **13**(b, c), can be applied in flexible devices. One of the critical issues of OLED devices is that the penetration of water vapor and oxygen may oxidize its metallic cathode and severely deteriorate its performance and lifetime. The current requirement for OLEDs demands coatings with exceedingly low OTR and WVTR of the order of 10^{-3} $cm^3/(m^2$ day) and 10^{-6} $g/(m^2$ day), respectively, for an OLED life time >10,000 h [78, 171]. The encapsulation structures of flexible OLED include the barrier-coated flexible lid and thin-film barrier coatings in intimate contact with the device, as respectively shown in Fig. **13**(b, c). Regarding these two structures, the thin-film direct-encapsulation technique has become a prevailing technique on account of its thinner structure and lesser concern for abrasion damage from the upper lid during in-flex use. In addition, the lowest-permeability epoxies are usually rigid and thereby fabricate an effective sealing structure by the laminated barrier-coated flexible lid technique could be a challenge [79].

As mentioned earlier, there are two approaches to increase the gas barrier performance: by increasing the packing density (or packing factor) of the barrier film and multilayer structures. The exploration on the increasing packing density for single-layer films is encouraging. WVTR rates can be decreased to the order of 10^{-3} to 5×10^{-5} $g/(m^2$ day) using advanced deposition techniques. Amorphous barrier films with very high density by sputtering technique have been produced by Symmorphix and the WVTR rates of the films were 8×10^{-5} $g/(m^2$ day) [172, 173]. Similarly, General Atomics has reported excellent permeation barrier performance from sputtered amorphous Al_2O_3, with a WVTR of 5×10^{-5} $g/(m^2$ day) [164]. Al_2O_3 films with a WVTR of 1×10^{-3} $g/(m^2$ day) have been produced by low temperature atomic layer deposition [174].

The barrier performance of multilayer structure comprised of alternating polymer and inorganic layers is much better than the single-layer films, where the WVTR rates can be generally lowered down to the order of 10^{-6} $g/(m^2$ day). Multilayer permeation barrier with graded junctions between multiple organic and inorganic layers by PECVD has been developed by GE with WVTR results in the 10^{-5}-10^{-6}

g/(m^2 day) range [175]. Multilayer barrier coated polymers from Vitex deposited by a hybrid process combining vacuum polymer deposition (VPD) and magnetron sputtering have demonstrated a WVTR estimated to be equivalent to 2×10^{-6} g/(m^2 day) at ambient conditions using the Ca test [176]. A cross-sectional micrograph of a multilayer sample is shown in Fig. **14**.

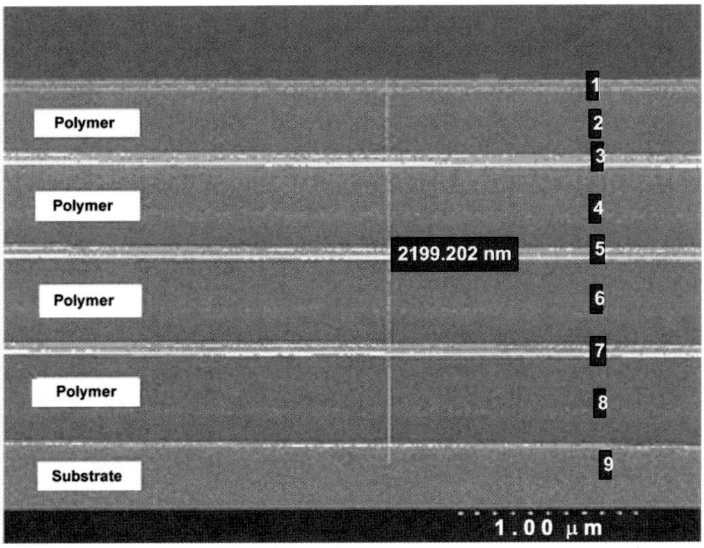

Figure 14: Layer structure of ultra-barrier coating, copied from reference [178] (courtesy of Dr. Peter Martin).

In addition, Vitex announced a combination of barrier-coated substrate and top encapsulation with a WVTR of 8×10^{-6} g/(m^2 day) at ambient conditions [177]. Recently, Wuu *et al.* [136] deposited SiN_x/parylene multilayer thin films onto flexible polyimide (PI) substrates using PECVD. They showed that the WVTR could be decreased to 7.41×10^{-6} g/(m^2 day) for the barrier with SiN_x films deposited at a relatively high growth temperature of 200°C. The progress demonstrated by multilayer structures is promising; however, barriers with satisfied performance at a cost that is compatible with large-scale, low-cost manufacturing have not yet been identified.

CONCLUSION

Oxynitrides have great potential to be applied on gas barriers. Although AlO_x and SiO_x are the most widely used oxides for gas barrier thin films, AlO_xN_y has been demonstrated to have better WVTR than its counterpart oxide. Recent research on transition metal oxynitrides ($TMeO_xN_y$) showed their promising potential to gas barrier applications; and the adjustable optical and electrical properties are especially attractive. To increase the gas barrier performance, there are two major approaches: by increasing the packing density of the barrier film and multilayer structures. For the applications requiring extremely low OTR and WVTR, such as flexible OLED, multilayer structures comprised of alternating polymer and inorganic layers are much better than the single-layer films, where the WVTR rates can be generally lowered down to the order of 10^{-6} g/(m^2 day). However, finding barriers with satisfying performance at a cost that is compatible with large-scale, low-cost manufacturing still remains a challenging issue.

ACKNOWLEDGEMENTS

Yung-Hsien Wu wishes to acknowledge the support from the National Science Council of Republic of China (Taiwan) under the contracts of NSC 98-2221-E-007-119 and NSC 98-2120-M-009-005. In addition, without the support of boost program 99N2520E1 from National Tsing Hua University, this work would not have been possible. Jia-Hong Huang would like to acknowledge the support of the National Science Council of Republic of China (Taiwan) under the contracts of NSC 99-2221-E-007-020-MY2 and NSC 99-2221-E-007-021. Mr. Han-Ming Chu is appreciated for the assistance of collecting references.

CONFLICT OF INTEREST

The author(s) confirm that this chapter content has no conflict of interest.

REFERENCES

[1] www.intel.com

[2] 2012 International Technology Roadmap for Semiconductors (http://www.itrs.net/)R. E. Matick and S. E Schuster, "Logic-based eDRAM: Origins and rationale for use", IBM J. Res. & Dev., vol. 49, pp. 145-165, 2005.

[3] M. L. Green, E. P. Gusev, R. Degraeve, and E. L. Garfunkel, "Ultrathin (<4 nm) SiO_2 and Si–O–N gate dielectric layers for silicon microelectronics: Understanding the processing, structure, and physical and electrical limits", *J. Appl. Phys.*, vol. 90, pp. 2057-2121, 2001.

[4] D. A. Buchanan, "Scaling the gate dielectric: materials, integration and reliability", *IBM J. Res. & Dev.*, vol. 43, pp. 245-264, 1999.

[5] D. A. Buchanan, "Scaling the gate dielectric: materials, integration and reliability", *IBM J. Res. & Dev.*, vol. 43, pp. 245-264, 1999.

[6] E. P. Gusev, H.-C. Lu, E. L. Garfunkel, T. Gustafsson, and M. L. Green, "Growth and characterization of ultrathin nitrided silicon oxide films", *IBM J. Res. Develop.,* vol. 43, pp. 265-286, 1999.

[7] Y. H. Wu, C. L. Ku, and J. F. Wang, "Boron-retarded gate dielectric formed by dry oxidation of thermal nitride", *J. Electrochem. Soc.*, vol. 151, pp. F1-F6, 2004.

[8] C. H. Chen, Y. K. Fang, C. W. Yang, S. F. Ting, Y. S. Tsair, M. C. Yu, T. H. Hou, M. F. Wang, S. C. Chen, C. H. Yu, and M. S. Liang, "Thermally-enhanced remote plasma nitrided ultrathin (1.65 nm) gate oxide with excellent performances in reduction of leakage current and boron diffusion", *IEEE Electron Device Lett.*, vol. 22, pp. 378-380, 2001.

[9] X. Guo and T. P. Ma, "Tunneling leakage current in oxynitride: dependence on oxygen/nitrogen content", *IEEE Electron Device Lett.*, vol. 19, pp. 207-209, 1998.

[10] I. Kingon, J. P. Maria, and S. K. Streiffer, "Alternative dielectrics to silicon dioxide for memory and logic devices", *Nature*, vol. 406, pp. 1032-1038, 2000.

[11] G. D. Wilk, R. M. Wallace, and J. M. Anthony, "High-k gate dielectrics: Current status and materials properties considerations", *J. Appl. Phys.*, vol. 89, pp. 5243-5275, 2001.

[12] Chin, C. C. Liao, C. H. Lu, W. J. Chen, and C. Tsai, "Device and reliability of high-k Al_2O_3 gate dielectric with good mobility and low D_{it}", in *VLSI Symp. Tech. Dig.*, 2006, pp. 135-136.

[13] H. Wong and H. Iwai, "On the scaling issues and high-k replacement of ultrathin gate dielectrics for nanoscale MOS transistors", *Microelectronic Eng.*, vol. 83, pp. 1867-1904, 2006.

[14] G. Shang, P. W. Peacock, and J. Robertson, "Stability and band offsets of nitrogenated high-dielectric-constant gate oxides", *Appl. Phys. Lett.*, vol. 84, pp. 106-108, 2004.

[15] T. Ino, Y. Kamimuta, M. Suzuki, M. Koyama, and A. Nishiyama, "Dielectric constant behavior of Hf–O–N system", *Jpn. J. Appl. Phys.*, vol. 45, pp. 2908-2913, 2004.

[16] S. Jeon, C. J. Choi, T. Y. Seong, and H. Hwang, "Electrical characteristics of ZrO_xN_y prepared by NH_3 annealing of ZrO_2", *Appl. Phys. Lett.*, vol. 79, pp. 245-247, 2001.

[17] M. Koyama, K. Suguro, M. Yoshiki, Y. Kamimuta, M. Koike, M.Ohse, C. Hongo, and A. Nishiyama, "Thermally stable ultra-thin nitrogen incorporated ZrO_2 gate dielectric prepared by low temperature oxidation of ZrN", in *IEDM Tech. Dig.*, 2001, pp. 20.3.1 - 20.3.4.

[18] R. E. Nieh, C. S. Kang, H. J. Cho, K. Onishi, R. Choi, S. Krishnan, J. H. Han, Y. H. Kim, M. S. Akbar, and J. C. Lee, "Electrical characterization and material evaluation of zirconium oxynitride gate dielectric in TaN-gated NMOSFETs with high-temperature forming gas annealing", *IEEE Trans. Electron Devices*, vol. 50, pp. 333-340, 2003.

[19] Chin, Y. H. Wu, S. B. Chen, C. C. Liao, and W. J. Chen, "High quality La_2O_3 and Al_2O_3 gate dielectrics with equivalent oxide thickness 5-10 Å", in *VLSI Symp. Tech. Dig.*, 2006, pp. 16-17.

[20] Y. H. Wu, M. Y. Yang, A. Chin, W. J. Chen, and C. M. Kwei, "Electrical characteristics of high quality La_2O_3 gate dielectric wit equivalent oxide thickness of 5 Å", *IEEE Electron Device Lett.*, vol. 21, pp. 341-343, 2000.

[21] K. Ohmori1, P. Ahmet, K. Shiraishi, K. Yamabe, H. Watanabe, Y. Akasaka, N. Umezawa, K. Nakajima, M. Yoshitake, T. Nakayama, K.-S. Chang, K. Kakushima, Y. Nara, M.L. Green, H. Iwai, K. Yamada, and T. Chikyow, "Controllability of flatband voltage in high-k

gate stack structures-remarkable advantages of La_2O_3 over HfO_2", in *IEDM Tech. Dig.*, 2006, pp. 376-379.

[22] S. J. Doh, H.-S. Jung, Y.-S. Kim, H.-J. Lim, J. P. Kim, J. H. Lee, J.-H. Lee, N.-I. Lee, H.-K. Kan, K.-P. Suh, S. G. Park, S. B. Kang, G. H. Choi, Y.-S. Chung, H.-S. Baikz, H.-S. Chang, M.-H. Cho, D.-W. Moon, H. B. Park, M. Cho, and C. S. Hwang, "Improvement of NBTI and electrical characteristics by ozone pre-treatment and PBTI issues in HfAlO(N) high-k gate dielectrics", in *IEDM Tech. Dig.*, 2003, pp. 943-946.

[23] Shanware, M.R. Visokay, J.J. Chambers, A.L.P. Rotondaro, J. McPherson, and L. Colombo, "Characterization and comparison of the charge trapping in HfSiON and HfO_2 gate dielectrics", in *IEDM Tech. Dig.*, 2003, pp. 938-942.

[24] X. B. Lu, Z.G. Liu, Y.P. Wang, Y. Yang, X.P. Wang, H.W. Zhou, and B.Y. Nguyen "Structure and dielectric properties of amorphous $LaAlO_3$ and $LaAlO_xN_y$ films as alternative gate dielectric materials", *J. Appl. Phys.*, vol. 94, pp. 1229- 1234, *2003*

[25] Y. Kamata, "High-k/Ge MOSFETs for future nanoelectronics", *Mater. Today,* vol. 11, pp. 30-38, 2008.

[26] C. O. Chui, F. Ito, and K. C. Saraswat, "Nanoscale germanium MOS dielectrics—Part I: Germanium oxynitrides", *IEEE Trans. Electron Devices.,* vol. 53, pp. 1501-1508, 2006.

[27] C. O. Chui, F. Ito, and K. C. Saraswat, "Scalability and electrical properties of germanium oxynitride MOS dielectrics", *IEEE Electron Device Lett.*, vol. 25, pp. 613-615, 2004.

[28] H. Shang, H. O. Schmidt, K. K. Chan, M. Copel, J. A. Ott, P. M. Kozlowski, S. E. Steen, S. A. Cordes, H. S. P. Wong, E. C. Jones, and W. E. Haensch, "High mobility p-channel germanium MOSFETs with a thin Ge oxynitride gate dielectric", in *IEDM tech. Dig.*, 2002, pp. 441-444.

[29] Ritenour, S. Yu, M. L. Lee, N. Lu, W. Bai, A. Pitera, E. A. Fitzgerald, D. L. Kwong, and D. A. Antoniadis, "Epitaxial strained germanium p-MOSFETs with HfO_2 gate dielectric and TaN gate electrode", in *IEDM tech. Dig.,* 2003, pp. 18.2.1-18.2.4.

[30] O. Chui, H. Kim, P. C. McIntryre, and K. C. Saraswat, "Atomic layer deposition of high-k dielectric for germanium MOS applications—substrate surface preparation", *IEEE Electron Device Lett.,* vol. 25, pp. 274-276, 2004.

[31] Y. H. Wu, J. R. Wu, and M. L. Wu, "Thermal gate SiO_2 for Ge metal-oxide-semiconductor capacitors fabricated on Si substrate", *Appl. Phys. Lett.*, vol. 91, pp. 093503, 2009.

[32] Y. H. Wu, M. L. Wu, Y. S. Lin, and J. R. Wu, "Electrical characteristics of thermal-SiON-gated Ge p-MOSFET formed on Si substrate", *IEEE Electron Device Lett.*, vol. 30, pp. 72-74, 2009.

[33] C. H. Tu, T. C. Chang, P. T. Liu, H. C. Liu, C. F. Weng, J. H, Shy, B. H Tseng, T. Y. Tseng, S. M. Sze, and C. Y. Chang, "A fabrication of germanium nanocrystal embedded in silicon-oxygen-nitride layer", *Electrochem. Solid-State Lett.,* vol. 9, pp. G358-G360, 2006.

[34] B. D. Salvo, C. Gerardi, R. V. Schaijk, S. A. Lombardo, D. Corso, C. Plantamura, S. Serafino, G. Ammendola, M. V. Duuren, P. Goarin, W. Y. Mei, K. V. d. Jeugd, T. Baron, M. Gély, P. Mur, and S. Deleonibus, "Performance and reliability features of advanced nonvolatile memories based on discrete traps (silicon nanocrystals, SONOS)", *IEEE Trans. Dev. Mat. Rel.*, vol. 4, pp.377-389, 2004.

[35] H. Aozasa, I. Fujiwara, K. Nomoto, H. Komatsu, K. Koyama, T. Kobayashi, and Tatsuji Oda, "Effects of nitridation on the electrical properties of MONOS nonvolatile memories", *J. Electrochem. Soc.,* vol. 154, pp. H798-H804, 2007.

[36] Y. H. Wu, C. K. Kao, C. Y. Wang, Y. S. Lin, C. M. Chang, C. H. Chuang, C. Y. Lee, C. M. Kuo, and A. Ku, "High-density silicon nanocrystal formed on nitrided tunnel oxide for nonvolatile memory application", *Electrochem. Solid-State Lett.,* vol. 11, pp. H131-H134, 2008.

[37] Polman, "Photonic materials: Teaching silicon new tricks", *Nature Materials.,* vol. 1, pp. 10-12, 2002.

[38] R. J. Walters, G. I. Bourianoff, H. A. Atwater, "Field-effect electroluminescence in silicon nanocrystals", *Nature Materials.*, vol. 4, pp. 143-146, 2004.

[39] S. Godefroo, M. Hayne, M. Jivanescu, A. Stesmans, M. Zacharias, O. I. Lebedev, G. Van Tendeloo, and V. V. Moshchalkov, "Classification and control of the origin of photoluminescence from Si nanocrystals", *Nature Nanotechnology.,* vol. 3, pp. 174-178, 2008.

[40] B. Govoreanu, P. Blomme, M. Rosmeulen, J. Van Houdt, and K. De Meyer, "VARIOT: A novel multilayer tunnel barrier concept for low-voltage nonvolatile memory devices", *IEEE Electron Device Lett.*, vol. 24, pp. 99-101, 2003.

[41] P. Xuan, M. She, B. Harteneck, A. Liddle, J. Bokor, and T. J. King, "FinFET SONOS flash memory for embedded applications", in *IEDM Tech. Dig.*, 2003, pp. 609–613.

[42] T. Sugizaki, M. Kobayashi, M. Ishidao, H. Minakata, M. Yamaguchi, Y. Tamura, Y. Sugiyama, T. Nakanishi, and H. Tanaka, "Novel multi-bit SONOS type flash memory using a high-k charge trapping layer", in *VLSI Symp. Tech. Dig.*, 2003, pp. 27–28.

[43] A. Chin, C. C. Liao, C. Chen, K. C. Chiang, D. S. Yu, W. J. Yoo, G. S. Samudra, T. Wang, I. J. Hsieh, S. P. McAlister, and C. C. Chi, "Low voltage high speed SiO_2/AlGaN/$AlLaO_3$/TaN memory with good retention", in *IEDM Tech. Dig.*, 2005, pp. 158–161.

[44] Y. N. Tan, W. K. Chim, W. K. Choi, M. S. Joo, and B. J. Cho, "Hafnium aluminum oxide as charge storage and blocking-oxide layers in SONOS-type nonvolatile memory for high-speed operation", *IEEE Trans. Electron Devices*, vol. 53, pp. 654–662, 2006.

[45] H. J. Yang, A. Chin, W. J. Chen, C. F. Cheng, W. L. Huang, I. J. Hsieh, and S. P. McAlister, "A program-erasable high-κ $Hf_{0.3}N_{0.2}O_{0.5}$ MIS capacitor with good retention", *IEEE Electron Device Lett.*, vol. 28, pp. 913-915, 2007.

[46] C. H. Lai, A. Chin, H. L. Kao, K. M. Chen, M. Hong, J. Kwo and C. C. Chi, "Very low voltage SiO_2/HfON/HfAlO/TaN memory with fast speed and good retention", in *VLSI Symp. Tech. Dig.*, 2006, pp. 54-55.

[47] X. Zhao and D. Vanderbilt, "Phonons and lattice dielectric properties of zirconia", *Phys. Rev. B, Condens. Matter*, vol. 68, pp. 075105, 2002.

[48] D. Fischer and A. Kersch, "The effect of dopants on the dielectric constant of HfO_2 and ZrO_2 from first principles", *Appl. Phys. Lett.*, vol. 92, pp. 012908, 2008.

[49] S. Bang, S. Lee, S. Jeon, S. Kwon, W. Jeong, S. Kim, and H. Jeon, "Physical and electrical properties of hafnium–zirconium–oxide films grown by atomic layer deposition", *J. Electrochem. Soc.*, vol. 155, pp. H633–H637, 2008.

[50] H. J. Cho, Y. D. Kim, D. S. Park, E. Lee, C. H. Park, J. S. Jang, K. B. Lee, H. W. Kim, Y. J. Ki, I. K. Han, and Y. W. Song, "New TIT capacitor with $ZrO_2/Al_2O_3/ZrO_2$ dielectrics for 60 nm and below DRAMs", *Solid-State Electronics*, vol. 51, pp. 1529-1533, 2007.

[51] J.-H. Kim, V. A. Ignatova, J. Heitmann, and L. Oberbeck, "Deposition temperature effect on electrical properties and interface of high-k ZrO_2 capacitor", *J. Phys. D: Appl. Phys.*, pp. 1–6, 2008.

[52] Y. H. Wu, C. K. Kao, B. Y. Chen, Y. S. Lin, M. Y. Li and H. C. Wu, "High density metal-insulator-metal capacitor based on $ZrO_2/Al_2O_3/ZrO_2$ laminate dielectric", *Appl. Phys. Lett.*, vol. 93, pp. 033511, 2008.

[53] Y. H. Wu, L. L. Chen, Y. S. Lin, M. Y. Li, and H. C. Wu, "Nitrided tetragonal ZrO_2 as the charge-trapping layer for nonvolatile memory application", *IEEE Electron Device Lett.*, vol. 30, pp. 1290-1292, 2009.

[54] C. H. Wu, F. Hung, A. Chin, S. J. Wang, W. J. Chen, X. P. Wang, M.-F. Li, C. Zhu, Y. Jin, H. J. Tao, S. C. Chen, and M. S. Liang, "High temperature stable [Ir_3Si-TaN]/HfLaON CMOS with large work-function difference", in *IEDM Tech. Dig.*, 2006, pp. 617-620.

[55] H. J. Yang, C. F. Cheng, W. B. Chen, S. H. Lin, F. S. Yeh, Sean P. McAlister and A. Chin, "Comparison of MONOS memory device integrity when using $Hf_{1-x-y}N_xO_y$ trapping layers with different N compositions", *IEEE Trans. Electron Devices*, vol. 55, pp. 1417–1422, 2008.

[56] N. Lee, K. Eup, Y.Gun, K. Do, "Perspectives on giga-bit scaled DRAM technology generation", *Microelectronic. Reliability*, vol. 40, pp. 191–206, 2000.

[57] J. Lutzen, A. Birner, M. Goldbach, M. Gutsche, T. Hecht, S. Jakschik, A. Orth, A. Sanger, U. Schroder, H. Seidl, B. Sell, and D. Schumann, "Integration of capacitor for sub-100-nm DRAM trench technology", in *Symp. VLSI Tech. Dig.*, 2002, pp. 178–179.

[58] M. Gutsche, H. Seidl, J. Luetzen, A. Birner, T. Hecht, S. Jakschik, M. Kerber, M. Leonhardt, P. Moll, T. Pompl, H. Reisinger, S. Rongen, A. Saenger, U. Schroeder, B. Sell, A. Wahl, and D. Schumann, "Capacitance enhancement techniques for sub-100 nm trench DRAMs", in *IEDM Tech. Dig.*, 2001, pp. 411–414.

[59] G. Aichmayr, A. Avellán, G. S. Duesberg, F. Kreupl, S. Kudelka, M. Liebau, A.Orth, A. Sänger, J. Schumann, O. Storbeck, "Carbon/high-k trench capacitor for the 40 nm DRAM generation", in *Symp. VLSI Tech. Dig.*, 2007, pp. 186–187.

[60] U. Schroeder, S. Jakschik, A. Avellan, E. Erben, B. Hintze, R. Duschl, M. Kerber, A. Link, and A. Kersch, "Recent developments in ALD technology for 50 nm trench DRAM applications", in *Proc. 208th Meeting Electrochem. Soc.*, 2005.

[61] Y. H. Wu, E. Hsieh, R. Kuo, S. Lai, and A. Ku, "Extending storage dielectric scaling limit by reoxidizing nitrided NO dielectric for trench DRAM", *IEEE Electron Device Lett.*, vol. 26, pp. 66-68, 2005.

[62] Y. H. Wu, I. Chang, C.-Y. Wang, T. Kao, I. Chang, C.-M. Kuo, and A. Ku, "Oxide-nitride storage dielectric formation in a single furnace process for trench DRAM", *IEEE Electron Device Lett.*, vol. 27, pp. 734–736, 2006.

[63] J. K. Park, J. H. Yang, and W. J. Cho, "Channel recessed one transistor dynamic random access memory with SiO2/Si3N4/SiO2 gate dielectric," *Jpn. J. Appl. Phys.*, vol. 51, p. 06FE08, 2011.

[64] T. Ando, N. Sato, S. Hiyama, T. Hirano, K. Nagaoka, H. Abe, A. Okuyama, H. Ugajin, K. Tai, S. Fujita, K. Watanabe, R. Katsumata, J. Idebuchi,T. Suzuki T. Hasegawa, H. Iwamoto and S. Kadomura, "Application of HfSiON to deep-trench capacitors of sub-45-nm-node embedded dynamic random-access memory", *Jpn.J. Appl. Phys.*, vol. 45, pp. 3165–3169, 2006.

[65] I.-S. Park, B. T. Lee, S. J. Choi, J. S. Im, S. H. Lee, K. Y. Park, J. W. Lee, Y. W. Hyung, Y. K. Kim, H. S. Park, Y. W. Park, S. I. Lee and M. Y. Lee, "Novel MIS Al_2O_3 Capacitor as a prospective technology for Gbit DRAMs", in *Symp. VLSI Tech. Dig.*, 2000, pp. 42–43.

[66] J. S. Lim, Y. K. Kim, S. J. Choi, J. H. Lee, Y. S. Kim, B. T. Lee, H. S. Park, Y. W. Park, S. I. Lee, "Novel Al_2O_3 capacitor for high density DRAMs", in *Int. Conf. on VLAI and CAD*, pp. 506-509, 1999.

[67] S. J. Won, Y. W. Hyung, K. J. Nam, Y. D. Kim, K. Y. Park, Y. W. Park, "Inner cylinder Ta_2O_5 capacitor process for 1 Gb DRAM and beyond", in *Symp. VLSI Tech. Dig.*, 1997.

[68] E. Gerritsen, N. Emonet, C. Caillat, N. Jourdan,M. Piazza, D. Fraboulet, B. Boeck, A. Berthelot,S. Smith, P. Mazoyer, "Evolution of materials technology for stacked-capacitors in 65nm embedded-DRAM", *Solid-State Electronics,* vol.49, pp.1767-1775, 2005.

[69] E. Atanassova, A. Paskaleva, "Challenges of Ta_2O_5 as high-k dielectric for nanoscale DRAMs", *Microelectronics Reliability,* vol. 47, pp.913–923, 2007.

[70] D.-S. Kil, H.-S. Song, K.-J. Lee, K. Hong, J.-H. Kim, K.-S. Park, S.-J. Yeom, J.-S. Roh, N.-J. Kwak, H.-C. Sohn, J.-W. Kim, and S.-W. Park, "Development of new $TiN/ZrO_2/Al_2O_3/ZrO_2/TiN$ capacitors extendable to 45nm generation DRAMs replacing HfO_2 based dielectrics", in *Symp. VLSI Tech. Dig.*, 2006, pp. 38–39.

[71] Y. K. Park, S. H. Lee, J. W. Lee, J. Y. Lee, S. H. Han, E. C. Lee, S. Y. Kim, J. Han, J. H. Sung, Y. J. Cho, J. Y. Jun, D. J. Lee, K. H. Kim, D. K. Kim, S. C. Yang, B. Y. Song, Y. S. Sung, H. S. Byun, W. S. Yang, K. H. Lee, S. H. Park, C. S. Hwang, T. Y. Chung and W. S. Lee, "Fully Integrated 56nm DRAM Technology for 1 Gb DRAM", in *Symp. VLSI Tech. Dig.*, 2007, pp. 190–191.

[72] P. Mazoyer, S. Blonkowski, F. Mondon, A. Farcy, J. Torres, G. Reimbold, F. Martin, J. F. Damlencourt, Y. Morand, N. Bicais, S. Descombes, "MIM HfO_2 low leakage capacitors for eDRAM integration at interconnect levels", in *Int. Interconnect Tech. Conf.*, 2003, pp. 117-119.

[73] S.-H. Oh, J.-H. Chung, J.-H. Choi, C.-Y. Yoo, Y. S. Kim, S. T. Kim, U.-I. Chung, J. T. Moon, "$TiN/HfO_2/TiN$ capacitor technology applicable to 70nm generation DRAMs", in *Symp. VLSI Tech. Dig.*, 2003, pp. 73–74.

[74] M. Lin, C. -Y. Chang, T.-Y. Huang and W. -Y. Shieh, "Leakage current reduction of chemical-vapor-deposited Ta_2O_5 films on rugged polycrystalline silicon electrode for dynamic random access memory application", *Jpn. J. Appl. Phys.*, vol. 38, pp. 1927–1931, 1999.

[75] H. Miki, Y. Shimamoto, and R. Furukawa, "Highly reliable MIS capacitors with plasma nitridation and doubled dielectric-constant tantalum pentoxide", *IEEE Trans. Electron Devices,* vol. 52, pp. 1832–1838, 2005.

[76] P. R. Johansen, " Large-area industrial vacuum coating in the 1990's", *J. Vac. Sci. Technol., A*, vol. 8, pp. 2798-2801, 1990.

[77] C. Charton, N. Schiller, M. Fahland, A. Hollander, A. Wedel, and K. Noller, "Development of high barrier films on flexible polymer substrates", *Thin Solid Films*, vol. 502, pp. 99-103, 2006.

[78] M. S. Weaver, J. J. Brown, R. H. Hewitt, S. Y. Mao, L. A. Michalski, T. Ngo, K. Rajan, M. A. Rothman, J. A. Silvernail, W. E. Bennet, C. Bonham, P. E. Burrows, G. L. Graff, M. E. Gross, M. Hall, E. Mast, and P. M. Martin, "Flexible organic LEDs: Plastics promise thin panels but present new problems," *Inform. Display*, vol. 17, p. 26, 2001.

[79] J. Lewis, "Material challenge for flexible organic devices", *Mater. Today*, vol. 9, pp. 38-45, 2006.

[80] http://www.materialsnet.com.tw, vol. 210, p.190, 2004.

[81] G. Odian, "*Principles of Polymerization*", 4th ed., New Jersey: John Wiley & Sons. Inc., 2004.

[82] N. Inagaki, S. Tasaka, and H. Hiramatsu, "Preparation of oxygen gas barrier poly(ethylene terephthalate) films by deposition of silicon oxide films plasma-polymerized from a mixture of tetramethoxysilane and oxygen", *J. Appl. Polym. Sci.*, vol. 71 pp. 2091-2100, 1999.

[83] A. S. da Silva Sobrinho, N. Schuhler, J. E. Klemberg-Sapieha, M. R. Wertheimer,M. Andrews, and S. C. Gujrathi, "Plasma-deposited silicon oxide and silicon nitride films on poly(ethylene terephthalate): a multitechnique study of the interphase regions", *J. Vac. Sci. Technol. A*, vol. 16, pp. 2021–2030, 1998.

[84] D. S. Wuu, W. C. Lo, C. C. Chiang, H. B. Lin, L. S. Chang, R. H. Horng, C. L. Huang, and Y. J. Gao, "Water and oxygen permeation of silicon nitride films prepared by plasma-enhanced chemical vapor deposition", *Surf. Coat. Technol.*, vol.198, pp. 114-117, 2005.

[85] T. N. Chen, D. S. Wuu, C. C. Wu, C. C. Chiang, H. B. Lin, Y. P. Chen, R. and H. Horng, "Effects of plasma pretreatment on silicon nitride barrier films on polycarbonate substrates", *Thin Solid Films*, vol. 514, pp. 188-192, 2006.

[86] A. G. Eralt, B. M. Henry, J. J. Ingram, D. B. Mountain, A. McGuigan, R. P. Howson, C. R. M. Grovenor, G. A. D. Briggs, and Y. Tsukahara, "Characterisation of aluminium oxynitride gas barrier films", *Thin Solid Films*, vol. 388, pp. 78-86, 2001.

[87] M. C. Lin, L. S. Chang, and H. C. Lin, "Gas barrier properties of titanium oxynitride films deposited on polyethylene terephthalate substrates by reactive magnetron sputtering" *Appl. Surf. Sci.*, vol. 254, pp. 3509-3516, 2008.

[88] M. C. Lin, C. H. Tseng, L. S. Chang, and D. S. Wuu, "Characterization of the silicon oxide thin films deposited on polyethylene terephthalate substrates by radio frequency reactive magnetron sputtering" *Thin Solid Films*, vol. 515, pp. 4596-4602, 2007.

[89] D. S. Wuu, W. C. Lo, L. S. Chang, and R. H. Horng, "Properties of SiO_2-like barrier layers on polyethersulfone substrates by low-temperature plasma-enhanced chemical vapor deposition", *Thin Solid Films*, vol. 468, pp. 105-108, 2004.

[90] D. S. Wuu, W. C. Lo, C. C. Chiang, H. B. Lin, L. S. Chang, R. H. Horng, C. L. Huang, and Y. J. Gao, "Plasma-deposited silicon oxide barrier films on polyethersulfone substrates: temperature and thickness effects", *Surf. Coat. Technol.*, vol. 197, pp. 253-259, 2005.

[91] C. C. Chiang, D. S. Wuu, H. B. Lin, Y. P. Chen, T. N. Chen, Y. C. Lin, C. C. Wu, W. C. Chen, T. H. Jaw, and R. H. Horng, "Deposition and permeation properties of SiNX/parylene multilayers on polymeric substrates", *Surf. Coat. Technol.*, vol. 200, 5843-5848, 2006.

[92] D.S. Wuu, T.N. Chen, C.C. Wu, C.C. Chiang, Y.P. Chen, R H. Horng, and F.S. Juang, "Transparent barrier coatings for flexible organic light-emitting diode applications", *Chem. Vap. Deposition*, vol. 12, pp. 220-224, 2006.

[93] T. N. Chen, D. S. Wuu, C. C. Wu, C. C. Chiang, Y. P. Chen, and R. H. Horng, "High-performance transparent barrier films of SiO_x/SiN_x stacks on flexible polymer substrates", *J. Electrochem. Soc.*, vol. 153, pp. F244-F248, 2006.

[94] Y. G. Tropsha and N. G. Harvey, "Activated rate theory treatment of oxygen and water transport through silicon oxide/ poly(ethylene terephthalate) composite barrier structures", *J. Phys. Chem. B*, vol. 101, pp.2259–2266, 1997.

[95] A. G. Erlat, R. J. Spontak, R. P. Clarke, T. C. Robinson, P. D. Haaland, Y. Tropsha, N. G. Harvey, and E. A. Vogler, "SiOx gas barrier coatings on polymer substrates: morphology and gas transport considerations", *J. Phys. Chem. B*, vol. 103, pp. 6047–6055, 1999.

[96] B. M. Henry, A. G. Erlat, A. McGuigan, C. R. M. Grovenor, G. A. D. Briggs, Y. Tsukahara, T. Miyamoto, N. Noguchi, and T. Niijima, "Characterization of transparent aluminium oxide and indium tin oxide layers on polymer substrates", *Thin Solid Films*, vol. 382, pp. 194-201, 2001.

[97] Satoru Iwamori, Yumi Gotoh, and Krzysztof Moorthi, "Characterization of silicon oxynitride gas barrier films", *Vacuum*, vol. 68, pp. 113-117, 2003.

[98] Takuya Oyaidu, Yohei Ogawa, Kazuhiko Tsurumaki, Keisuke Ohdaira, and Hideki Matsumura, "Formation of gas barrier films by Cat-CVD method using organic silicon compounds", *Thin Solid Films*, vol. 516, pp. 604-606, 2008.

[99] K. Lau, J. Weber, H. Bartzsch, and P. Frach, "Reactive pulse magnetron sputtered SiO_xN_y coatings on polymers", *Thin Solid Films*, vol. 517, pp. 3110-3114, 2009.

[100] T. Krug, R. Ludwig, and G. Steiniger, "New developments in transparent barrier coatings", in *Proc. Soc. Vacuum Coaters 36th Annu. Tech. Conf.*, 1993, p. 302-306.

[101] N. Inagaki, S. Tasaka, and M. Makino, "Plasma polymer deposition from mixture of tetramethoxysilane and oxygen on PET films and their oxygen gas barrier properties", *J. Appl. Polym. Sci.*, vol. 64, pp. 1031-1039, 1997.

[102] June Hee Lee, Chang Hyun Jeong, Jong Tae Lim, Viktor A. Zavaleyev, Se Jin Kyung and Geun Young Yeom, "Characteristics of a Multilayer $SiO_x(CH)_yN_z$ Film Deposited by Low Temperature Plasma Enhanced Chemical Vapor Deposition Using Hexamethyldisilazane/Ar/N_2O", *Jpn. J. Appl. Phys*, vol. 45, pp. 8430-8434, 2006.

[103] Juno Shim, Ho Gyu Yoon, Sang-Hyun Na, Insun Kim, and Soonjong Kwak, "Silicon oxynitride gas barrier coatings on poly(ether sulfone) by plasma-enhanced chemical vapor deposition", *Surf. Coat. Technol.*, vol. 202, pp. 2844-2849, 2008.

[104] J.H. Lee, C.H. Jeong, J.T. Lim, V.A. Zavaleyev, S.J. Kyung, and G.Y. Yeom, "SiO_xN_y thin film deposited by plasma enhanced chemical vapor deposition at low temperature using HMDS–O_2–NH_3–Ar gas mixtures", *Surf. Coat. Technol.*, vol. 201, pp. 4957-4960, 2007.

[105] A. Heya, T. Minamikawa, T. Niki, S. Minami, A. Masuda, H. Umemoto, N. Matsuo, and H. Matsumura, "Cat-CVD SiN passivation films for OLEDs and packaging", *Thin Solid Films*, vol. 516, pp. 553-557, 2008.

[106] J. H. Lee, C. H. Jeong, H. B. Kim, J. T. Lim, S. J. Kyung, G. Y. Yeom, "Characteristics of SiO_xN_y films deposited by inductively coupled plasma enhanced chemical vapor deposition using HMDS/NH_3/O_2/Ar for water vapor diffusion barrier", *Thin Solid Films*, vol. 515, pp. 917-921, 2006.

[107] Tsai-Ning Chen, Dong-Sing Wuu, Chia-Cheng Wu, Cheng-Chung Chiang, Yung-Pei Chen, and Ray-Hua Horng, "Improvements of permeation barrier coatings using encapsulated parylene interlayers for flexible electronic applications", *Plasma Process. Polym.* vol. 4, pp. 180–185, 2007.

[108] Dong Sing Wuu, W.C. Lo, C.C. Chiang, H.B. Lin, L.S. Chang, R.H. Horng, C.L. Huang, and Ying Jun Gao "Transparent Barrier Coatings on Flexible Polyethersulfone Substrates for Moisture-Resistant Applications", *Mater. Sci. Forum.*, vol. 475-479, pp. 4017-4020, 2005.

[109] K. Inagawa, T. Zeniya, and N. Hibino, "Characterization of diamond-like carbon films formed by magnetically enhanced plasma chemical vapor deposition", *Surf. Coat. Technol.* vol. 169/170, pp. 344-347, 2003.

[110] S. Yamamoto, H. Kodama, T. Hasebe, A. Shirakura, and T. Suzuki, "Oxygen transmission of transparent diamond-like carbon films", *Diamond Relat. Mater.* vol. 14, pp. 1112-1115, 2005.

[111] M. Yoshida, S. Watanabe, T. Tanaka, T. Takagi, M. Shinohara, and J.W. Lee, "Investigation of diamond-like carbon formed on PET film by plasma-source ion implantation using C_2H_2 and CH_4", *Nucl. Instrum. Methods Phys. Res.* B, vol. 206, pp. 712-716, 2003.

[112] M. Yoshida, T. Tanaka, S. Watanabe, M. Shinohara, J.W. Lee, and T. Takagi, "Improvement of oxygen barrier of PET film with diamond-like carbon film by plasma-source ion implantation", *Surf. Coat. Technol.*, vol. 174/175, pp.1033-1037, 2003.

[113] [113] S.Watanabe,M. Shinohara, H. Kodama, T. Tanaka, M. Yoshida, and T. Takagi, "Amorphous carbon layer deposition on plastic film by PSII", *Thin Solid Films*, vol. 420/421, pp.253-258, 2002.

[114] H. Kodama, A. Shinohara, A. Hotta, and T. Suzuki, "Gas barrier properties of carbon films synthesized by atmospheric pressure glow plasma", *Surf. Coat. Technol.*, vol. 201, pp. 913-917, 2006.

[115] F. Vaz, P. Cerqueira, L. Rebouta, S.M.C. Nascimento, E. Alves, Ph. Goudeau, J.P. Riviere, K. Psichow, and J. de Rijk, "Structural, optical and mechanical properties of coloured TiN_xO_y thin films", *Thin Solid Films*, vol. 447/448, pp. 449-454, 2004.

[116] N. Martin, O. Banakh, A.M.E. Santo, S. Springer, R. Sanjines, J. Takadoum, and F. Levy, "Correlation between processing and properties of TiO_xN_y thin films sputter deposited by the reactive gas pulsing technique", *Appl. Surf. Sci.*, vol. 185, pp.123-133, 2001.

[117] F. Vaz, P. Cerqueira, L. Rebouta, S.M.C. Nascimento, E. Alves, Ph. Goudeau, and J.P. Riviere, "Preparation of magnetron sputtered TiN_xO_y thin films", *Surf. Coat. Technol.* vol. 174/175, pp.197-203, 2003.

[118] Jia-Hong Huang, Zhang-En Tsai, and Ge-Ping Yu, "Mechanical properties and corrosion resistance of nanocrystalline ZrN_xO_y coatings on AISI 304 stainless steel by ion plating", *Surf. Coat. Technol.*, vol. 202, pp. 4992-5000, 2008.

[119] J. L. Lin. "TiN/Ti thin film on flexible PET substrate deposited by RF magnetron sputtering", M.S. thesis, National Tsing Hua University, Hsinchu, Taiwan, 2008.

[120] H.-M. Chu, "ZrN/Zr Thin Film Gas Barrier on Flexible PET Substrate Deposited by Magnetron Sputtering" M.S. thesis, National Tsing Hua University, Hsinchu, Taiwan, 2009.

[121] C.K. Hawangbo, L.J. Lingg, J.P. Lehan, H.A. Macleod, and F. Suits, "Reactive ion assisted deposition of aluminum oxynitride thin films", *Appl. Optics*, vol. 28, pp. 2779-2784, 1989.

[122] G.A. Al-Jumaily, T.A. Mooney, W.A. Spurgeon, and H.M. Dauplaise, "Ion assisted deposition of oxynitrides of aluminum and silicon", *J. Vac. Sci. Tech. A*, vol. 7, pp. 2280-2285, 1989.

[123] E. Budke, J. Krempel-Hesse, H. Maidhof, and H. Schussler, „Decorative hard coatings with improved corrosion resistance", *Surf. Coat. Technol.*, vol. 112, pp.108-113, 1999.

[124] E. J. Harju, I. M. Penttinen, and A. S. Korhonen, "Optimization of wear and corrosion resistance of triode-ion-plated nitride coatings", *Surf. Coat. Technol.*, vol. 41, pp. 157-166, 1990.

[125] P. C. Johnson and H. Randhawa, "Zirconium nitride films prepared by cathodic arc plasma deposition process", *Surf. Coat. Technol.*, vol. 33, pp. 53-62, 1987.

[126] E. Kelesoglu, C. Mitterer, M. K. Kazmanli, and M.Urgen, Microstructure and properties of nitride and diboride hard coatings deposited under intense mild-energy ion bombardment", *Surf. Coat. Technol.*, vol. 116, pp. 133-140, 1999.

[127] A. F. Jankowski, J. P. Hayes, T. E. Felter, C. Evans, and A. J. Nelson, "Sputter deposition of silicon–oxide coatings", *Thin Solid Films*, vol. 420, pp. 43-46, 2002.

[128] B. M. Henry, F. Dinelli, K. Y. Zhao, C. R. M. Grovenor, O. V. Kolosov, G. A. D. Briggs, A. P. Roberts, R. S. Kumar, and R. P. Howson, "A microstructural study of transparent metal oxide gas barrier films", *Thin Solid Films*, vol. 355, pp. 500-505, 2002.

[129] C. J. Tavares, S. M. Marques, S. Lanceros-Mendez, V. Sencadas, V. Teixeira, J. O. Carneiro, A. J. Martins, and A. J. Fernandes, "Strain analysis of photocatalytic TiO_2 thin films on polymer substrates", *Thin Solid Films*, vol. 516, pp. 1434-1438, 2008.

[130] A. G. Erlat, B. M. Henry, C. R. M. Grovenor, and A. G. D. Briggs, "Mechanism of water vapor transport through PET/AlO_xN_y gas barrier films", *J. Phys. Chem. B*, vol. 108, pp. 883-890, 2004.

[131] B.M. Henry, F. Dinelli, K.-Y. Zhao, C.R.M. Grovenor, O.V. Kolosov, G.A.D. Briggs, A.P. Roberts, R.S. Kumar, and R.P. Howson, "A microstructural study of transparent metal oxide gas barrier films", *Thin Solid Films*, vol. 355/356, pp. 500-505, 1999.

[132] K. Tadanaga, K. Iwashita, and T. Minami, "Coating and water permeation properties of SiO_2 thin films prepared by the sol-gel method on nylon-6 substrates", *J. Sol-Gel Sci. Technol.*, vol. 6, pp. 107-111, 1996.

[133] F. Casey, A. W. Smith, and G. Eills, "Clear Barrier Coatings", in *Proc. Soc. Vacuum Coaters 42nd Annu. Tech. Conf.*, 1999, pp.415-419.

[134] G. Garcia-Ayuso, L. Vazquez, and J. M. Martinez-Duart, "Atomic force microscopy (AFM) morphological surface characterization of transparent gas barrier coatings on plastic films," *Surf. Coat. Technol.*,vol. 80, pp. 203–206, 1996.

[135] A. S. da Silva Sobrinho, G. Czeremuszkin, M. Latreche, and M. R. Wertheimer, "Defect-permeation correlation for ultrathin transparent barrier coatings on polymers," *J. Vac. Sci. Technol. A*, vol. 18, pp.149–157, 2000.

[136] Dong-Sing Wuu, Tsai-Ning Chen, Eddy Lay, Chia-Hsiang Liu, Chia-Hao Chang, Hsiao-Fen Wei, Liang-You Jiang, Hung-Uang Lee, and Yu-Yang Chang, "Transparent barrier coatings on high temperature resisting polymer substrates for flexible electronic applications", *J. Electrochem. Soc.*, vol. 157, pp. C47-C51, 2010.

[137] L. Martinu, J. E. Klemberg-Sapieha, O. M. Kuttel, A. Raveh, and M. R. Wertheimer, "Critical ion energy and ion flux in the growth of films by plasma-enhanced chemical-vapor deposition", *J. Vac. Sci. Technol., A*, vol. 12, pp.1360-1364, 1994.

[138] B. Window, "Recent advances in sputter deposition", *Surf. Coat. Technol.*, vol.71, pp. 93-97, 1995.

[139] D.G. Teer, "Magnetron sputter ion plating", US Patent No. 5,556,519, September 17, 1996.

[140] P.J. Kelly and R.D. Arnell, "Magnetron sputtering: a review of recent developments and applications", *Vacuum*, vol. 56, pp. 159-172, 2000.

[141] C. S. Deng, H. E. Assender, F. Dinelli, V. Kolosov, G. A. D. Briggs, T. Miyamoto, and Y. Tsukahara, "Nucleation and growth of gas barrier aluminium oxide on surfaces of poly(ethylene terephthalate) and polypropylene: effects of the polymer surface properties", *J. Polym. Sci., Part B: Polym. Phys.*, vol. 38, pp. 3151-3162, 2000.

[142] M. Vogt and R. Hauptmann, "Plasma-deposited passivation layers for moisture and water protection," *Surf. Coat. Technol.*, vol. 74–75, pp.676–681, 1995.

[143] H. Lin, L. Xu, X. Chen, X. Wang, M. Sheng, F. Stubhan, K.-H. Merkel, and J. Wilde, "Moisture-resistant properties of SiN_x films prepared by PECVD", *Thin Solid Films*, vol. 333, pp. 71-76, 1998.

[144] A. P. Roberts, B. M. Henry, A. P. Sutton, C. R. M. Grovenor, G. A. D. Briggs, T. Miyamoto, M. Kano, Y. Tsukahara., and M. Yanaka, "Gas permeation in silicon-oxide/polymer (SiO_x/PET) barrier films: role of the oxide lattice, nano-defects and macro-defects", *J. Membr. Sci.*, vol. 208, pp. 75-88, 2002.

[145] H. Chatham, “Oxygen diffusion barrier properties of transparent oxide coatings on polymeric substrates,” *Surf. Coat. Technol.*, vol. 78, pp. 1–9, 1996.
[146] J. E. Shelby, *Handbook of Gas Diffusion in Solids and Melts*, Ohio: ASM International, 1996.
[147] R. Cueff, G. Baud, M. Benmalek, J. P. Besse, J. R. Butruille, and M. Jacquet, “X-ray photoelectron spectroscopy studies of plasma-modified PET surface and alumina/PET interface”, *Appl. Surf. Sci.*, vol. 115, pp. 292-298, 1997.
[148] M. Neuhauser, S. Barwulf, H. Hilgers, E. Lugscheider, and M. Riester, “Optical emission spectroscopy studies of titanium nitride sputtering on thermoplastic polymers“, *Surf. Coat. Technol.*, vol. 116, pp. 981-985, 1999.
[149] J. C. Rotger, J. J. Pireaux, R. Caudano, N. A. Thorne, H. M. Dunlop, and M. Benmalek, “Deposition of silicon oxide onto polyethylene and polyethyleneterephthalate: An x-ray photoelectron spectroscopy interfacial study”, *J. Vac. Sci. Technol. A*, vol. 13, pp. 260–266, 1995.
[150] L. Martinu, M. R. Wertheimer, and J. E. Klemberg-Sapieha, “Recent advances in plasma deposition of functional coatings on polymers”, in *Proc. Mater. Res. Soc. Symp.*, vol. 544, 1999, pp. 251–256.
[151] R. M. Barrer, *Diffusion in and through solids*, New York: Cambridge University Press, 1941.
[152] J. T. Felts and A. D. Grubb, “Commercial-scale application of plasma processing for polymeric substrates: From laboratory to production”, *J. Vac. Sci. Technol., A*, vol. 10, pp. 1675-1681, 1992.
[153] B. F. Chen, W. L. Pan, G. P. Yu, J. H. Hwang, and J. H. Huang, “On the Corrosion Behavior of TiN-Coated AISI D2 Steel”, *Surf. Coat. Technol.*, vol. 111, pp. 16-21, 1999.
[154] J. H. Huang, F. Y. Ouyang, and G. P. Yu, " Effect of Film Thickness and Ti Interlayer on the Structure and Properties of Nanocrystalline TiN Thin Films on AISI D2 Steel", *Surf. Coat. Technol.*, vol. 201, pp. 7043-7053, 2007.
[155] Y. Leterrier, “Durability of nanosized oxygen-barrier coatings on polymers,” *Prog. in Mater. Sci.*, vol. 48, pp. 1–55, 2003.
[156] Z. Suo and J. W. Hutchinson, “Steady-state cracking in brittle substrates beneath adherent films”, *Int. J. Solids Struct.*, vol. 25, pp. 1337-1353, 1989.
[157] J. W. Hutchinson and Z. Suo, “Mixed Mode Cracking in Layered Materials”, *Adv. Appl. Mech.*, vol. 29, pp. 63–191, 1992.
[158] M. Yanaka, Y. Tsukahara, N. Nakaso, and N. Takeda, “Cracking phenomena of brittle films in nanostructure composites analysed by a modified shear lag model with residual strain,” *J. Mater. Sci.*, vol. 33, pp. 2111–2119, 1998.
[159] U. Welzel, J. Ligot, P. Lamparter, A. C. Vermeulen and E. J. Mittemeijer, “Stress analysis of polycrystalline thin films and surface regions by X-ray diffraction”, *J. Appl. Cryst.*, vol. 38, pp.1-29, 2005.
[160] J. D. Affinito, M. E. Gross, C. A. Coronado, G. L. Graff, E. N. Greenwell, and P. M. Martin, “Anewmethod for fabricating transparent barrier layers,” *Thin Solid Films*, vol. 290–291, pp. 63–67, 1996.
[161] J. D. Affinito, S. Eufinger, M. E. Gross, G. L. Graff, and P. M. Martin, “PML/oxide/PML barrier layer performance differences arising from use of UV or electron beam polymerization of the PML layers,” *Thin Solid Films*, vol. 308–309, pp. 19–25, 1997.

[162] M. Scheapkens, T.W. Kim, A.G. Erlet, M. Yan, K.W. Flanagan, C.M. Heller, and P.A. McConnelee, "Ultrahigh barrier coating deposition on polycarbonate substrates", *J. Vac. Sci. Technol., A*, vol. 22, pp. 1716-1722, 2004.

[163] G.L. Graff, R.E. Williford, and P.E. Burrows, "Mechanisms of vapor permeation through multilayer barrier films: Lag time *vs.* equilibrium permeation", *J. Appl. Phys.*, vol. 96, pp. 1840-1849, 2004.

[164] http://displayproducts.ga.com/thinfilm.html

[165] J. Yun, S. Lee, Y. Jeong, H. R. Lee, J. D. Kwon, and G. H. Lee," Reduction of Defects in SiO_x Vapor Permeation Barriers on Polymer Substrates by Introducing a Sputtered Interlayer", *Jpn. J. Appl. Phys., Part 1*, vol. 48, 055503, 2009.

[166] W. L. Pan, G. P. Yu, and J. H. Huang, "Mechanical Properties of Ion-Plated TiN Films on AISI D2 Steel", *Surf. Coat. Technol.*, vol. 110, pp. 111-119, 1998.

[167] J. Y. Chen, G. P. Yu, and J. H. Huang, "Corrosion Behavior and Adhesion of Ion-Plated TiN Film on AISI 304 Stainless Steel", *Mater. Chem. Phys.*, vol. 65, pp. 310-315, 2000.

[168] W. J. Chou, G. P. Yu, and J. H. Huang, "Corrosion Resistance of ZrN Films on 304 Stainless Steel Substrates", *Surf. Coat. Technol.*, vol. 167, pp. 59-67, 2003.

[169] Jia-Hong Huang, Cheng-Hsin Ma, and Haydn Chen, "Effect of Interlayer on the Residual Stress and Texture Development of TiN Thin Film by Unbalanced Magnetron Sputtering", *Surf. Coat. Technol.*, vol. 201, pp.3199-3204, 2006.

[170] J. S. Lewis and M. S. Weaver, "Thin-Film Permeation-Barrier Technology for Flexible Organic Light-Emitting Devices", *IEEE J. Sel. Top. Quantum Electron.*, vol. 10, pp. 45-57, 2004.

[171] P. E. Burrows, G. L. Graff, M. E. Gross, P. M. Martin, M. Hall, E. Mast, C. Bonham, W. Bennett, L. Michalski, M. Weaver, J. J. Brown, D. Fogarty, and L. S. Sapochak, "Gas permeation and lifetime tests on polymer-based barrier coatings," *Proc. SPIE*, vol. 4105, pp. 75–83, 2001.

[172] M. Narasimhan, "Progress on barrier on plastic substrates", Presented at *USDC Flexible Displays Microelectron. Conf.*, Phoenix, AZ, 2005.

[173] http://www.symmorphix.com

[174] M.D. Groner and S. M. George, R.S. McLean, and P.F. Carcia, "Gas diffusion barriers on polymers using Al_2O_3 atomic layer deposition", in *Proc. Soc. Vacuum Coaters 48th Annu. Tech. Conf.*, 2005, p. 169-172.

[175] A. G. Erlat, A.G., M. Yan, T.W. Kim, M. Schaepkens, J. Liu, C.M. Heller, P. McConnelee, T. Feist, and A.R. Duggal, "Ultra-high barrier coatings for flexible organic electronics", in *Proc. Soc. Vacuum Coaters 48th Annu. Tech. Conf.*, 2005, p. 116-120.

[176] G. Nisato, P. C. P. Bouten, P. J. Slikkerveer, W. D. Bennett, G. L. Graff, N. Rutherford, and L. Wiese, "Evaluating high performance diffusion barriers: The calcium test," in *Proc. Int. DisplayWorkshop/Asia Display*, 2001, pp. 1435–1438.

[177] N. Rutherford, "Flexible displays– a low cost substrate/encapsulation packaging solution", Presented at *USDC Flex. Displ. Microelectron. Conf.*, Phoenix, AZ, 2005.

[178] P. M. Martin, "Hybrid Deposition Process 1:Vacuum Polymer Deposition", *Vac. Technol. Coat.*, vol. 10, p.40-45, February 2009.

Author Index

A

B

C

D

E

F

G

Filipe Vaz, Nicolas Martin and Martin Fenker (Eds)

Subject Index

A

B

C

D

Filipe Vaz, Nicolas Martin and Martin Fenker (Eds)

R

S

T

W

Z

Made in the USA
Las Vegas, NV
09 November 2024

11349669R00217